Use of Modern Materials in Technological Processes Accompanied by Frictional Heating

Use of Modern Materials in Technological Processes Accompanied by Frictional Heating

Editors

Aleksander Yevtushenko
Michal Kuciej

MDPI

Basel • Beijing • Wuhan • Barcelona • Belgrade • Novi Sad • Cluj • Manchester

Editors

Aleksander Yevtushenko
Department of Mechanics and Applied Computer Science
Białystok University of Technology
Białystok
Poland

Michal Kuciej
Department of Mechanics and Applied Computer Science
Białystok University of Technology
Białystok
Poland

Editorial Office
MDPI
St. Alban-Anlage 66
4052 Basel, Switzerland

This is a reprint of articles from the Special Issue published online in the open access journal *Materials* (ISSN 1996-1944) (available at: www.mdpi.com/journal/materials/special_issues/materials_frictional_heating).

For citation purposes, cite each article independently as indicated on the article page online and as indicated below:

Lastname, A.A.; Lastname, B.B. Article Title. *Journal Name* **Year**, *Volume Number*, Page Range.

ISBN 978-3-0365-8463-8 (Hbk)
ISBN 978-3-0365-8462-1 (PDF)
doi.org/10.3390/books978-3-0365-8462-1

Contents

About the Editors

Aleksander Yevtushenko

Aleksander Yevtushenko is a professor in the Department of Mechanics and Applied Computer Science at Bialystok University of Technology (BUT), Poland. He received his PhD in mechanics from the Institute for Applied Problems of Mechanics and Mathematics of the National Academy of Sciences of Ukraine in Lviv in 1979. He became a full professor in 2002. He is the director of doctoral studies at BUT and has supervised eight doctoral degree theses. He is a specialist in the field of the thermomechanics and mechanics of damage and cracking. He is the author and co-author of numerous books and scientific papers in international peer-reviewed journals.

Michal Kuciej

Michał Kuciej is a professor working at Bialystok University of Technology (BUT) in Poland. He obtained his doctoral degree in mechanics in 2008, and became a full professor in 2020. He is a specialist in the field of computer methods in mechanics, especially in thermomechanics. He has participated in numerous national and international research projects and different scientific conferences and congresses. He has published many articles in scientific journals. He became a dean of the Faculty of Mechanical Engineering at BUT in 2023.

Preface

This reprint contains 14 articles published in the Special Issue, "Use of Modern Materials in Technological Processes Accompanied by Frictional Heating".

Analytical one-dimensional models of unsteady heat generation due to friction in the tribosystem of two semi-bounded bodies made of FGM have been developed [1-6]. The key assumptions of the models relate to how the thermal conductivity exponentially increases as the distance from the contact surface increases; meanwhile, these models meet the conditions for perfect frictional thermal contact. The exact solutions for the corresponding boundary-value problems of heat conduction were obtained using the mathematical apparatus of the Laplace integral transform. Cases of slipping with a constant specific power of friction [1] and its time profile, taking into consideration the time of the contact pressure increase [2], were considered. The influence of the thermal sensitivity of the FGM components on the temperature field, which resulted from friction during single braking with constant deceleration, was investigated [3]. An analytical model was proposed to determine the temperature of a friction system containing one homogeneous element and the other made of FGM [4]. A methodology for determining the heat partition ratio in a friction couple made of FGM was developed [5]. This methodology was adapted to determine the maximum temperature of the disc brake system operating in a repetitive short-term (RST) mode of operation [6]. The influence of the flash temperature on the maximum temperature during RST braking in a system that consists of two elements made of homogeneous materials was investigated [7].

A comparative analysis of the railway brake temperature modes was carried out using axisymmetric (2D) and spatial (3D) numerical models with FEM [8].

Appropriate numerical models for railway solid and ventilated brake discs have also been developed [9].

The effect of adding a carbon-containing additive to the friction material of the wet clutch on the temperature generated by friction has been studied [10]. For the same system, the effect of ceramic and intermetallic powder additives on the tribological properties and temperature field was also investigated [11].

The problem of the identification of the unknown temperature for frictional heat generation in a cylindrical tribo-couple was analyzed [12].

An approach to determine the stress state of bi-material structures with interfacial homogeneous ribbon-like deformable inclusions has been proposed [13]. The influence of FGM on the stress–strain state of ribbon-reinforced composites was investigated as well [14].

Aleksander Yevtushenko and Michal Kuciej
Editors

materials MDPI

Article

The Effect of Functionally Graded Materials on Temperature during Frictional Heating: Under Uniform Sliding

Aleksander Yevtushenko, Katarzyna Topczewska * and Przemysław Zamojski

Faculty of Mechanical Engineering, Bialystok University of Technology (BUT), 45C Wiejska Street, 15-351 Białystok, Poland; a.yevtushenko@pb.edu.pl (A.Y.); zamojski.przemyslaw@gmail.com (P.Z.)
* Correspondence: k.topczewska@pb.edu.pl

Abstract: The mathematical model of heating process for a friction system made of functionally graded materials (FGMs) was proposed. For this purpose, the boundary-value problem of heat conduction was formulated for two semi-spaces under uniform sliding taking into consideration heating due to friction. Assuming an exponential change in thermal conductivities of the materials, the exact, as well as asymptotic (for small values of time), solutions to this problem were obtained. A numerical analysis was performed for two elements made of ZrO_2–Ti-6Al-4V and Al_3O_2–TiC composites. The influence of the gradient parameters of both materials on the evolution and spatial distributions of the temperature were investigated. The temperatures of the elements made of FGMs were compared with the temperatures found for the homogeneous ceramic materials.

Keywords: frictional heating; functionally gradient materials; temperature; composite; ceramic

Citation: Yevtushenko, A.; Topczewska, K.; Zamojski, P. The Effect of Functionally Graded Materials on Temperature during Frictional Heating: Under Uniform Sliding. *Materials* **2021**, *14*, 4285. https://doi.org/10.3390/ma14154285

Academic Editor: Jae-il Jang

Received: 9 June 2021
Accepted: 27 July 2021
Published: 31 July 2021

1. Introduction

A new class of composite materials with non-homogeneous spatial distribution of properties has emerged in recent years in the field of materials science [1]. Such properties are intentionally obtained during manufacturing by grading the internal structure of a material. Depending on the fabrication process, they are designed as stepwise-graded or continuous-graded materials [2]. The typical representatives of stepwise-graded composites are the laminates. The defect of such materials is the discontinuity of stress on the interfaces between adjacent discrete layers [3]. Materials with a continuous change in properties, known as functionally graded materials (FGMs) are devoid of this drawback. Nowadays, FGMs are usually a mixture of two distinct materials with continuously varying volume fractions of the constituents that, in effect, possess smooth properties which change along a certain direction [4]. Functionally graded materials possess a number of advantages that make them attractive in potential applications [5]. For example, a significant reduction of thermal stress in a heated element has been be achieved by introducing a thermal conductivity gradient in the material [6]. The results of studies indicate that a controlled continuous change in material properties can lead to a significant improvement in resistance to contact deformation and damage [4,7]. Thus, functionally graded coatings have been proposed as an alternative to replace conventional homogeneous coatings of frictional elements [8,9]. It has been proven that FGM coatings subjected to thermal shocks may suffer less damage than conventional ceramic coatings [6].

Usually, functionally graded materials are made of ceramic-metal composites and have superior characteristics of both components, i.e., heat and corrosion resistance of the ceramic and mechanical strength of the metal, at the same time [10]. Therefore, FGMs are considered to be advanced materials resistant to wear and elevated temperature conditions, and therefore they have great potential for use in heavy loaded sliding systems. One such application is brake discs exposed to intensive heating due to friction. At the core of an FGM disc, the material is steel to maintain structural rigidity, which gradually changes along the thickness and approaches purely ceramic at the friction surfaces to resist the severe thermal loading [5,11]. This significantly improves the thermomechanical behavior of the brake system as a whole [12].

Investigations associated with the development of frictional heating models for FGMs to determine the distributions of temperature and thermal stresses in brake systems, have received a great deal of attention from many researchers. The most common investigations have simulated the temperature regime in FGM brakes using numerical methods, in particular, by means of the finite element method (FEM). The finite element analysis of axisymmetric thermoelastic contact problems for a functionally graded disc with material property changes in the radial direction was performed by Shahzamanian et al. [13,14]. In [5], the corresponding problem was analyzed for a disc with properties dependent on the depth, along a normal direction to the friction surface of the disc. It was established that with the same operating parameters, the temperature gradient in a functionally graded disc was significantly lower than in a conventional steel disc. In a study by [9], the finite element methodology was used to compute the subsurface stresses in functionally graded coatings subjected to frictional contact with heat generation.

In addition to the well-established finite element method, there are other numerical methods for solving the corresponding heat problems of friction for functionally gradient materials. An advanced computational method for transient heat conduction analysis in a non-homogeneous FGM, based on local boundary integral equations, was proposed by Sladek et al. [15]. The Green's functions for the three-dimensional FGM transient heat conduction equation was derived using an exponential variation transform by [16]. The boundary integral equation based upon this approach has been solved numerically using a Galerkin approximation. The hybrid numerical method, based on the weighed residual and Fourier transform methods, to investigate the temperature distribution in the FGM plates under the exponential heat source load, was adopted by Tian and Jiang [17].

However, the closed-form analytical solutions to the thermal problems of friction for FGMs have higher accuracy and require less computational time than other methods. In general, the problems of thermomechanical contact with frictional heating for material with non-homogeneous properties are difficult to solve analytically due to the high mathematical complexity. For such materials, the equations of thermal conductivity and thermoelasticity contain coefficients that depend on the spatial coordinate [18]. Thus, the exact solutions of these equations and the determination of temperature distributions on their basis, require some special assumptions [19]. It is known that the superb performance of a functionally graded brake disc is achieved by introducing the appropriate gradient of thermomechanical properties by adjusting the gradient index [4]. The distribution of material properties in the FGM models is usually limited to unidirectional changes in the constituents of the composite [5]. There are two main distinctive ways to approximate the distribution of material properties through the graded direction, i.e., by means of an exponential and a power function. Note that the actual variations of properties depend on the material manufacturing process, which is neither exponential nor power law, therefore, in both cases, some level of curve fitting is implied [20]. However, both of these functions have a parameter that can be regulated to improve the fit and to adjust the gradation of the material. This role is played by the exponential decay rate in the exponential and the power exponent in the power law. The selection of these functions is also crucial from the point of view of the difficulty solving the thermal problems for FGMs by analytical methods.

The one-dimensional transient heat conduction problem for an axisymmetric FGM cylindrical shell with nonlinear thermal conductivity distributed according to the power law has been solved by the methods of separation of variables and Bessel functions [21]. The analytical formulas for calculating the thermal and mechanical stresses in a hollow cylinder made of FGM with properties modeled by the power law, using the direct method of solution to the Navier equation were obtained by [16]. Steady-state and unsteady temperature and thermal stress distributions in a plate, a hollow circular cylinder, and a hollow sphere made of functionally gradient material have been studied [22–24]. They proposed the original analytical method for solution to the one-dimensional heat conductivity problem for heterogeneous FGMs, which was performed with proper displacement of variables, Laplace transform, and the perturbation method. It should be noted that the perturbation method may be employed for the study of all classes of thermoelastic problems for functionally graded materials, even with consideration of the thermal sensitivity of material [11]. In [10], an analytical solution of the heat conduction problem for FGM cylinders subjected to non-uniform heat flux was obtained by using the method of matched asymptotic expansion in the perturbation technique. In [25], the Hankel transform method was used to obtain an analytical solution of the axisymmetric stationary problem of heat conduction for an FGM layer with thermal conductivity dependent on the depth from the heated boundary surface. The same technique has been applied to solve the steady axisymmetric boundary problem of thermoelasticity for non-homogeneous semi-space with thermomechanical properties that depend exponentially on the distance from the heated surface [26]. This approach can be used for modeling layered composites with stepwise gradation of the properties, and also for approximate modeling of the materials with a functional change in properties. In this last case, the functionally graded coatings were replaced by a package of layers, whose material properties were assumed to be constant. This simplification of material property gradation allows one to implement analytical methods for each sublayer, using the known solutions to the thermoelastic problems of friction for isotropic bodies. This approach is known as the multi-layered model of functionally graded materials. It has been shown that the results obtained for the FGM layer, divided into a sufficient number of the sublayers, are in good agreement with the data found using the corresponding exact solutions [26]. The same approach has been used to simulate FGM with sinusoidal and cosinusoidal power and exponential distribution of properties for a cylinder subjected to non-uniform heat flux [27]. Thermoelastic frictional contact of the FGMs with arbitrarily varying properties has been investigated using the multi-layered model by Liu et al. [4,7]. On the basis of the same approach, the non-stationary temperature field in a functional gradient layer with continuous and piecewise change in material properties has been determined by means of the Laplace transform, asymptotic analysis and integration technique [6]. The multi-layered model has been developed for analysis of the two-dimensional sliding frictional contact problem with a functionally graded coating [28]. This model has been used to solve the transient heat conduction and thermal stress problems for the FGM plate taking into consideration temperature-dependent material properties [17].

Most of the above-mentioned studies have considered the problems for a heated FGM layer on homogeneous substrate or cylinder, which can successfully simulate thermoelastic behavior of a brake disc with FGM coating. The temperature mode of a pad-disc tribosystem has been simulated using the thermal problem of friction for a functionally graded coated half-space (a disc) sliding against a homogeneous body (a pad) in [4,7–9,27,29]. While modern materials for friction pads in brake systems are usually composites, the proportion of individual components can also be changed along with the distance from the friction surface. Experimental investigations have shown that functional variations in the properties of the pad material significantly improve their braking characteristics [12,30]. In particular, the results have indicated that the wear resistance of a specimen made of a functionally graded material is higher than the wear resistance of its analogue made of a homogeneous material [12]. Therefore, FGMs are real candidates for the role of automotive brake pads [29,31]. In connection with this potential application, we see the need for the development of mathematical models of frictional heating of two element systems of the pad-disc type, both made of functionally graded materials. The development of such models is also associated with the possibility of their use in the study of thermoelastic instability (TEI) due to frictional heating. It is known that the system exhibits TEI in brakes when the sliding speed exceeds a critical value [13]. The emergence of instability is accompanied by the concentration of frictional heating over regions much smaller than the nominal contact region, thus, leading to high localized temperature and contact pressure. The appearance of these so-called hot spots results in various undesirable effects such as material transformations, thermal cracking, and brake fade [20]. The studies concerning the effect of material non-homogeneity on thermal instability in brakes have shown that an FGM disc reduces the susceptibility towards TEI by increasing the critical speed of sliding [8,20,27].

2. Statement of the Problem

Consider a heat-conduction problem for two semi-infinite FGM bodies (Figure 1).

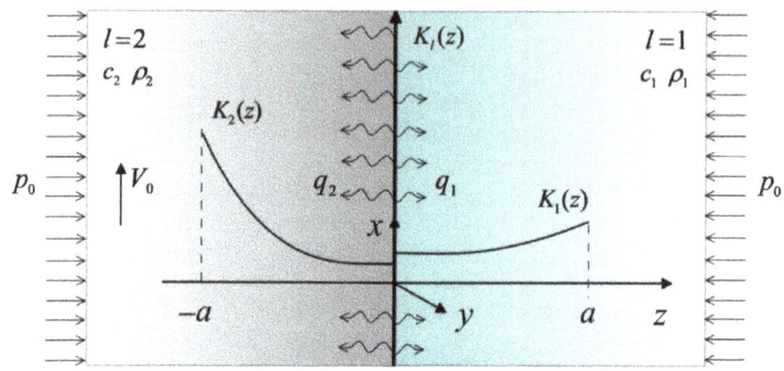

Figure 1. Scheme of the problem.

It is assumed that:

1. The bodies are related to the coordinate Cartesian system $Oxyz$, and their initial temperature distribution is homogeneous and equal to the ambient temperature T_a;
2. At the initial time moment $t = 0$, the bodies are pressed to each other with uniform pressure p_0 acting parallel to the z axis and simultaneously start sliding with constant relative speed V_0 in the positive direction of the x axis;
3. Due to friction, on the contact surface $z = 0$ heat is generated, which is absorbed by the elements of friction pair in the form of heat fluxes, causing an increase in their temperature $T(z,t)$ over the initial value T_a;
4. During frictional heating, the sum $q_1 + q_2$ of intensities of heat fluxes directed from the contact surface $z = 0$ along the normal to the insides of the bodies, is equal to the specific power of friction $q_0 = fp_0V_0$, where f is the coefficient of friction. At the same time, the temperatures T on the friction surfaces of both bodies are equal [32,33];
5. Changes in the temperature gradients in the directions x and y are negligible and the gradient in the direction z decreases, along with the distance from the contact surface;
6. Thermal conductivity of materials K_l are exponential functions of variable z, and their specific heat c_l and density ρ_l, $l = 1,2$ are constant [34]. Here and further, the lower index $l = 1$ indicates the parameters and quantities relating to the first element, and $l = 2$ to the second element.

On the basis of the above assumptions, the temperature rise $\Theta(z,t) = T(z,t) - T_a$ of the friction pair was found as the solution to the following boundary-value problem of heat conduction:

$$\frac{\partial}{\partial z}\left[K_1(z)\frac{\partial\Theta(z,t)}{\partial z}\right] = c_1\rho_1\frac{\partial\Theta(z,t)}{\partial t},\ z > 0, t > 0, \tag{1}$$

$$\frac{\partial}{\partial z}\left[K_2(z)\frac{\partial\Theta(z,t)}{\partial z}\right] = c_2\rho_2\frac{\partial\Theta(z,t)}{\partial t},\ z < 0, t > 0, \tag{2}$$

$$K_2(z)\frac{\partial\Theta(z,t)}{\partial z}\bigg|_{z=0^-} - K_1(z)\frac{\partial\Theta(z,t)}{\partial z}\bigg|_{z=0^+} = q_0,\ t > 0, \tag{3}$$

$$\Theta(0^-,t) = \Theta(0^+,t),\ t > 0, \tag{4}$$

$$\Theta(z,t) \to 0\,,\ |z| \to \infty,\ t > 0, \tag{5}$$

$$\Theta(z,0) = 0, |z| < \infty. \tag{6}$$

Taking into consideration the dependencies:

$$K_l(z) = K_{l,0}e^{\gamma_l|z|},\ |z| < \infty,\ K_{l,0} \equiv K_l(0),\ \gamma_l \geq 0, l = 1,2, \tag{7}$$

the problem Equations (1)–(6) can be written in the form:

$$\frac{\partial^2\Theta(z,t)}{\partial z^2} + \gamma_1\frac{\partial\Theta(z,t)}{\partial z} = \frac{e^{-\gamma_1 z}}{k_{1,0}}\frac{\partial\Theta(z,t)}{\partial t},\ z > 0, t > 0, \tag{8}$$

$$\frac{\partial^2\Theta(z,t)}{\partial z^2} - \gamma_2\frac{\partial\Theta(z,t)}{\partial z} = \frac{e^{\gamma_2 z}}{k_{2,0}}\frac{\partial\Theta(z,t)}{\partial t},\ z < 0, t > 0, \tag{9}$$

$$K_{2,0}\frac{\partial\Theta(z,t)}{\partial z}\bigg|_{z=0^-} - K_{1,0}\frac{\partial\Theta(z,t)}{\partial z}\bigg|_{z=0^+} = q_0,\ t > 0, \tag{10}$$

$$\Theta(0^-,t) = \Theta(0^+,t),\ t > 0 \tag{11}$$

$$\Theta(z,t) \to 0\ ,\ |z| \to \infty,\ t > 0 \tag{12}$$

$$\Theta(z,0) = 0, |z| < \infty \tag{13}$$

where

$$k_{l,0} = \frac{K_{l,0}}{c_l\rho_l},\ l = 1,2. \tag{14}$$

are the coefficients of thermal diffusivity of materials on the surface of friction $z = 0$.

3. Solution to the Problem

The Laplace transform [35]:

$$L[\,\Theta(z,t);p] \equiv \overline{\Theta}(z,p) = \int_0^\infty \Theta(z,t)e^{pt}dt,\ \mathrm{Re}p > 0. \tag{15}$$

application to the problem Equations (8)–(13), gives:

$$\frac{d^2\overline{\Theta}(z,p)}{dz^2} + \gamma_1\frac{d\overline{\Theta}(z,p)}{dz} - \frac{p}{k_{1,0}}e^{-\gamma_1 z}\overline{\Theta}(z,p) = 0,\ z > 0, \tag{16}$$

$$\frac{d^2\overline{\Theta}(z,p)}{dz^2} - \gamma_2\frac{d\overline{\Theta}(z,p)}{dz} - \frac{p}{k_{2,0}}e^{\gamma_2 z}\overline{\Theta}(z,p) = 0,\ z < 0, \tag{17}$$

$$K_{2,0}\frac{d\overline{\Theta}(z,p)}{dz}\bigg|_{z=0^-} - K_{1,0}\frac{d\overline{\Theta}(z,p)}{dz}\bigg|_{z=0^+} = \frac{q_0}{p}, \tag{18}$$

$$\overline{\Theta}(0^-,p) = \overline{\Theta}(0^+,p), \tag{19}$$

$$\overline{\Theta}(z,p) \to 0\ ,\ |z| \to \infty. \tag{20}$$

Introducing the new variables and dimensionless parameters:

$$\xi_1 = \xi e^{-\gamma_1 z/2},\ z \geq 0,\ \xi_2 = \gamma_\varepsilon \xi e^{\gamma_2 z/2},\ z \leq 0,\ \gamma_\varepsilon = \gamma^*\sqrt{k_0^*}, \tag{21}$$

$$\xi = \frac{2}{\gamma_1}\sqrt{\frac{p}{k_{1,0}}},\ \gamma^* = \frac{\gamma_1}{\gamma_2},\ k_0^* = \frac{k_{1,0}}{k_{2,0}}, \tag{22}$$

the following derivatives can be found:

$$\frac{d\overline{\Theta}(z,p)}{dz} = (-1)^l\frac{1}{2}\gamma_l\xi_l\frac{d\overline{\Theta}(\xi_l,p)}{d\xi_l},\ \frac{d^2\overline{\Theta}(z,p)}{dz^2} = \frac{1}{4}\gamma_l^2\xi_l^2\frac{d^2\overline{\Theta}(\xi_l,p)}{d\xi_l{}^2} + \frac{1}{4}\gamma_l^2\xi_l\frac{d\overline{\Theta}(\xi_l,p)}{d\xi_l}, l = 1,2. \tag{23}$$

Taking into consideration the relations (23), Equations (16) and (17) are brought to the form:

$$\frac{d^2\overline{\Theta}(\xi_l,p)}{d\xi_l^2} - \frac{1}{\xi_l}\frac{d\overline{\Theta}(\xi_l,p)}{d\xi_l} - \overline{\Theta}(\xi_l,p) = 0,\ \xi_l > 0,\ l = 1,2 \tag{24}$$

The general solution to Equation (24), satisfying the boundary condition (20), has the form:

$$\overline{\Theta}(\xi_l,p) = A_l(p)\xi_l \mathrm{I}_1(\xi_l),\ l = 1,2 \tag{25}$$

where $\mathrm{I}_k(\cdot)$ are the modified Bessel functions of the first kind of the kth order, and $A_l(p)$ are the unknown functions. After differentiating the solution (25), and taking into consideration the relations (21), (22) and derivative $[x\mathrm{I}_1(x)]' = x\mathrm{I}_0(x)$ [36] (here and further, the symbol $'$ denotes the ordinary derivative), the following is found:

$$\left.\frac{d\overline{\Theta}}{dz}\right|_{z=0^+} = -\frac{\gamma_1}{2}A_1(p)\xi^2\mathrm{I}_0(\xi),\ \left.\frac{d\overline{\Theta}}{dz}\right|_{z=0^-} = \frac{\gamma_2}{2}A_2(p)(\gamma_\varepsilon\xi)^2\mathrm{I}_0(\gamma_\varepsilon\xi). \tag{26}$$

Substituting the derivatives (26) into the boundary conditions (18) and (19), the system of two linear algebraic equations is obtained with respect to the unknown functions $A_l(p)$, $l = 1,2$, the solution of which, has the form:

$$A_1(p) = 2\Lambda\frac{\mathrm{I}_1(\gamma_\varepsilon\xi)}{p\gamma_\varepsilon\xi^2\psi(p)},\ A_2(p) = 2\Lambda\frac{\mathrm{I}_1(\xi)}{p(\gamma_\varepsilon\xi)^2\psi(p)} \tag{27}$$

where

$$\psi(p) = \mathrm{I}_0(\gamma_\varepsilon\xi)\mathrm{I}_1(\xi) + K_\varepsilon\mathrm{I}_0(\xi)\mathrm{I}_1(\gamma_\varepsilon\xi), \tag{28}$$

$$K_\varepsilon = \frac{K_0^*}{\sqrt{k_0^*}},\ K_0^* = \frac{K_{1,0}}{K_{2,0}},\ \Lambda = \frac{q_0}{\gamma_2 K_{2,0}}. \tag{29}$$

Taking into consideration the forms of variables $\xi_l, l = 1,2$ (21), (22), and functions $A_l(p), l = 1,2$ (27)–(29) the solutions (25) are given as:

$$\overline{\Theta}(z,p) = 2\Lambda e^{-\gamma_1 z/2}\frac{\varphi_1(z,p)}{\Psi(p)},\ z \geq 0,\ \overline{\Theta}(z,p) = 2\Lambda e^{\gamma_2 z/2}\frac{\varphi_2(z,p)}{\Psi(p)},\ z \leq 0, \tag{30}$$

$$\begin{gathered}\varphi_1(z,p) = \mathrm{I}_1(\gamma_\varepsilon\xi)\mathrm{I}_1(\xi e^{-\gamma_1 z/2}),\ \varphi_2(z,p) = \mathrm{I}_1(\xi)\mathrm{I}_1(\gamma_\varepsilon\xi e^{\gamma_2 z/2}),\\ \Psi(p) = p\gamma_\varepsilon\xi\psi(p).\end{gathered} \tag{31}$$

Using the Vashchenko–Zakharchenko theorem [37,38], the inverse Laplace transform of the solutions (30) and (31) can be written in the form:

$$\Theta(z,t) = 2\Lambda e^{-\frac{1}{2}\gamma_1 z}\left[\frac{\varphi_1(z,0)}{\Psi'(0)} + \sum_{n=1}^{\infty}\frac{\varphi_1(z,p_n)}{\Psi'(p_n)}e^{-p_n t}\right],\ z \geq 0,\ t \geq 0, \tag{32}$$

$$\Theta(z,t) = 2\Lambda e^{\frac{1}{2}\gamma_2 z}\left[\frac{\varphi_2(z,0)}{\Psi'(0)} + \sum_{n=1}^{\infty}\frac{\varphi_2(z,p_n)}{\Psi'(p_n)}e^{-p_n t}\right],\ z \leq 0, t \geq 0. \tag{33}$$

where $p_n > 0, n = 1,2,\ldots$ are the real roots of the transcendental equation $\psi(p) = 0$ with function $\psi(p)$ (28).

With consideration of the expansions [36]:

$$\mathrm{I}_0(x) = 1 + \frac{x^2}{4} + \frac{x^4}{64} + \ldots, \mathrm{I}_1(x) = \frac{x}{2} + \frac{x^3}{16} + \ldots, \tag{34}$$

from Equation (31), it can be found that:

$$\varphi_l(z,p) \cong \xi^2\tilde{\varphi}_l(z,p),\ l = 1,2,\ \Psi(z,p) \cong \xi^2\tilde{\Psi}(z,p), \tag{35}$$

$$\widetilde{\varphi}_1(z,p) = \frac{1}{4}\gamma_\varepsilon e^{-\gamma_1 z/2}\left[1+\frac{1}{8}(\gamma_\varepsilon\xi)^2\right],\ z \geq 0,\ \varphi_2(z,p) = \frac{1}{4}\gamma_\varepsilon e^{\gamma_2 z/2}\left(1+\frac{1}{8}\xi^2\right),\ z \leq 0 \quad (36)$$

$$\widetilde{\Psi}(p) = p\gamma_\varepsilon\left[\frac{1}{2}(1+\gamma_\varepsilon K_\varepsilon) + \frac{1}{16}(1+2\gamma_\varepsilon K_\varepsilon + 2\gamma_\varepsilon^2 + K_\varepsilon\gamma_\varepsilon^3)\xi^2\right],\ \xi^2 = \frac{4p}{\gamma_1^2 k_{1,0}}, \quad (37)$$

At $p \to 0$, Equations (35)–(37) lead to:

$$\frac{\varphi_1(0)}{\Psi'(0)} = \frac{e^{-\gamma_1 z/2}}{2(1+\gamma_\varepsilon K_\varepsilon)},\ z \geq 0,\ \frac{\varphi_2(0)}{\Psi'(0)} = \frac{e^{\gamma_2 z/2}}{2(1+\gamma_\varepsilon K_\varepsilon)},\ z \leq 0, \quad (38)$$

Using the relation [36]:

$$\mathrm{I}_0(x) = J_0(ix),\ \mathrm{I}_1(x) = -iJ_1(ix),\ J'_0(x) = -J_1(x),\ J'_1(x) = J_0(x) - x^{-1}J_1(x),\ i \equiv \sqrt{-1}, \quad (39)$$

where $J_k(\cdot)$ are the Bessel functions of the first kind of the kth order, and denoted $\mu \equiv i\xi$, the temperature rise (32), (33), with consideration of Equations (22) and (38), can be written in the form:

$$\Theta(z,t) = \Lambda e^{-\gamma_1 z/2}\left[\frac{e^{-\gamma_1 z/2}}{(1+\gamma_\varepsilon K_\varepsilon)} + \frac{4}{\gamma_\varepsilon}\sum_{n=1}^{\infty}\frac{\hat{\varphi}_1(z,\mu_n)}{\hat{\Psi}'(\mu_n)}e^{-p_n t}\right],\ z \geq 0,\ t \geq 0, \quad (40)$$

$$\Theta(z,t) = \Lambda e^{\gamma_2 z/2}\left[\frac{e^{\gamma_2 z/2}}{(1+\gamma_\varepsilon K_\varepsilon)} + \frac{4}{\gamma_\varepsilon}\sum_{n=1}^{\infty}\frac{\hat{\varphi}_2(z,\mu_n)}{\hat{\Psi}'(\mu_n)}e^{-p_n t}\right],\ z \leq 0,\ t \geq 0, \quad (41)$$

where

$$\hat{\varphi}_1(z,\mu_n) = J_1(\mu_n)J_1(\gamma_\varepsilon\mu_n e^{-\gamma_1 z/2}),\ \hat{\varphi}_2(z,\mu_n) = J_1(\mu_n)J_1(\gamma_\varepsilon\mu_n e^{\gamma_2 z/2}), \quad (42)$$

$$\hat{\Psi}'(\mu_n) = \mu_n^2[(1+\gamma_\varepsilon K_\varepsilon)J_0(\mu_n)J_0(\gamma_\varepsilon\mu_n) - (\gamma_\varepsilon + K_\varepsilon)J_1(\mu_n)J_1(\gamma_\varepsilon\mu_n)], \quad (43)$$

$$p_n = 0.25k_{1,0}\gamma_1^2\mu_n^2, \quad (44)$$

$\mu_n > 0, n = 1,2,3,\ldots$, are the real roots of the functional equation:

$$J_0(\gamma_\varepsilon\mu)J_1(\mu) + K_\varepsilon J_0(\mu)J_1(\gamma_\varepsilon\mu) = 0. \quad (45)$$

On the contact surface $z = 0$ from Equations (40)–(42), we achieve:

$$\Theta(t) \equiv \Theta(0,t) = \Lambda\left[\frac{1}{(1+\gamma_\varepsilon K_\varepsilon)} + \frac{4}{\gamma_\varepsilon}\sum_{n=1}^{\infty}\frac{\hat{\varphi}(\mu_n)}{\hat{\Psi}'(\mu_n)}e^{-p_n t}\right],\ t \geq 0 \quad (46)$$

$$\hat{\varphi}(\mu_n) \equiv \hat{\varphi}_1(0,\mu_n) = \hat{\varphi}_2(0,\mu_n) = J_1(\gamma_\varepsilon\mu_n)J_1(\mu_n), \quad (47)$$

Additionally, assuming that the materials of the friction pair are the same ($K_{1,0} = K_{2,0} \equiv K_0$, $k_{1,0} = k_{2,0} \equiv k_0$, $\gamma_1 = \gamma_2 \equiv \gamma$), then, from Equations (21), (22), and (29), it follows that $K_\varepsilon = \gamma_\varepsilon = 1$ and solution (46) and (47) take the form:

$$\Theta(t) = 2\Lambda\left(\frac{1}{4} - \sum_{n=1}^{\infty}\frac{e^{-p_n t}}{\mu_n^2}\right),\ t \geq 0, \quad (48)$$

where $J_0(\mu_n) \equiv 0, p_n = 0.25k_0\gamma^2\mu_n^2$.

Introducing the following dimensionless variables and parameters:

$$\zeta = \frac{z}{a},\ \tau = \frac{k_{1,0}t}{a^2},\ \gamma_l = \frac{\gamma_l^*}{a},\ l = 1,2,\ \Theta_0 = \frac{q_0 a}{K_{1,0}},\ \Theta^* = \frac{\Theta}{\Theta_0} \quad (49)$$

where a is the thickness of the friction pair elements participating in heat absorption. These parameters are closely related to the concept of effective thickness, i.e., the distance from the friction surface where the temperature is equal to 5% of the maximum value [39].

Taking into consideration the notations (49) in Equations (40)–(45), the dimensionless temperature rise can be written as:

$$\Theta^*(\zeta,\tau)=\frac{K_0^*}{\gamma_2^*}e^{-\gamma_1^*\zeta/2}\left[\frac{e^{-\gamma_1^*\zeta/2}}{(1+\gamma_\varepsilon K_\varepsilon)}+\frac{4}{\gamma_\varepsilon}\sum_{n=1}^{\infty}\frac{\varphi_1^*(\zeta,\mu_n)}{\hat{\Psi}'(\mu_n)}e^{-\lambda_n^2\tau}\right],\ \zeta\geq 0,\ \tau\geq 0, \tag{50}$$

$$\Theta^*(\zeta,\tau)=\frac{K_0^*}{\gamma_2^*}e^{\gamma_2^*\zeta/2}\left[\frac{e^{\gamma_2^*\zeta/2}}{(1+\gamma_\varepsilon K_\varepsilon)}+\frac{4}{\gamma_\varepsilon}\sum_{n=1}^{\infty}\frac{\varphi_2^*(\zeta,\mu_n)}{\hat{\Psi}'(\mu_n)}e^{-\lambda_n^2\tau}\right],\ \zeta\leq 0,\ \tau\geq 0, \tag{51}$$

where

$$\varphi_1^*(\zeta,\mu_n)=J_1(\gamma_\varepsilon\mu_n)J_1(\mu_n e^{-\gamma_1^*\zeta/2}),\ \varphi_2^*(\zeta,\mu_n)=J_1(\mu_n)J_1(\gamma_\varepsilon\mu_n e^{\gamma_2^*\zeta/2}), \tag{52}$$

$$\lambda_n=0.5\gamma_1^*\mu_n,\ n=1,2,\ldots \tag{53}$$

On the contact surface $\zeta=0$ from Equations (50)–(52), it follows that:

$$\Theta^*(\tau)\equiv\Theta^*(0,\tau)=\frac{K_0^*}{\gamma_2^*}\left[\frac{1}{(1+\gamma_\varepsilon K_\varepsilon)}+\frac{4}{\gamma_\varepsilon}\sum_{n=1}^{\infty}\frac{\varphi^*(\mu_n)}{\hat{\Psi}'(\mu_n)}e^{-\lambda_n^2\tau}\right],\ \tau\geq 0, \tag{54}$$

where

$$\varphi^*(\mu_n)\equiv\varphi_1^*(0,\mu_n)=\varphi_2^*(0,\mu_n)=J_1(\gamma_\varepsilon\mu_n)J_1(\mu_n). \tag{55}$$

4. An Asymptotic Solution at the Initial Stage of Sliding

At large values of the parameter p of the Laplace integral transform (15), and taking into consideration the asymptotic behavior of the functions [36]:

$$\mathrm{I}_k(x)\cong\frac{e^x}{\sqrt{2\pi x}},\ k=0,1, \tag{56}$$

from Equations (28) and (31) it can be found that:

$$\varphi_1(z,p)\cong\frac{e^{(1+\gamma_\varepsilon-\gamma_1 z/2)\xi}}{2\pi\xi\sqrt{\gamma_\varepsilon}}e^{\gamma_1 z/4},\ z\geq 0,\ \varphi_2(z,p)\cong\frac{e^{(1+\gamma_\varepsilon+\gamma_2 z/2)\xi}}{2\pi\xi\sqrt{\gamma_\varepsilon}}e^{-\gamma_2 z/4},\ z\leq 0, \tag{57}$$

$$\Psi(p)\cong(1+K_\varepsilon)\frac{pe^{(1+\gamma_\varepsilon)\xi}}{2\pi\sqrt{\gamma_\varepsilon}}. \tag{58}$$

Substituting Equations (57) and (58) into Equation (30), the transforms of the temperature rise can be presented as:

$$\overline{\Theta}(z,p)=\frac{2\Lambda e^{-(1+2\xi)\gamma_1 z/4}}{(1+K_\varepsilon)p\xi},\ z\geq 0,\ \overline{\Theta}(z,p)=\frac{2\Lambda e^{(1+2\xi)\gamma_2 z/4}}{(1+K_\varepsilon)p\xi},\ z\leq 0, \tag{59}$$

In view of the notation ξ (22), the transformed solutions (59) can be written in the form:

$$\overline{\Theta}(z,p)=\frac{\Lambda\gamma_1}{(1+K_\varepsilon)}e^{-\gamma_1 z/4}\frac{e^{-\sqrt{\frac{p}{k_{1,0}}}z}}{p\sqrt{\frac{p}{k_{1,0}}}},\ z\geq 0,\ \overline{\Theta}(z,p)=\frac{\Lambda\gamma_1}{(1+K_\varepsilon)}e^{\gamma_2 z/4}\frac{e^{-\sqrt{\frac{p}{k_{2,0}}}z}}{p\sqrt{\frac{p}{k_{1,0}}}},\ z\leq 0, \tag{60}$$

Using the relation [40]:

$$L^{-1}[p^{-3/2}e^{-\alpha\sqrt{p}};t]=2\sqrt{t}\,\mathrm{ierfc}\left(\frac{\alpha}{2\sqrt{t}}\right),\ \alpha\geq 0, \tag{61}$$

Taking into consideration notations (29) and (49), the dimensionless temperature rise for small values of the Fourier number τ was received as:

$$\Theta^*(\zeta,\tau) = \frac{2\gamma_\varepsilon K_\varepsilon}{(1+K_\varepsilon)} e^{-\gamma_1^*\zeta/4}\sqrt{\tau}\,\mathrm{ierfc}\left(\frac{\zeta}{2\sqrt{\tau}}\right),\ \zeta \geq 0,\ 0 \leq \tau << 1, \quad (62)$$

$$\Theta^*(\zeta,\tau) = \frac{2\gamma_\varepsilon K_\varepsilon}{(1+K_\varepsilon)} e^{\gamma_2^*\zeta/4}\sqrt{\tau}\,\mathrm{ierfc}\left(-\frac{\zeta}{2}\sqrt{\frac{k_0^*}{\tau}}\right),\ \zeta \leq 0,\ 0 \leq \tau << 1, \quad (63)$$

where $\mathrm{ierfc}(x) = \pi^{-1/2}e^{-x^2} - x\mathrm{erfc}(x), \mathrm{erfc}(x) = 1 - \mathrm{erf}(x)$, and the $\mathrm{erf}(x)$ is the Gaussian error function [36]. On the contact surface $\zeta = 0$ from Equations (63) and (62), it can be obtained that:

$$\Theta^*(\tau) = \frac{2\gamma_\varepsilon K_\varepsilon}{(1+K_\varepsilon)}\sqrt{\frac{\tau}{\pi}},\ 0 \leq \tau << 1, \quad (64)$$

Substituting $\gamma_1 = \gamma_2 = 0$ and $\gamma^* = 1$ into Equations (62)–(64), the known solutions can be found to determine the dimensionless temperature increase in the homogeneous bodies [41]:

$$\Theta^*(\zeta,\tau) = \frac{2K_0^*}{(1+K_\varepsilon)}\sqrt{\tau}\,\mathrm{ierfc}\left(\frac{\zeta}{2\sqrt{\tau}}\right),\ \zeta \geq 0,\ 0 \leq \tau << 1, \quad (65)$$

$$\Theta^*(\zeta,\tau) = \frac{2K_0^*}{(1+K_\varepsilon)}\sqrt{\tau}\,\mathrm{ierfc}\left(-\frac{\zeta}{2}\sqrt{\frac{k_0^*}{\tau}}\right),\ \zeta \leq 0, 0 \leq \tau << 1, \quad (66)$$

$$\Theta^*(\tau) = \frac{2K_0^*}{(1+K_\varepsilon)}\sqrt{\frac{\tau}{\pi}},\ 0 \leq \tau << 1. \quad (67)$$

5. Numerical Analysis

The numerical analysis was performed based on the exact solutions (50)–(55) and the asymptotic Equations (62)–(64). The elements are both made of functionally graded materials in such a way that their friction surfaces $z = 0$ are purely ceramic ZrO_2 and Al_3O_2 and, along the thickness of the elements, they approach the core materials Ti-6Al-4V and TiC, respectively. The thermal properties of component materials are presented in Table 1.

Table 1. Thermal properties of the FGMs components [17,42].

Element Number.	Material	Thermal Conductivity $K[\mathbf{Wm^{-1}K^{-1}}]$	Thermal Diffusivity $k \times 10^6[\mathbf{m^2s^{-1}}]$
$l = 1$	ZrO_2	2.09	0.86
	Ti-6Al-4V	7.5	3.16
$l = 2$	Al_3O_2	1.5	4.98
	TiC	33.9	9.59

In view of notations (49), Equation (7), describing the change in thermal conductivity of materials with distance from the surface of friction, becomes:

$$K_l(z) = K_{l,0}K_l^*(\zeta), K_l^*(\zeta) = e^{\gamma_l^*|\zeta|},\ |\zeta| < \infty,\ l = 1,2 \quad (68)$$

where the values of the dimensionless gradient parameters can be calculated from the following relation [8]:

$$\gamma_l^* = \ln(K_{l,1}/K_{l,0}),\ K_{l,0} \equiv K_l^*(0),\ K_{l,1} \equiv K_l^*(1),\ l = 1,2. \quad (69)$$

The formula (69) provides that thermal conductivity changes in a manner suitable for the FGM composition variations from pure ceramic on the friction surface, achieving the pure core material on the effective thickness a ($|\zeta| = 1$) inside elements. The effective thicknesses 3.2 mm and 7.7 mm for the first ($l = 1$) and second ($l = 2$) elements, respectively were calculated in accordance with the methodology [39]. Hence, it can be assumed that

$a = 7.7$ mm. Then, from Table 1, the following data are taken: $K_{1,0} = 2.09$ Wm^{-1}K^{-1}, $K_{1,1} = 7.5$ Wm^{-1}K^{-1} for the FGM ZrO_2–Ti-6Al-4V (l = 1) and $K_{2,0} = 1.5$ Wm^{-1}K^{-1}, $K_{2,1} = 33.9$ Wm^{-1}K^{-1} for the FGM Al_3O_2–TiC (l = 2). Substituting these coefficients into the formula (69) we obtain the dimensionless gradient parameters values $\gamma_1^* = 1.28$, $\gamma_2^* = 3.12$. Distribution of the thermal conductivity along the distance from the friction surface, for considered tribosystem is presented in the Figure 2. The positive roots of the nonlinear functional Equation (45) were found by means of the bisection method [43]. It was necessary to take at least 70 roots of Equation (45) in order to perform calculations according to Equations (50)–(55) with a relative accuracy of 10^{-3}.

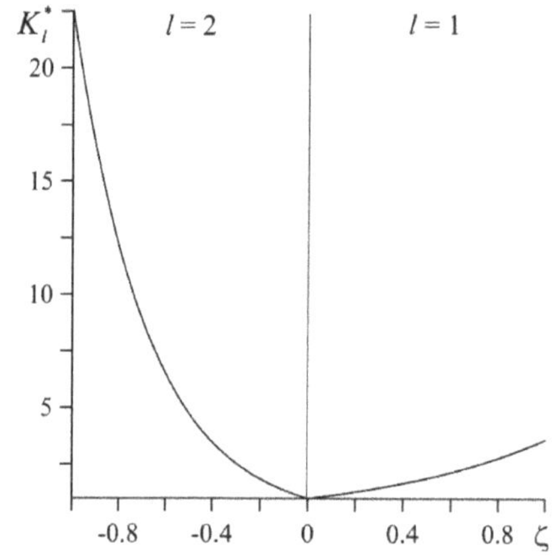

Figure 2. Distributions of the dimensionless thermal conductivities K_l^*, of FGM ZrO_2–Ti-6Al-4V (l = 1) and Al_3O_2–TiC (l = 2) along the dimensionless distance ζ from the friction surface.

Variations of the dimensionless temperature rise $\Theta^*(\zeta, \tau)$ (50)–(55) in the friction elements ZrO_2–Ti-6Al-4V ($l = 1$) and Al_3O_2–TiC ($l = 2$) during the sliding, are shown by the continuous curves in Figure 3, while the dashed lines in this figure illustrate the corresponding results obtained from the solutions (65)–(67) for the friction pair elements made of homogeneous materials ZrO_2 ($l = 1$) and Al_3O_2 ($l = 2$). At a certain distance ζ, the temperature monotonically increases over time (Fourier number τ). The highest temperature is achieved on the contact surface $\zeta = 0$. It can be seen that the elements of tribocouple made of homogeneous materials are heated more intensively during the sliding than the FGMs. Differences between the compared results increase over the time of heating. Taking into consideration notations (49), it can be established that, the maximum temperature rises are $\Theta_{max} = 604$ °C and $\Theta_{max} = 765$ °C achieved at the end of the process, for the friction pairs made of functionally graded and homogeneous materials, respectively.

Distribution of dimensionless maximum temperature Θ^*_{max}, achieved at the end of the process, along the distance from the contact surface is presented in Figure 4. With the distance from the contact surface in the element $l = 1$, the difference between continuous and dashed lines decreases. Unlike in the element $l = 2$, where this difference remains almost unchanged along the thickness, and even slightly increases (to the 238°C at distance $|z| = 1.85$ mm). A much higher temperature level is reached in the homogeneous element $l = 2$ made of ceramic (Al_3O_2) as compared with the temperature achieved in the FGM element Al_3O_2–TiC, which is caused by application of the core material (TiC) with high thermal conductivity and diffusivity.

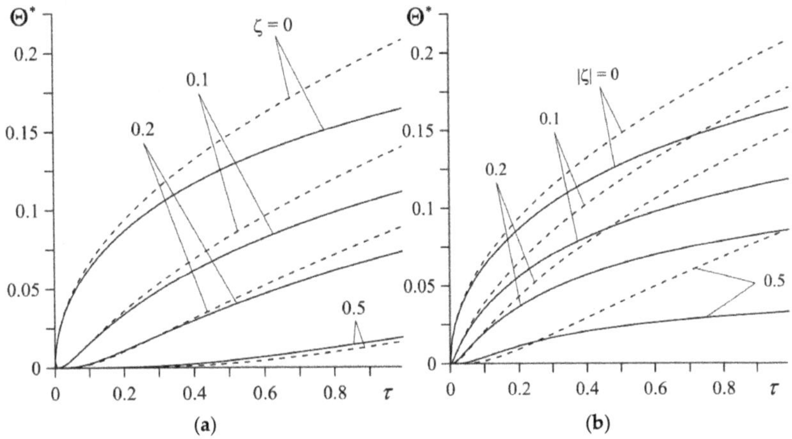

Figure 3. Evolution of the dimensionless temperature $\Theta^*(\zeta,\tau)$ during sliding at different distances from the friction surface. Continuous curves represent FGMs: (**a**) ZrO_2–Ti-6Al-4V; (**b**) Al_3O_2–TiC and the dashed curves represent homogeneous materials: (**a**) ZrO_2; (**b**) Al_3O_2.

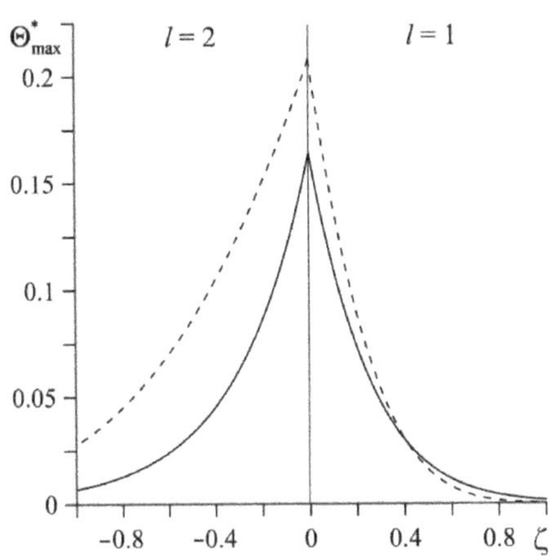

Figure 4. Distribution of the dimensionless maximum temperature rise Θ^*_{max} reached at the end of friction process, along the distance ζ from the friction surface. Continuous curves represent FGMs ZrO_2–Ti-6Al-4V (l = 1) and Al_3O_2–TiC (l = 2); the dashed curves represent homogeneous materials ZrO_2 (l = 1) and Al_3O_2 (l = 2).

The time profiles of dimensionless temperature rise Θ^* on the contact surface $\zeta = 0$ for different values of the parameter $\gamma_l^*, l = 1, 2$, are demonstrated in Figure 5. At a certain moment of time, the temperature of the friction surface increases with a decrease in the material gradient parameter (the continuous curves), approaching the temperature values obtained for a friction pair made of homogeneous materials (the dashed curves). The differences between the individual curves obtained for FGMs and homogeneous materials grow with the time of sliding.

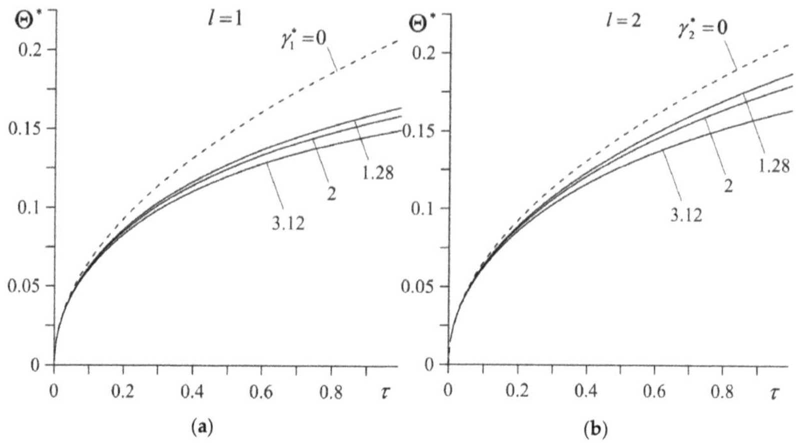

Figure 5. Evolutions of dimensionless temperature rise Θ^* on the contact surface of the friction pair for various values of parameter: (**a**) γ_1^* for $\gamma_2^* = 3.12$; (**b**) γ_2^* for $\gamma_1^* = 1.28$. Continuous curves represent FGMs ZrO_2–Ti-6Al-4V (l = 1) and Al_3O_2–TiC (l = 2), the dashed curves represent homogeneous materials ZrO_2 (l = 1) and Al_3O_2 (l = 2).

The influence of dimensionless gradients of materials $\gamma_l^*, l = 1,2$ (69) on the dimensionless maximum temperature $\Theta_{\max}^*$ of the contact surface is shown in Figure 6 (the continuous curves). The dashed lines in this figure present the corresponding data calculated on the basis of the solution found for the homogeneous materials. The highest values $\Theta_{\max}^*$ are achieved for the elements of the friction pair made of the homogeneous ceramic materials. Increasing the gradient parameters of the material reduces the maximum temperature of the friction system. The highest reduction of $\Theta_{\max}^*$ takes place while increasing the parameter γ_1^* in the element made of ZrO_2–Ti-6Al-4V, while the parameter γ_2^* value of the element Al_3O_2–TiC remains constant (Figure 6a).

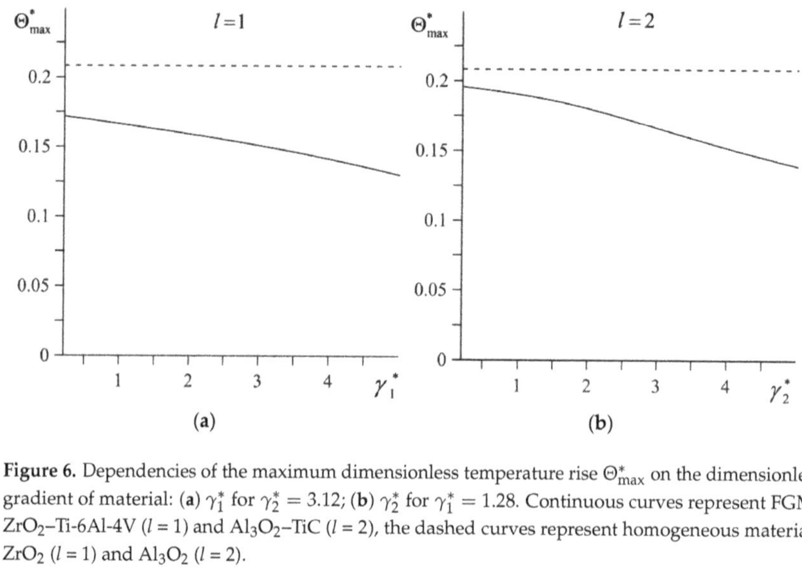

Figure 6. Dependencies of the maximum dimensionless temperature rise $\Theta_{\max}^*$ on the dimensionless gradient of material: (**a**) γ_1^* for $\gamma_2^* = 3.12$; (**b**) γ_2^* for $\gamma_1^* = 1.28$. Continuous curves represent FGMs ZrO_2–Ti-6Al-4V (l = 1) and Al_3O_2–TiC (l = 2), the dashed curves represent homogeneous materials ZrO_2 (l = 1) and Al_3O_2 (l = 2).

Distributions of dimensionless temperature Θ^* at the end of the sliding, along the distance $0 \leq |\zeta| \leq 1$ from the friction surface is demonstrated in Figure 7. The highest distance value $|\zeta| = 1$ corresponds to the previously established maximum effective thickness of heating $a = 7$ mm. As agreed, so far, on the contact surface $\zeta = 0$, the temperature of the friction pair made of FGMs is lower than that in the case of the tribocouple with homogeneous materials. Increasing the distance from the contact surface reduces the temperature in both cases, the system made of FGMs (continuous curves) and the friction pair made of ceramic homogeneous materials (dashed curves).

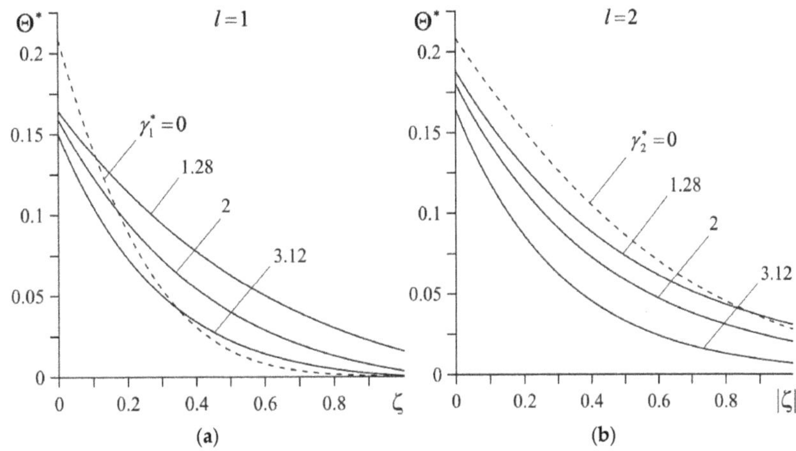

Figure 7. Dependencies of dimensionless temperature rise Θ^* at the end of heating, on the dimensionless distance $|\zeta|$ from the contact surface for different values of parameters: (**a**) γ_1^* for $\gamma_2^* = 3.12$; (**b**) γ_2^* for $\gamma_1^* = 1.28$. Continuous curves represent FGMs ZrO_2–Ti-6Al-4V (l = 1) and Al_3O_2–TiC (l = 2) and the dashed curves represent homogeneous materials ZrO_2 (l = 1) and Al_3O_2 (l = 2).

The temperature in the first element ($l = 1$) decreases faster than that in the case of the homogeneous material ZrO_2, while in the second element ($l = 2$), the temperature of the homogeneous material Al_3O_2 remains higher than the temperature of the element made of FGM Al_3O_2–TiC, throughout the whole effective thickness. At a certain distance from the friction surface, increasing the material gradient parameters (enhancement of the volume fraction of the core material in the composite structure) causes a drop of the temperature in both FGMs used.

6. Conclusions

According to the obtained solutions, the numerical analysis of the temperature mode was performed for friction pair elements made of functionally graded materials, under uniform sliding. The friction surfaces of these elements are ceramic materials, i.e., zirconium dioxide ZrO_2 and aluminum oxide Al_3O_2. The volume fraction of ceramics in the materials decreases with the depth, in favor of the core materials.

The composites are the high-class titanium-aluminum-vanadium alloy Ti-6Al-4V and titanium carbide TiC, with higher thermal conductivities than ceramics. On the basis of the results of the calculations, the influences of the values of the friction material gradient parameters on the time-space temperature distributions in the tribological system were investigated. The obtained data show that the use of selected composites with a continuous (exponential) change of thermal conductivity, improves the friction conditions, causing a significant decrease in the temperature level reached on the friction surface, especially the maximum value at the end of the sliding.

Despite its purely theoretical importance, the determined analytical solution also has practical significance. On the basis of this closed-formed expression, it is possible to quickly estimate the temperature mode of a friction system made of FGMs with an exponential gradient under uniform sliding. Furthermore, the exact solutions play the role of a template for testing the approximate numerical methods. It should be noted that the solution of formulated thermal problem of friction was obtained assuming an exponential change in thermal conductivity. Thus, the developed model is oriented only to the FGM class with just such a gradient. In this sense, it is a natural limitation of the solution. Other application limitations of this model (unidirectional heating process, ideal thermal contact of bodies, etc.) are presented in the assumptions.

In the next report, we plan to present the results concerning a study of the impact of functional gradient structure of friction materials on the temperature in a disc brake system during braking.

Author Contributions: Conceptualization and methodology, A.Y.; software, K.T. and P.Z.; validation, P.Z.; formal analysis, K.T. and P.Z.; investigation, A.Y., K.T. and P.Z.; writing—original draft preparation, K.T. and P.Z.; writing—review and editing, A.Y., K.T. and P.Z.; visualization and figures preparation, K.T. and P.Z.; supervision, A.Y.; project administration, K.T.; funding acquisition, K.T. All authors have read and agreed to the published version of the manuscript.

Funding: This study was performed within the framework of research project no. 2017/27/B/ST8/01249, funded by the National Science Centre, Poland and project financing through the program of the Minister of Education and Science of Poland named "Regional Initiative of Excellence" in 2019–2022, project no. 011/RID/2018/19, amount financing 12,000,000 PLN.

Institutional Review Board Statement: Not applicable.

Informed Consent Statement: Not applicable.

Data Availability Statement: No new data were created or analyzed in this study. Data sharing is not applicable to this article.

Conflicts of Interest: The authors declare no conflict of interest.

Nomenclature

a	Effective depth of heat penetration (m)
c_l	Specific heat (Jkg^{-1}K^{-1})
f	Coefficient of friction (dimensionless)
$I_k(\cdot)$	The modified Bessel functions of the first kind of the kth order
$J_k(\cdot)$	The Bessel functions of the first kind of the kth order
k_l	Thermal diffusivity (m^{2}s^{-1})
K_l	Thermal conductivity (Wm^{-1}K^{-1})
p	Parameter of the Laplace transform (dimensionless)
p_0	Contact pressure (Pa)
q_l	Intensity of the frictional heat flux (Wm^{-2})
q_0	Specific power of friction (Wm^{-2})
t	Time (s)
T	Temperature (°C)
T_a	Initial (ambient) temperature (°C)
V_0	Sliding velocity (ms^{-1})
x, y, z	Spatial coordinates (m)

Glossary

γ_l	Parameter of material gradient (m^{-1})
γ_l^*	Parameter of material gradient (dimensionless)
ζ	Thickness (dimensionless)
Θ_l	Temperature rise (°C)
Θ_l^*	Temperature rise (dimensionless)
Θ_0	Temperature scaling factor (°C)
ρ_l	Density (kgm^{-3})
τ	Time (dimensionless)
lower l	Number of the main ($l = 1$) and frictional ($l = 2$) elements of the friction pair

References

1. Udupa, G.; Rao, S.S.; Gangadharan, K.V. Functionally graded composite materials: An overview. *Procedia Mater. Sci.* **2014**, *5*, 1291–1299. [CrossRef]
2. Mahamood, R.M.; Akinlabi, E.T. *Functionally Graded Materials*; Springer: Cham, Switzerland, 2017.
3. Birman, V.; Byrd, L.W. Modelling and analysis of functionally graded materials and structures. *Appl. Mech. Rev.* **2007**, *60*, 195–216. [CrossRef]
4. Liu, J.; Ke, L.L.; Wang, Y.S. Two-dimensional thermoelastic contact problem of functionally graded materials involving frictional heating. *Int. J. Solids Struct.* **2011**, *48*, 2536–2548. [CrossRef]
5. Hosseini, T.P.; Talebi, M. Stress and temperature distribution study in a functionally graded brake disk. *Int. J. Automot. Eng.* **2012**, *2*, 172–179.
6. Jin, Z.H. An asymptotic solution of temperature field in a strip of a functionally graded material. *Int. Commun. Heat Mass Transf.* **2002**, *29*, 887–895. [CrossRef]
7. Liu, J.; Ke, L.L.; Wang, Y.S.; Yang, J.; Alam, F. Thermoelastic frictional contact of functionally graded materials with arbitrarily varying properties. *Int. J. Mech. Sci.* **2012**, *63*, 86–98. [CrossRef]
8. Mao, J.J.; Ke, L.L.; Yang, J.; Kitipornchai, S.; Wang, Y.S. The coupled thermoelastic instability of FGM coatings with arbitrarily varying properties: In-plane sliding. *Acta Mech.* **2018**, *229*, 2979–2995. [CrossRef]
9. Balci, M.N.; Dag, S.; Yildirim, B. Subsurface stresses in graded coatings subjected to frictional contact with heat generation. *J. Therm. Stresses* **2017**, *40*, 517–534. [CrossRef]
10. Yaghoobi, M.P.; Ghannad, M. An analytical solution for heat conduction of FGM cylinders with varying thickness subjected to non-uniform heat flux using a first-order temperature theory and perturbation technique. *Int. Commun. Heat Mass Transf.* **2020**, *116*, 104684. [CrossRef]
11. Noda, N. Thermal stresses in functionally graded materials. *J. Therm. Stresses* **1999**, *22*, 477–512. [CrossRef]
12. Govindaraju, M.; Megalingam, A.; Murugasan, J.; Vignesh, V.R.; Kota, K.P.; Ram, S.A.; Lakshana, P.; Kumar, N.V. Investigations on the tribological behavior of functionally gradient iron-based brake pad material. *Proc. Inst. Mech. Eng. Part C J. Mech. Eng. Sci.* **2020**, *234*, 2474–2486. [CrossRef]
13. Shahzamanian, M.M.; Sahari, B.B.; Bayat, M.; Ismarrubie, Z.N.; Mustapha, F. Transient and thermal contact analysis for the elastic behaviour of functionally graded brake disks due to mechanical and thermal loads. *Mater. Des.* **2010**, *31*, 4655–4665. [CrossRef]
14. Shahzamanian, M.M.; Sahari, B.B.; Bayat, M.; Mustapha, F.; Ismarrubie, Z.N. Finite element analysis of thermoelastic contact problem in functionally graded axisymmetric brake disks. *Compos. Struct.* **2010**, *92*, 1591–1602. [CrossRef]
15. Sladek, J.; Sladek, V.; Zhang, C. Transient heat conduction analysis in functionally graded materials by the meshless local boundary integral equation method. *Comput. Mater. Sci.* **2003**, *28*, 494–504. [CrossRef]
16. Sutradhar, A.; Paulino, G.H.; Gray, L.J. Transient heat conduction in homogeneous and non-homogeneous materials by the Laplace transform Galerkin boundary element method. *Eng. Anal. Bound. Elem.* **2002**, *26*, 119–132. [CrossRef]
17. Tian, J.H.; Jiang, K. Heat conduction investigation of the functionally graded materials plates with variable gradient parameters under exponential heat source load. *Int. J. Heat Mass Trans.* **2018**, *122*, 22–30. [CrossRef]
18. Jabbari, M.; Sohrabpour, S.; Eslami, M.R. Mechanical and thermal stresses in a functionally graded hollow cylinder due to radially symmetric loads. *Int. J. Press. Vessel. Pip.* **2002**, *79*, 493–497. [CrossRef]
19. Tanigawa, Y.; Akai, T.; Kawamura, R.; Oka, N. Transient heat conduction and thermal stress problems of a nonhomogeneous plate with temperature-dependent material properties. *J. Therm. Stresses* **1996**, *19*, 77–102. [CrossRef]
20. Lee, S.W.; Jang, Y.H. Frictionally excited thermoelastic instability in a thin layer of functionally graded material sliding between two half-planes. *Wear* **2009**, *267*, 1715–1722. [CrossRef]
21. Hosseini, S.M.; Akhlaghi, M.; Shakeri, M. Transient heat conduction in functionally graded thick hollow cylinders by analytical method. *Heat Mass Transf.* **2007**, *43*, 669–675. [CrossRef]
22. Obata, Y.; Noda, N. Unsteady Thermal Stresses in a Functionally Gradient Material Plates Analysis of One-Dimensional Unsteady Heat Transfer Problem. *Jpn. Soc. Mech. Eng. Ser. A* **1993**, *59*, 1090–1096. [CrossRef]
23. Obata, Y.; Noda, N. Steady thermal stresses in a hollow circular cylinder and a hollow sphere of a functionally gradient material. *J. Therm. Stresses* **1994**, *17*, 471–487. [CrossRef]

24. Kim, K.S.; Noda, N. Green's function approach to unsteady thermal stresses in an infinite hollow cylinder of functionally graded material. *Acta Mech.* **2002**, *156*, 145–161. [CrossRef]
25. Perkowski, D.M. On axisymmetric heat conduction problem for FGM layer on homogeneous substrate. *Int. Commun. Heat Mass Transf.* **2014**, *57*, 157–162. [CrossRef]
26. Kulchytsky-Zhyhailo, R.; Bajkowski, A.S. Axisymmetrical problem of thermoelasticity for half-space with gradient coating. *Int. J. Mech. Sci.* **2016**, *106*, 62–71. [CrossRef]
27. Mao, J.J.; Ke, L.L.; Yang, J.; Kitipornchai, S.; Wang, Y.S. Thermoelastic instability of functionally graded coating with arbitrarily varying properties considering contact resistance and frictional heat. *Appl. Math. Model.* **2017**, *43*, 521–537. [CrossRef]
28. Ke, L.L.; Wang, Y.S. Two-dimensional sliding frictional contact of functionally graded materials. *Eur. J. Mech. A Solids* **2007**, *26*, 171–188. [CrossRef]
29. Lee, S.W.; Jang, Y.H. Effect of functionally graded material on frictionally excited thermoelastic instability. *Wear* **2009**, *266*, 139–146. [CrossRef]
30. Polajnar, M.; Kalin, M.; Thorbjornsson, I.; Thorgrimsson, J.T.; Valle, N.; Botor-Probierz, A. Friction and wear performance of functionally graded ductile iron for brake pads. *Wear* **2017**, *382*, 85–94. [CrossRef]
31. Soundararajan, R.; Karthik, S.; Sivaraman, S. Automotive Brake Pad by Using, Functionally Graded Hybrid Composites and Their Behaviour. *Int. J. Mech. Eng. Technol.* **2018**, *9*, 318–328.
32. Ling, F.F. *Surface Mechanics*; John Wiley & Sons: New York, NY, USA, 1973.
33. Yevtushenko, A.A.; Kuciej, M. One-dimensional thermal problem of friction during braking: The history of development and actual state. *Int. J. Heat Mass Transf.* **2012**, *55*, 4148–4153. [CrossRef]
34. Suresh, S.; Mortensen, A. *Fundamentals of Functionally Graded Materials*; The Institute of Materials, IOM Communications Ltd.: London, UK, 1998.
35. Sneddon, I.N. *The Use of Integral Transforms*; McGraw-Hill: New York, NY, USA, 1972.
36. Abramowitz, M.; Stegun, I. *Handbook of Mathematical Functions with Formulas, Graphs, and Mathematical Tables*; United States Department of Commerce, National Bureau of Standards (NBS): Washington, DC, USA, 1964.
37. Vashchenko-Zakharchenko, M.E. *Symbolic Calculus and Its Application to Integration of Linear Differential Equations*; University Press: Kiev, Ukraine, 1862.
38. Luikov, A.V. *Analitycal Heat Diffusion Theory*; Academic Press: New York, NT, USA, 1968.
39. Chichinadze, A.V. *Polymers in Friction Assembles of Machines and Devices: A Handbook*; Allerton Press Inc.: New York, NY, USA, 1984.
40. Bateman, H.; Erdelyi, A. *Tables of Integrals Transforms*; McGraw-Hill: New York, NY, USA, 1954; Volume 1.
41. Carslaw, H.S.; Jaeger, J.C. *Conduction of Heat in Solids*; Oxford University Press: Oxford, UK, 1959.
42. Chen, L.; Kny, E. Reaction hot-pressed sub-micro Al_2O_3+TiC ceramic composite. *Int. J. Refract. Met. Hard Mater.* **2000**, *18*, 163–167. [CrossRef]
43. Press, W.; Flannery, B.P.; Teukolsky, S.A.; Vetterling, W.T. *Numerical Recipes*; Cambridge University Press: Cambridge, UK, 2007.

materials

Article

The Effect of Functionally Graded Materials on Temperature during Frictional Heating at Single Braking

Aleksander Yevtushenko, Katarzyna Topczewska * and Przemysław Zamojski

Faculty of Mechanical Engineering, Bialystok University of Technology (BUT), 45C Wiejska Street, 15-351 Bialystok, Poland; a.yevtushenko@pb.edu.pl (A.Y.); zamojski.przemyslaw@gmail.com (P.Z.)
* Correspondence: k.topczewska@pb.edu.pl

Abstract: A mathematical model for evaluation of the temperature mode of the disc–pad system during single braking is proposed. The model is based on the thermal problem of friction formulated for two semi-infinite bodies, compressed with pressure increasing over time while reducing the sliding velocity from the initial value to zero at the stop. The exact solution to this problem was obtained by means of Duhamel's theorem. Validation of the solution was performed by achieving in special cases parameters of known solution to this problem with constant pressure and velocity (under uniform sliding). The results of the numerical calculations are presented for a selected friction pair, made of functionally graded materials with titanium alloy (disc) and aluminum alloy (pad) cores coated with ceramics graded toward friction surfaces. For the established values of the parameters such as the rise time in pressure and the FGM gradients, the ability to quickly obtain spatiotemporal temperature distributions in the disc and pad was presented. The influence of the variability of these parameters on the maximum temperature of the brake system was also investigated.

Keywords: functionally graded materials; braking; frictional heating; temperature

Citation: Yevtushenko, A.; Topczewska, K.; Zamojski, P. The Effect of Functionally Graded Materials on Temperature during Frictional Heating at Single Braking. *Materials* **2021**, *14*, 6241. https://doi.org/10.3390/ma14216241

Academic Editor: Liaoliang Ke

Received: 24 September 2021
Accepted: 19 October 2021
Published: 20 October 2021

1. Introduction

Friction elements of braking systems are subjected to severe conditions such as high temperature and intensive wear. During braking, performance of these components in terms of efficiency, service life, and dissipation of heat from the contact surface depends on the operating conditions and material properties. It has been shown that the operating characteristics of the entire assembly of the braking system can be significantly improved by introducing a smooth gradient in the microstructure of the friction materials [1]. Such functionally graded materials (FGMs) are a class of heterogeneous materials with continuous variation of properties over their volume. Generally, these materials are composites formed by smooth gradation of two or more constituent phases along certain dimensions of a structure. This gradation can be regulated by changing the volume fraction distribution of component of material from one to another in a controlled manner [2]. As a result, the thermophysical properties of material continuously vary as a function of position along a certain direction. This allows designing a functionally graded material in order to obtain optimized friction characteristics of a brake.

In general, statements of thermal problems of friction contain partial differential equations with variable coefficients. Therefore, the application of analytical methods to their solution is difficult or even impossible. Hence, numerical methods are often used to consider such problems. An FGM disc subjected to thermal load due to frictional heating while taking into account the inertial force due to the rotation of the disc was studied by Afsar and Go [3]. A 2D finite element analysis (FEA) for a circular disc with exponential variations in thermophysical properties in the radial direction was performed. An axisymmetric FEA of a brake disc, with properties distributed according to the power-law function of radial position, was executed by Shahzamanian et al. [4,5]. It was found that the gradation index of the disc material has a crucial influence on the thermomechanical behavior of the entire braking system.

A finite element thermal contact analysis of a functionally graded disc under dry friction was performed by Hosseini and Talebi [6]. The core part of the considered disc was steel, and it gradually changed through the thickness of the disc, according to a power law, approaching pure ceramic at the outer surface. It was shown that the temperature and the corresponding thermal displacements in the FGM disc are much lower than in the conventional steel disc. Furthermore, it was established that the use of an FGM brake disc may eliminate thermal cracking and wear. In particular, functionally graded materials composed of ceramic and metal perform very well in contact problems involving friction, since they combine the advantages of both components [7]. These elements mostly have a metal core in order to maintain strength and rigidity, whereas ceramic is present on the outer surfaces to resist intensive wear and elevated temperature conditions.

Separately, the study of the phenomenon of thermoelastic instability (TEI) of brake systems with FGM should be mentioned. The solution of a 2D thermal contact problem of friction for a functionally graded cermet brake disc was obtained by means of FEA [8]. They investigated TEI caused by the coupled interaction of the mechanical and thermal loads in the sliding system. Generally, this leads to the establishment of localized high-temperature zones on the contact surface, known as hot spots, which are directly attributable to the premature failure of the friction system. This instability is often called the frictionally excited TEI and occurs in tribosystems when the sliding velocity exceeds a certain critical value. It was shown that the value of the critical velocity for a functionally graded brake disc is higher than that for a conventional homogeneous disc. This conclusion was confirmed by further research investigating TEI in an FGM strip sliding uniformly against two homogeneous semi-spaces [9,10]. Assuming an exponential variation of the thermophysical properties along the thickness of an FG strip permitted obtaining an exact solution using the analytical perturbation method. Using the same methodology, the TEI of the brake modeled as an FGM semi-infinite body sliding against a homogeneous semi-space under uniform pressure taking into account the frictional heating and thermal contact resistance was investigated by Mao et al. [11]. As a result, they determined the stability boundaries of thermoelastic instability in the considered sliding system. The effect of the arbitrarily varying thermoelastic properties of the FGM on the TEI was considered by Mao et al. [12]. To simulate the distribution of the FGM properties, a homogeneous multilayered model was employed. This approach is a replacement of the continuous FGM material with a package of homogeneous layers with constant properties. The gradient was simulated by assigning different properties values to each sublayer. It was proven that the results received for the FGM strip, divided into a sufficient number of layers, were close to the results found using the corresponding exact solutions [13]. It should be noted that this conclusion is dependent on the problem under consideration, and the differences between the obtained results may be significant in some cases [14]. This is particularly true for FGM with temperature-dependent properties. A multilayered model was used in [15] in order to establish the coupled effect of the frictional heat and the thermal contact resistance. Since the homogeneous multilayered model deals with the arbitrarily varying properties of FGM, the power-law, exponential, sinusoidal, and cosinusoidal distributions of the brake disc properties were considered. The perturbation and transfer matrix methods were used to deduce the characteristic equation of the TEI problem, to obtain the relationship between the critical sliding velocity and the critical heat flux [15]. The formulated conclusions confirm that the application of ceramic-based FGM in a brake disc, consisting of ceramics at the sliding interface and steel in the middle layer, reduces the susceptibility of braking system toward TEI [9–11,15].

However, FGMs are increasingly finding applications in braking systems, in the fabrication of not only discs, but also brake pads. Experimental investigations revealed that FGMs could successfully fulfill the demands for brake pads and improve their characteristics [1,16]. The novel functionally graded ductile iron for brake pads was investigated in a well-controlled model sliding test and a disc-brake machine in [16]. The results of the tribological tests revealed a positive effect of the functional gradient of properties on the wear of the pads and the improved stability of the friction coefficient. Govindaraju et al. [1] developed and investigated Fe-based material on brake pads with graded composition. The FG specimens were subjected to a dry sliding test for studying their tribological behavior. The results were compared with the conventional brake pad specimen. It was found that the wear resistance of the functionally gradient specimen is much greater compared to the conventional pad material [1].

We note that a more comprehensive review of the literature on thermoelastic contact problems with frictional heating for functionally graded materials was provided in our previous article [17]. This article is a continuation of the research cycle started in [17], in which the case of the uniform sliding of an FGM tribosystem was considered. The present article concerns the transient thermal problem of friction during braking, which takes into account the time-dependent specific friction power due to the exponential increase in contact pressure.

2. Statement to the Problem

The frictional heating in a brake disc system during a single braking process is considered. Frictional elements of the system are two identical pads, located symmetrically to the brake disc. At the initial time moment $t = 0$, pads are pressed to the friction surfaces of the disc with uniformly distributed on the contact area and time-dependent pressure [18].

$$p(t) = p_0 p^*(t), \quad p^*(t) = 1 - e^{-t/t_i}, \quad 0 \le t \le t_s, \tag{1}$$

where $t_i \ge 0$ is the rise time in contact pressure from zero to the nominal value p_0, and t_s is the time of stop. Due to the interaction of friction forces, the linear velocity of vehicle V is reduced from the initial value $V_0 \equiv V(0)$ to zero at the stop time moment $t = t_s$ according to the following law [19,20]:

$$V(t) = V_0 V^*(t),\ V^*(t) = 1 - \frac{t}{t_s^0} + \frac{t_i}{t_s^0} p^*(t),\ t_s^0 = \frac{W_0}{f p_0 A_a V_0},\ 0 \le t \le t_s, \tag{2}$$

where W_0 is the initial kinetic energy of the system, f is the friction coefficient, A_a is the nominal contact area between the pad and disc, and t_s^0 is the braking time with constant deceleration ($t_i \to 0$). The braking time, taking into account the temporal profile of the velocity (Equation (2)), is determined from the stop condition $V^*(t_s) = 0$. For $0 < t_i \le 0.3 t_s^0$ it was established [19] that $t_s \cong t_s^0 + 0.99 t_i$.

The sliding velocity reduction during braking is accompanied by the generation of frictional heat on the contact surface of the friction pair. In order to determine the temperature field generated in this way, the corresponding thermal problem of friction is formulated on the basis of the following assumptions:

1. The materials of the pads and the disc are functionally graded with an exponential decrease in thermal conductivity along their thickness, with invariant specific heat and density;
2. The initial temperature of all elements is the same and equal to the ambient temperature T_a;
3. The whole work of friction goes to heating the bodies, while the wear of the friction surfaces is neglected;
4. The free surfaces of the pads and the disc are adiabatic;
5. The thermal and mechanical properties and coefficient of friction are independent of the temperature T;
6. Only the change in the temperature gradient in the direction perpendicular to the friction surface is taken into account;
7. The thermal contact of friction between the pads and the disc is perfect; the temperatures of their friction surfaces during braking are the same, and the sum of the

intensity of the heat fluxes directed to both elements along the normal to the contact surface is equal to the specific friction power:

$$q(t) = q_0 q^*(t),\ q_0 = f p_0 V_0,\ q^*(t) = p^*(t) V^*(t),\ 0 \le t \le t_s, \tag{3}$$

where the temporal profiles of pressure $p^*(t)$ and velocity $V^*(t)$ have the forms expressed in Equations (1) and (2), respectively;

8. Due to the symmetry with respect to the center plane of the disc, to establish the temperature of the braking system, it is sufficient to consider the contact scheme of one pad with a disc of half of its thickness.

With such assumptions, a contact scheme of two sliding semi-infinite bodies (semi-spaces) related to the Cartesian system $0xyz$ (Figure 1) was adopted to describe the process of frictional heating in the disc–pad system.

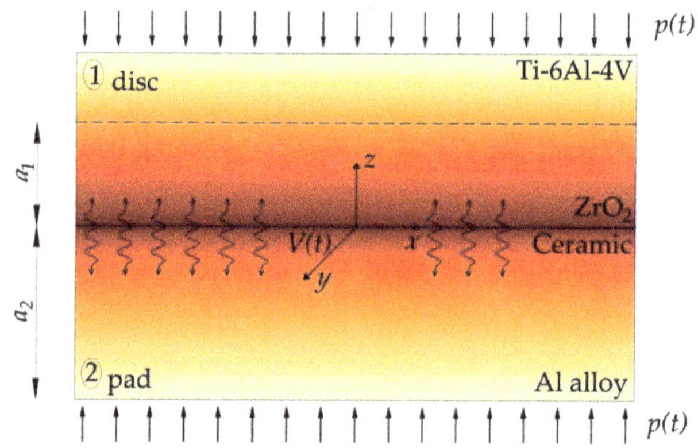

Figure 1. Scheme of the frictional heating in the disc–pad system.

The temperature rise $\Theta = T - T_a$ was determined from solution to the following one-dimensional boundary value problem of heat conduction taking into account the generation of heat due to friction:

$$\frac{\partial}{\partial z}\left[K_1(z)\frac{\partial \Theta(z,t)}{\partial z}\right] = c_1\rho_1 \frac{\partial \Theta(z,t)}{\partial t}, z > 0,\ 0 < t \le t_s, \tag{4}$$

$$\frac{\partial}{\partial z}\left[K_2(z)\frac{\partial \Theta(z,t)}{\partial z}\right] = c_2\rho_2 \frac{\partial \Theta(z,t)}{\partial t}, z < 0,\ 0 < t \le t_s, \tag{5}$$

$$K_2(z)\frac{\partial \Theta(z,t)}{\partial z}\bigg|_{z=0^-} - K_1(z)\frac{\partial \Theta(z,t)}{\partial z}\bigg|_{z=0^+} = q(t), 0 < t \le t_s, \tag{6}$$

$$\Theta(0^-,t) = \Theta(0^+,t), 0 < t \le t_s, \tag{7}$$

$$\Theta(z,t) \to 0\ ,\ |z| \to \infty, 0 < t \le t_s, \tag{8}$$

$$\Theta(z,0) = 0, |z| < \infty, \tag{9}$$

where

$$K_l(z) = K_{l,0} e^{\gamma_l |z|}, |z| < \infty, K_{l,0} \equiv K_l(0), \gamma_l \ge 0,\ l = 1,\ 2, \tag{10}$$

and function $q(t)$ has the form expressed in Equation (3). Here and further, the subscript l indicates the parameters and quantities related to the certain element—$l = 1$ for the disc, and $l = 2$ for the pad. Taking into account the relations in Equation (10), the problem in Equations (6)–(9) was written in the following form:

$$\frac{\partial^2\Theta(z,t)}{\partial z^2} + \gamma_1 \frac{\partial\Theta(z,t)}{\partial z} = \frac{e^{-\gamma_1 z}}{k_{1,0}} \frac{\partial\Theta(z,t)}{\partial t}, z > 0,\ 0 < t \le t_s, \quad (11)$$

$$\frac{\partial^2\Theta(z,t)}{\partial z^2} - \gamma_2 \frac{\partial\Theta(z,t)}{\partial z} = \frac{e^{\gamma_2 z}}{k_{2,0}} \frac{\partial\Theta(z,t)}{\partial t}, z < 0,\ 0 < t \le t_s, \quad (12)$$

$$K_{2,0} \left.\frac{\partial\Theta(z,t)}{\partial z}\right|_{z=0^-} - K_{1,0} \left.\frac{\partial\Theta(z,t)}{\partial z}\right|_{z=0^+} = q(t),\ 0 < t \le t_s, \quad (13)$$

$$\Theta(0^-,t) = \Theta(0^+,t),\ 0 < t \le t_s, \quad (14)$$

$$\Theta(z,t) \to 0\ ,\ |z| \to \infty, 0 < t \le t_s, \quad (15)$$

$$\Theta(z,0) = 0, |z| < \infty, \quad (16)$$

where

$$k_{l,0} = \frac{K_{l,0}}{c_l \rho_l},\ l = 1,\ 2 \quad (17)$$

are the coefficients of thermal diffusivity of the materials on their contact surfaces; $z = 0$.

3. Solution to the Problem

In the case of a uniform slip with a constant specific power of friction $q(t) = q_0, t \ge 0$, the solution to the problem in Equations (11)–(16) can be written in the following form [17]:

$$\hat{\Theta}(z,t) = \Lambda e^{-\gamma_1 z/2} \left[\frac{e^{-\gamma_1 z/2}}{(1+\gamma_\varepsilon K_\varepsilon)} + \frac{4}{\gamma_\varepsilon} \sum_{n=1}^{\infty} \frac{\phi_1(z,\mu_n)}{\Psi(\mu_n)} e^{-p_n t} \right], z \ge 0, t \ge 0, \quad (18)$$

$$\hat{\Theta}(z,t) = \Lambda e^{\gamma_2 z/2} \left[\frac{e^{\gamma_2 z/2}}{(1+\gamma_\varepsilon K_\varepsilon)} + \frac{4}{\gamma_\varepsilon} \sum_{n=1}^{\infty} \frac{\phi_{2,n}(z,)}{\Psi(\mu_n)} e^{-p_n t} \right], z \le 0, t \ge 0, \quad (19)$$

where

$$\phi_1(z,\mu_n) = J_1(\mu_n) J_1(\gamma_\varepsilon \mu_n e^{-\gamma_1 z/2}),\ \phi_2(z,\mu_n) = J_1(\mu_n) J_1(\gamma_\varepsilon \mu_n e^{\gamma_2 z/2}), \quad (20)$$

$$\Psi(\mu_n) = \mu_n^2 [(1+\gamma_\varepsilon K_\varepsilon) J_0(\mu_n) J_0(\gamma_\varepsilon \mu_n) - (\gamma_\varepsilon + K_\varepsilon) J_1(\mu_n) J_1(\gamma_\varepsilon \mu_n)], \quad (21)$$

$$K_\varepsilon = \frac{K_0^*}{\sqrt{k_0^*}},\ \gamma_\varepsilon = \gamma^* \sqrt{k_0^*},\ K_0^* = \frac{K_{1,0}}{K_{2,0}},\ k_0^* = \frac{k_{1,0}}{k_{2,0}},\ \gamma^* = \frac{\gamma_1}{\gamma_2},\ \Lambda = \frac{q_0}{\gamma_2 K_{2,0}}, \quad (22)$$

$$p_n = 0.25 k_{1,0} \gamma_1^2 \mu_n^2, \quad (23)$$

$\mu_n > 0,\ n = 1,2,3,\ldots$, are the real roots of the following functional equation:

$$J_0(\gamma_\varepsilon \mu) J_1(\mu) + K_\varepsilon J_0(\mu) J_1(\gamma_\varepsilon \mu) = 0, \quad (24)$$

where $J_k(x)$ denotes the Bessel functions of the first kind of the k-th order [21].

The temperature rise $\Theta(z,t)$ corresponding to the specific friction power $q(t)$ in Equation (3) is searched on the basis of Duhamel's formula [22].

$$\Theta(z,t) = \frac{\partial}{\partial t} \int_0^t q^*(t-s) \hat{\Theta}(z,s) ds,\ 0 < t \le t_s, \quad (25)$$

where $\hat{\Theta}(z,t)$ is the temperature rise in Equations (18)–(24) for constant specific friction power $q(t) = q_0$. Taking into account the solutions to Equations (18) and (19) in Duhamel's integral (Equation (25)), it was achieved that

$$\Theta(z,t) = \Lambda e^{-\gamma_1 z/2}\left[\frac{e^{-\gamma_1 z/2}}{(1+\gamma_\varepsilon K_\varepsilon)}q^*(t) + \frac{4}{\gamma_\varepsilon}\sum_{n=1}^{\infty}\frac{\phi_1(z,\mu_n)}{\Psi(\mu_n)}G'_n(t)\right], z \geq 0, 0 \leq t \leq t_s, \quad (26)$$

$$\Theta(z,t) = \Lambda e^{\gamma_2 z/2}\left[\frac{e^{\gamma_2 z/2}}{(1+\gamma_\varepsilon K_\varepsilon)}q^*(t) + \frac{4}{\gamma_\varepsilon}\sum_{n=1}^{\infty}\frac{\phi_2(z,\mu_n)}{\Psi(\mu_n)}G'_n(t)\right], z \leq 0, 0 \leq t \leq t_s, \quad (27)$$

where $G'_n(t)$ is a derivative of the function $G_n(t)$, which is determined as

$$G_n(t) = \int_0^t q^*(t-s)e^{-p_n t}dt, n = 1,2,3,\ldots \quad (28)$$

Substituting the temporal profile of the specific power of friction $q^*(t)$ in Equation (3) into Equation (28) yielded the following equation:

$$G_n(t) = G_{n,1}(t) - \frac{1}{t_s^0}G_{n,2}(t) + \frac{t_i}{t_s^0}G_{n,3}(t), n = 1,2,3,\ldots, \quad (29)$$

where

$$G_{n,1}(t) = \int_0^t p^*(t-s)e^{-p_n t}dt,\ G_{n,2}(t) = \int_0^t (t-s)p^*(t-s)e^{-p_n t}dt,$$
$$G_{n,3}(t) = \int_0^t [p^*(t-s)]^2 e^{-p_n t}dt. \quad (30)$$

The calculations of integrals in Equation (30) taking into account the time profile of contact pressure $p^*(t)$ (1), give

$$G_{n,1}(t) = p_n^{-1}(1-e^{-p_n t}) + a_n^{-1}(e^{-p_n t} - e^{-t/t_i}), \quad (31)$$

$$G_{n,2}(t) = t(p_n^{-1} - a_n^{-1}e^{-t/t_i}) - p_n^{-2}(1-e^{-p_n t}) - a_n^{-2}(e^{-p_n t} - e^{-t/t_i}), \quad (32)$$

$$G_{n,3}(t) = p_n^{-1}(1-e^{-p_n t}) + 2a_n^{-1}(e^{-p_n t} - e^{-t/t_i}) - b_n^{-1}(e^{-p_n t} - e^{-2t/t_i}), \quad (33)$$

where

$$a_n = p_n - t_i^{-1} \neq 0,\ b_n = p_n - 2t_i^{-1} \neq 0 n = 1,2,3,\ldots \quad (34)$$

If for any $n = k,\ k = 1,2,\ldots$, the equality $p_k = t_i^{-1}$ ($a_k = 0,\ b_k = -t_i^{-1}$) is true, then the integration of the Equation (30) gives

$$G_{k,1}(t) = t_i(1-e^{-t/t_i}) - te^{-t/t_i}, \quad (35)$$

$$G_{k,2}(t) = t_i[t - t_i(1-e^{-t/t_i})] - 0.5t^2 e^{-t/t_i}, \quad (36)$$

$$G_{k,3}(t) = t_i(1-e^{-2t/t_i}) - 2te^{-t/t_i}. \quad (37)$$

On the other hand, for $p_k = 2t_i^{-1}$ ($a_k = t_i^{-1},\ b_k = 0$) it was obtained that

$$G_{k,1}(t) = 0.5t_i(1-e^{-t/t_i})^2, \quad (38)$$

$$G_{k,2}(t) = 0.5t_i[t - 0.5t_i(1-e^{-2t/t_i})] - t_i[t - t_i(1-e^{-t/t_i})]e^{-t/t_i}, \quad (39)$$

$$G_{k,3}(t) = 0.5t_i(1-e^{-2t/t_i}) - 2t_i(1-e^{-t/t_i})e^{-t/t_i} + te^{-2t/t_i}. \quad (40)$$

Substituting the function $G_{n,i}(t)$, $i = 1,2,3$ in Equations (31)–(33) into the right side of Equation (29) yields

$$\begin{aligned} G_n(t) &= \left(1+\frac{t_i}{t_s^0}+\frac{1}{t_s^0 p_n}\right)\frac{(1-e^{-p_n t})}{p_n}-\left(1+\frac{2t_i}{t_s^0}+\frac{1}{t_s^0 a_n}\right)\frac{(e^{-t/t_i}-e^{-p_n t})}{a_n}+ \\ &+\frac{t_i(e^{-2t/t_i}-e^{-p_n t})}{t_s^0 b_n}-\frac{t}{t_s^0}\left(\frac{1}{p_n}-\frac{e^{-t/t_i}}{a_n}\right),\ 0\le t\le t_s,\ n=1,2,\ldots. \end{aligned} \tag{41}$$

The searched derivative of the function $G_n(t)$ in Equation (41), meeting the conditions in Equation (34), has the following form:

$$\begin{aligned} G'_n(t) &= \left(1+\frac{t_i}{t_s^0}\right)e^{-p_n t}-\frac{(1-e^{-p_n t})}{t_s^0 p_n}+\left(1+\frac{2t_i}{t_s^0}+\frac{1}{t_s^0 a_n}\right)\frac{(t_i^{-1}e^{-t/t_i}-p_n e^{-p_n t})}{a_n}+ \\ &+\frac{1}{t_s^0 a_n}\left(1-\frac{t}{t_i}\right)e^{-t/t_i}-\frac{t_i(2t_i^{-1}e^{-2t/t_i}-p_n e^{-p_n t})}{t_s^0 b_n},\ 0\le t\le t_s,\ n=1,2,\ldots. \end{aligned} \tag{42}$$

Proceeding in a similar manner, from Equations (29) and (35)–(40), the derivative for $p_k = t_i^{-1}$ was found.

$$G'_k(t)=\frac{t}{t_s^0}\left(3+\frac{t_s^0}{t_i}-\frac{t}{2t_i}\right)e^{-t/t_i}+\frac{t_i}{t_s^0}\left(2e^{-2t/t_i}-e^{-t/t_i}-1\right), \tag{43}$$

That for $p_k = 2t_i^{-1}$ was also found.

$$G'_k(t)=\left(1+4\frac{t_i}{t_s^0}\right)\left(e^{-t/t_i}-e^{-2/t_i}\right)-\frac{t_i}{2t_s^0}\left(1-e^{-2t/t_i}\right)-\frac{t}{t_s^0}\left(e^{-t/t_i}+2e^{-2t/t_i}\right). \tag{44}$$

Approaching $p_n \to 0$ $(a_n \to -t_i^{-1},\ b_n \to -2t_i^{-1})$, the limit of Equation (42) was found.

$$\begin{aligned} \lim_{p_n\to 0} G'_n(t) &= 1+\frac{t_i}{t_s^0}-\frac{t}{t_s^0}-\left(1+\frac{t_i}{t_s^0}\right)e^{-t/t_i}+\frac{t_i}{t_s^0}e^{-2t/t_i}-\frac{t_i}{t_s^0}\left(1-\frac{t}{t_i}\right)e^{-t/t_i}= \\ &=\left(1-e^{-t/t_i}\right)\left[1-\frac{t}{t_s^0}+\frac{t_i}{t_s^0}\left(1-e^{-t/t_i}\right)\right]=p^*(t)\left[1-\frac{t}{t_s^0}+\frac{t_i}{t_s^0}p^*(t)\right]=q^*(t), \\ &\qquad 0\le t\le t_s, \end{aligned} \tag{45}$$

where $p^*(t)$ and $q^*(t)$ are the dimensionless temporal profiles of pressure (Equation (1)) and specific friction power (Equation (3)), respectively, where the function $q^*(t)$ occurs beyond the sign of the sum in the solutions in Equations (26)–(28).

It should be noted that, at the initial time moment, from Equation (3), it follows $q^*(0) = 0$ and, from Equation (42), taking into account Equation (34), it was found that

$$\begin{aligned} G'_n(0) &= 1+\frac{t_i}{t_s^0}+\left(1+\frac{2t_i}{t_s^0}+\frac{1}{t_s^0 a_n}\right)\frac{(t_i^{-1}-p_n)}{a_n}-\frac{t_i(2t_i^{-1}-p_n)}{t_s^0 b_n}+\frac{1}{t_s^0 a_n}= \\ &=1+\frac{t_i}{t_s^0}-1-\frac{2t_i}{t_s^0}-\frac{1}{t_s^0 a_n}+\frac{t_i}{t_s^0}+\frac{1}{t_s^0 a_n}=0,\ p_n\ne t_i^{-1}\vee p_n\ne 2t_i^{-1},\ n=1,2,\ldots. \end{aligned} \tag{46}$$

If $p_k = t_i^{-1}$ or $p_k = 2t_i^{-1}$, then, from Equations (43) and (44), it follows that $G'_k(0) = 0$. In this way, it was shown that the solution in Equations (26) and (27) meets the initial condition in Equation (16).

In the special case for $t_i \to 0$, when the pressure $p(t)$ in Equation (1) attains the nominal value p_0 immediately, and the velocity $V(t)$ in Equation (2) reduces linearly (braking with constant deceleration), the dimensionless temporal profile of the specific friction power $q^*(t)$ and function $G'_n(t)$ in Equation (42) takes the following form:

$$q^*(t)=1-\frac{t}{t_s^0},\ G'_n(t)=e^{-p_n t}-\frac{(1-e^{-p_n t})}{t_s^0 p_n},\ 0\le t\le t_s^0. \tag{47}$$

From Equations (47) and (48), it follows that $q^*(0) = 1$, $G'_n(0) = 1$. This means that fulfillment of the initial condition in Equation (16) in this case is possible when the following equality is satisfied:

$$\sum_{n=1}^{\infty} \frac{\phi_1(z,\mu_n)}{\Psi(\mu_n)} = \frac{0.25\gamma_\varepsilon}{1+\gamma_\varepsilon K_\varepsilon} e^{-\gamma_1 z/2}, z \geq 0, \; \sum_{n=1}^{\infty} \frac{\phi_2(z,\mu_n)}{\Psi(\mu_n)} = \frac{0.25\gamma_\varepsilon}{1+\gamma_\varepsilon K_\varepsilon} e^{\gamma_2 z/2}, z \leq 0, \tag{48}$$

where functions $\phi_l(z,\mu_n)$, $l = 1,2$, and $\Psi(\mu_n)$ have the form in Equations (20)–(22). The validation of the summation of functional series in Equation (48) was performed numerically.

4. Dimensionless Form of Solution

The following denotes are introduced:

$$\begin{gathered} \zeta = \frac{z}{a},\; \tau = \frac{k_{1,0}t}{a^2},\; \tau_s = \frac{k_{1,0}t_s}{a^2},\; \tau_s^0 = \frac{k_{1,0}t_s^0}{a^2},\; \tau_i = \frac{k_{1,0}t_i}{a^2},\; \gamma_l = \frac{\gamma_l^*}{a}, l = 1,2, \\ \Theta_0 = \frac{q_0 a}{K_{1,0}},\; \Theta^* = \frac{\Theta}{\Theta_0}, \end{gathered} \tag{49}$$

where $a = \max\{a_1,\ a_2\}$, a_l, and $l = 1,2$ is the thickness of the friction pair element, which actively participates in the absorption of heat. This is the distance from the friction surface, on which the temperature is 5% of maximum values achieved on this surface [23].

$$a_l = \sqrt{3k_{l,0}t_s},\; l = 1,2. \tag{50}$$

Taking into account the denotes in Equation (49) in Equations (1)–(3), (20), and (42), and the solutions in Equations (26) and (27), the dimensionless temperature rise can be written in the following form:

$$\Theta^*(\zeta,\tau) = \frac{K_0^*}{\gamma_2^*} e^{-\gamma_1^*\zeta/2}\left[\frac{e^{-\gamma_1^*\zeta/2}}{(1+\gamma_\varepsilon K_\varepsilon)} q^*(\tau) + \frac{4}{\gamma_\varepsilon}\sum_{n=1}^{\infty}\frac{\phi_1^*(\zeta,\mu_n)}{\Psi(\mu_n)} G'(\tau,\mu_n)\right],\; \zeta \geq 0, 0 \leq \tau \leq \tau_s, \tag{51}$$

$$\Theta^*(\zeta,\tau) = \frac{K_0^*}{\gamma_2^*} e^{\gamma_2^*\zeta/2}\left[\frac{e^{\gamma_2^*\zeta/2}}{(1+\gamma_\varepsilon K_\varepsilon)} q^*(\tau) + \frac{4}{\gamma_\varepsilon}\sum_{n=1}^{\infty}\frac{\phi_2^*(\zeta,\mu_n)}{\Psi(\mu_n)} G'(\tau,\mu_n)\right],\; \zeta \leq 0, 0 \leq \tau \leq \tau_s, \tag{52}$$

where

$$q^*(\tau) = p^*(\tau)\left[1 - \frac{\tau}{\tau_s^0} + \frac{\tau_i}{\tau_s^0} p^*(\tau)\right],\; p^*(\tau) = 1 - e^{-\tau/\tau_i}, \tag{53}$$

$$\phi_1^*(\zeta,\mu_n) = J_1(\gamma_\varepsilon\mu_n) J_1(\mu_n e^{-\gamma_1^*\zeta/2}),\; \phi_2^*(\zeta,\mu_n) = J_1(\mu_n) J_1(\gamma_\varepsilon\mu_n e^{\gamma_2^*\zeta/2}), \tag{54}$$

$$\begin{gathered} G'_n(\tau) = \left(1 + \frac{\tau_i}{\tau_s^0}\right) e^{-\lambda_n\tau} - \frac{(1-e^{-\lambda_n\tau})}{\tau_s^0\lambda_n} + \left(1 + \frac{2\tau_i}{\lambda_s^0} + \frac{1}{\tau_s^0\alpha_n}\right)\frac{(\tau_i^{-1}e^{-\tau/\tau_i} - \lambda_n e^{-\lambda_n\tau})}{\alpha_n} - \\ - \frac{\tau_i(2\tau_i^{-1}e^{-2\tau/\tau_i} - \lambda_n e^{-\lambda_n\tau})}{\tau_s^0\beta_n} + \frac{1}{\tau_s^0\alpha_n}\left(1 - \frac{\tau}{\tau_i}\right) e^{-\tau/\tau_i}, \\ \alpha_n = \lambda_n - \tau_i^{-1} \neq 0,\; \beta_n = \lambda_n - 2\tau_i^{-1} \neq 0, \end{gathered} \tag{55}$$

$$\lambda_n = (0.5\gamma_1^*\mu_n)^2,\; n = 1,2,\ldots, \tag{56}$$

$$\tau_s \cong \tau_s^0 + 0.99\tau_i,\; 0 < \tau_i \leq 0.3\tau_s^0. \tag{57}$$

Function $\Psi(\mu_n)$ is given by Equation (21), and numbers $\mu_n > 0$ are the real roots of the functional Equation (24). From Equations (43) and (44), it follows that

$$G'_k(\tau) = \frac{\tau}{\tau_s^0}\left(3 + \frac{\tau_s^0}{\tau_i} - \frac{\tau}{2\tau_i}\right) e^{-\tau/\tau_i} + \frac{\tau_i}{\tau_s^0}\left(2e^{-2\tau/\tau_i} - e^{-\tau/\tau_i} - 1\right),\; \lambda_k = \tau_i^{-1}, \tag{58}$$

$$G'_k(\tau) = \left(1 + 4\frac{\tau_i}{\tau_s^0}\right)\left(e^{-t/t_i} - e^{-2\tau/\tau_i}\right) - \frac{\tau_i}{2\tau_s^0}\left(1 - e^{-2\tau/\tau_i}\right) - \frac{\tau}{\tau_s^0}\left(e^{-\tau/\tau_i} + 2e^{-2\tau/\tau_i}\right),\; \lambda_k = 2\tau_i^{-1}. \tag{59}$$

Substituting $\zeta = 0$ into Equations (51), (52), and (54), the dimensionless temperature rise on the contact surface can be written in the following form:

$$G'_k(\tau) = \left(1 + 4\frac{\tau_i}{\tau_s^0}\right)\left(e^{-t/t_i} - e^{-2\tau/\tau_i}\right) - \frac{\tau_i}{2\tau_s^0}\left(1 - e^{-2\tau/\tau_i}\right) - \frac{\tau}{\tau_s^0}\left(e^{-\tau/\tau_i} + 2e^{-2\tau/\tau_i}\right),\ \lambda_k = 2\tau_i^{-1}, \quad (60)$$

where

$$\phi^*(\mu_n) \equiv \phi_1^*(0, \mu_n) = \phi_2^*(0, \mu_n) = J_1(\gamma_\varepsilon \mu_n) J_1(\mu_n). \quad (61)$$

In case of braking with constant deceleration ($\tau_i \to 0$), from Equation (47), it can be obtained that

$$q^*(\tau) = 1 - \frac{\tau}{\tau_s^0},\ G'(\tau, \mu_n) = e^{-\lambda_n \tau} - \frac{1}{\tau_s^0 \lambda_n}(1 - e^{-\lambda_n \tau}),\ 0 \le \tau \le \tau_s^0. \quad (62)$$

It should be noted that the exact solution to the problem considering the contact scheme of friction for two semi-infinite bodies, made of homogeneous materials ($\gamma_1 = \gamma_2 = 0$), with account of the time of contact pressure increase, was achieved in [19]. A special case of this solution—braking with constant deceleration—was investigated in [24].

5. Numerical Analysis

On the basis of the obtained exact solutions in Equations (51), (52), and (60), the calculations of the temperature generated due to friction in the disc–pad system during single braking were performed. Materials of the friction surfaces of elements were zirconium dioxide ($l = 1$) and the other ceramic ($l = 2$). With the distance from these surfaces deeper into the bodies, their thermal conductivity coefficients increased exponentially in accordance with Equation (10), reaching at the effective depths a_l, $l = 1, 2$ values corresponding to titanium and aluminum alloys, respectively. The thermal properties of the abovementioned materials are listed in Table 1.

Table 1. Thermophysical properties of the FGM components [15,25].

Element Subscript	Material	Thermal Conductivity K [$Wm^{-1}K^{-1}$]	Thermal Diffusivity $k \times 10^6$ [m^2s^{-1}]
$l = 1$	ZrO_2	2.09	0.86
	Ti-6Al-4V	7.5	3.16
$l = 2$	ceramic	3	1.15
	aluminum alloy	173	67.16

Values of the remaining input parameters were as follows: $A_a = 0.442 \cdot 10^{-2} m^2$, $f = 0.27$, $p_0 = 0.607$ MPa, $T_0 = 20$ °C, $V_0 = 23.8$ m s^{-1}, and $W_0 = 103.54$ kJ [26]. From Equation (2), the braking time with constant deceleration was found $t_s^0 = 12$ s and, next, the stop time $t_s = 12.49$ s. This allowed determining from Equation (50) the effective depths of heat penetration $a_1 = 5.556$ mm and $a_2 = 6.435$ mm, as well as the value of the scaling parameter $a = a_2$. According to the methodology, described in detail in [17], the dimensionless parameters of the material gradient were also established as $\gamma_1^* = 1.28$ and $\gamma_2^* = 4.05$.

Isotherms of the temperature rise $\Theta(z, t)$ inside the elements of the friction pair are illustrated in Figure 2. The most heated ($\Theta = 800 \div 943$ °C) was a narrow, approximately 0.5 mm thick, near-surface area that appeared ≈ 3 s after start of braking. The lifetime of such a high-temperature area is ≈ 3 s. The friction surfaces of both elements were cooled down until the stop time moment. At the stop moment, the distance from the friction surface, where the noticeable temperature occurs in the disc was greater than in the pad.

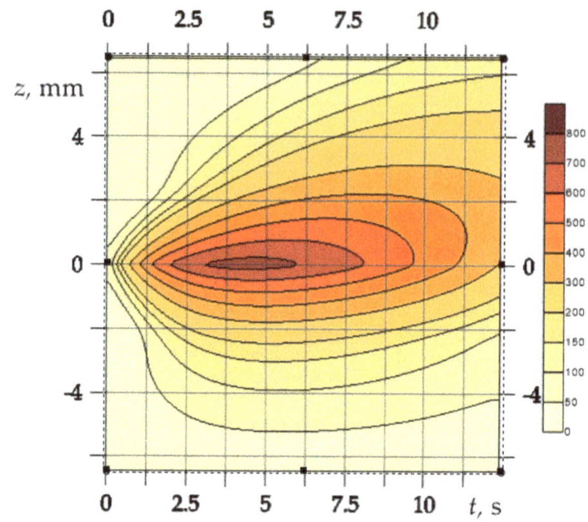

Figure 2. Isotherms of the temperature rise $\Theta(z,t)$ in the disc and the pad at $t_i = 0.5$ s.

Evolutions of the temperature rise $\Theta(z,t)$ during braking on the contact surface and inside the friction elements on different depths are presented in Figure 3. At the beginning of braking, the temperature on the friction surfaces $z = 0$ rapidly increases over time, achieving the maximum value $\Theta_{max} = 943°C$ at the moment $t_{max} = 5$ s. This is followed by a period of cooling of these surfaces until it stops. The temporal profiles of the temperature inside the disc and the pad also have a similar shape. However, the known "delay" effect is visible in the disc, which is that the time to reach the maximum temperature increases with the distance from the contact surface. At the same time, this effect is almost imperceptible. Noteworthy is also the process of rapid cooling on the friction surface of disc after reaching maximum Θ_{max}; at the stop moment, the temperature inside the disc is higher than on the surface. Again, this effect does not occur in the pad material.

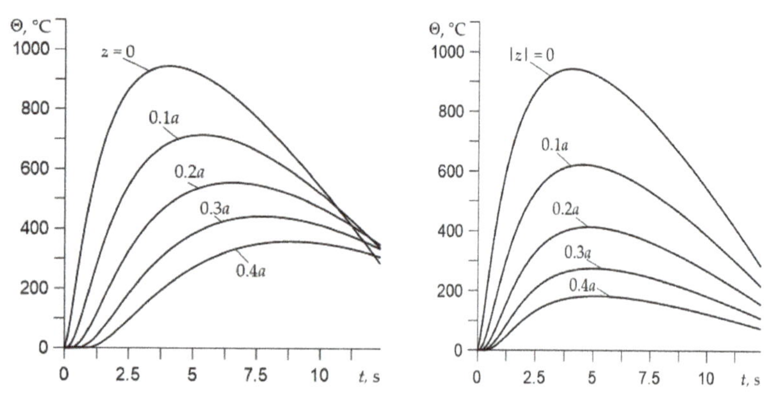

(a) (b)

Figure 3. Evolutions of the temperature rise $\Theta(z,t)$ during braking at $t_i = 0.5$ s for different distances from the friction surface: (**a**) the disc; (**b**) the pad.

Variations of the temperature during braking on the friction surfaces of disc and pad for different times of contact pressure increase are demonstrated in Figure 4. Extending the time of achieving the nominal value of pressure causes a drop of maximum temperature on the contact surface, while increasing the braking time. The effect of temperature drop with the growth of time of pressure increase is also presented in Figure 5.

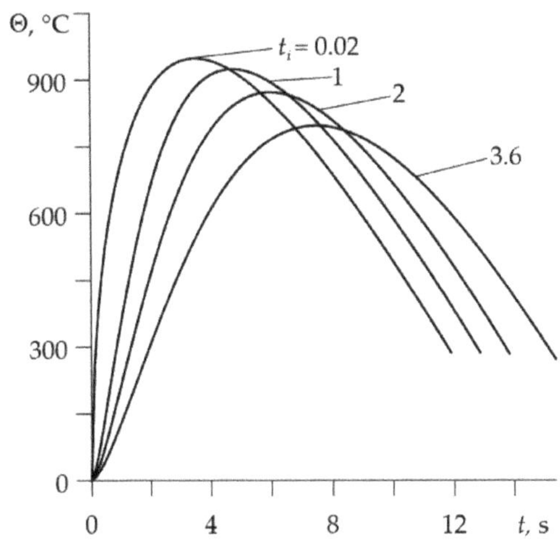

Figure 4. Evolutions of the temperature rise $\Theta(0, t)$ during braking for different values of the time t_i of contact pressure increase.

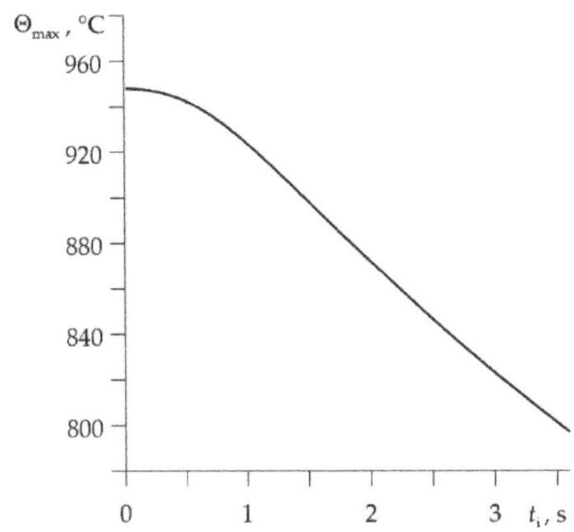

Figure 5. Dependence of the maximum temperature rise Θ_{max} on the time t_i of contact pressure increase.

The influence of dimensionless parameters of material gradients $\gamma_l^*, l = 1, 2$ on the dimensionless maximum temperature Θ_{max}^* on the contact surface is illustrated in Figure 6. It shows that an increase in the core material volume fraction in selected FGMs (Ti-6Al-4V for disc and aluminum alloy for pad) causes a decrease in the maximum temperature in the brake. The biggest drop in Θ_{max}^* occurs when the gradient of the pad material γ_2^* is increased (Figure 6b). However, the highest values of Θ_{max}^* are reached for the friction pair in which one of the elements is entirely made of homogeneous material. These are zirconium dioxide ZrO_2 for the disc ($\Theta^*_{max} = 995$ °C at $\gamma^*_1 = 0$, in Figure 6a) and the other ceramic for the pad ($\Theta^*_{max} = 1340$ °C at $\gamma^*_2 = 0$, in Figure 6b).

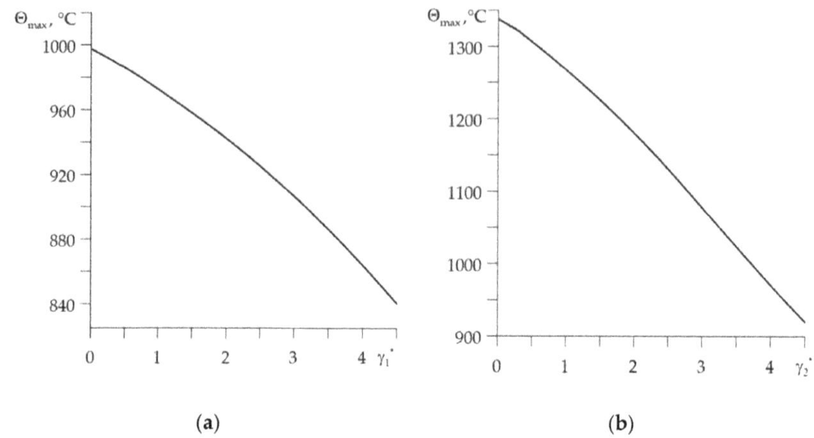

Figure 6. Dependence of the maximum temperature rise Θ_{max} at $t_i = 0.5$ s on the dimensionless gradient of material: (**a**) γ^*_1 for $\gamma^*_2 = 4.05$; (**b**) γ^*_2 for $\gamma^*_1 = 1.28$.

6. Conclusions

The presented results are the continuation of an investigation from a previous article of the authors [17], in which, in the dimensionless form, a comparative, qualitative analysis was performed in order to study the influence of gradient of FGMs on the temperature during frictional heating under uniform sliding. However, in this paper the mathematical model was derived to determine the temperature field in a disc–pad system during single braking. An important and unique feature of this model was its taking into account of the time-dependent pressure and velocity for friction elements, made of functionally graded materials with exponentially changing conductivity coefficients with thickness. The proposed model allows for a quick assessment of the brake temperature mode depending on the operational parameters, such as the time of contact pressure increase and the value of the gradient of the friction materials. The analysis was performed in the dimensional form. The friction surfaces of the materials were ceramic, and their cores were titanium alloys (disc) and aluminum alloys (pad). It was established that extending the time of pressure increase causes significant extending of the braking time and, thus, extending of the braking distance. The maximum temperature reached on the friction surfaces drops when the parameters of material gradients are increased.

Application of the proposed model has some limitations, resulting from the simplifying assumptions made, especially the use of only an exponential function to describe the thermal conductivity changes in FGMs. In further research, it is planned to include in the formulation of the boundary value problem of heat conduction, as well as the thermal resistance on the contact surface of the disc and the pad (imperfect thermal contact of friction), and to adapt the obtained exact solution to determine the temperature of the brake during a repeated short-term mode of braking.

As shown in the results of the numerical analysis presented in this article, the maximum temperature achieved even with a single braking is quite high. With such a temperature, the necessary problem is to develop a model that takes into account the thermal sensitivity of the materials. Some steps toward implementing the exact solutions of linear problems for homogeneous materials to take into account their thermal sensitivity have already been made for a single [27] and a repetitive short-term [28] braking modes. On the basis of this methodology, the development of appropriate models for FGM has begun.

Another problem caused by high temperatures is a reduction in the strength of the material, especially when the temperature exceeds the melting point of the aluminum alloy. Investigations of the strength were not the subject of this article, but they should also be considered in the future.

Author Contributions: Conceptualization and methodology, A.Y.; software, K.T. and P.Z.; validation, P.Z.; formal analysis, K.T. and P.Z.; investigation, A.Y., K.T., and P.Z.; writing—original draft preparation, P.Z.; writing—review and editing, A.Y. and K.T.; visualization and figure preparation, P.Z.; supervision, A.Y.; project administration, K.T.; funding acquisition, K.T. All authors read and agreed to the published version of the manuscript.

Funding: This study was performed within the framework of research project No. 2017/27/B/ST8/01249, funded by the National Science Center, Poland and project financing through the program of the Minister of Education and Science of Poland "Regional Initiative of Excellence" in 2019–2022, project No. 011/RID/2018/19 (amount of financing: 12,000,000 PLN).

Institutional Review Board Statement: Not applicable.

Informed Consent Statement: Not applicable.

Data Availability Statement: No new data were created or analyzed in this study. Data sharing is not applicable to this article.

Conflicts of Interest: The authors declare no conflict of interest.

Nomenclature

a_l	Effective depth of heat penetration (m)
A_a	Area of the nominal contact region (m^2)
c_l	Specific heat (J kg^{-1}K^{-1})
f	Coefficient of friction (dimensionless)
$J_k(\cdot)$	Bessel functions of the first kind of the k-th order
k_l	Thermal diffusivity (m^{2}s^{-1})
K_l	Thermal conductivity (W m^{-1}K^{-1})
p	Contact pressure (Pa)
p_0	Nominal value of the contact pressure (Pa)
q	Specific power of friction (W m^{-2})
q_0	Nominal value of the specific power of friction (W m^{-2})
t	Time (s)
t_i	Time of the contact pressure increase (s)
t_s^0	Stop time at braking with constant deceleration (s)
t_s	Stop time (s)
T	Temperature (°C)
T_a	Initial (ambient) temperature (°C)
V	Velocity (m s^{-1})
V_0	Initial velocity (m s^{-1})
W_0	Initial kinetic energy of the system (J)
x, y, z	Spatial coordinates (m)
lower l	Number of the main (l = 1) and frictional (l = 2) elements of the friction pair
γ_l	Parameter of material gradient (m^{-1})
γ_l^*	Parameter of material gradient (dimensionless)
Θ_l	Temperature rise (°C)
Θ_l^*	Temperature rise (dimensionless)
Θ_0	Temperature scaling factor (°C)
ρ_l	Density (kg m^{-3})
τ	Time (dimensionless)
τ_i	Time of contact pressure increase (dimensionless)
τ_s^0	Braking time at constant deceleration (dimensionless)
τ_s	Braking time (dimensionless)
ζ	Spatial coordinate in axial direction (dimensionless)

References

1. Govindaraju, M.; Megalingam, A.; Murugasan, J.; Vignesh, R.V.; Kota, P.K.; Ram, A.S.; Lakshana, P.; Kumar, V.N. Investigations on the tribological behavior of functionally gradient iron-based brake pad material. *Proc. Inst. Mech. Eng. Part C J. Mech. Eng. Sci.* **2020**, *234*, 2474–2486. [CrossRef]
2. Swaminathan, K.; Sangeetha, D. Thermal analysis of FGM plates—A critical review of various modeling techniques and solution methods. *Compos. Struct.* **2017**, *160*, 43–60. [CrossRef]
3. Afsar, A.; Go, J. Finite element analysis of thermoelastic field in a rotating FGM circular disk. *Appl. Math. Model.* **2010**, *34*, 3309–3320. [CrossRef]
4. Shahzamanian, M.; Bin Sahari, B.; Bayat, M.; Mustapha, F.; Ismarrubie, Z.N. Finite element analysis of thermoelastic contact problem in functionally graded axisymmetric brake disks. *Compos. Struct.* **2010**, *92*, 1591–1602. [CrossRef]
5. Shahzamanian, M.; Bin Sahari, B.; Bayat, M.; Ismarrubie, Z.N.; Mustapha, F. Transient and thermal contact analysis for the elastic behavior of functionally graded brake disks due to mechanical and thermal loads. *Mater. Des.* **2010**, *31*, 4655–4665. [CrossRef]
6. Hosseini, T.P.; Talebi, M. Stress and temperature distribution study in a functionally graded brake disk. *Int. J. Automot. Eng.* **2012**, *2*, 172–179.
7. Yaghoobi, M.P.; Ghannad, M. An analytical solution for heat conduction of FGM cylinders with varying thickness subjected to non-uniform heat flux using a first-order temperature theory and perturbation technique. *Int. Commun. Heat Mass Transf.* **2020**, *116*, 104684. [CrossRef]
8. Jang, Y.H.; Ahn, S.-H. Frictionally-excited thermoelastic instability in functionally graded material. *Wear* **2007**, *262*, 1102–1112. [CrossRef]
9. Lee, S.W.; Jang, Y.H. Frictionally excited thermoelastic instability in a thin layer of functionally graded material sliding between two half-planes. *Wear* **2009**, *267*, 1715–1722. [CrossRef]
10. Lee, S.W.; Jang, Y.H. Effect of functionally graded material on frictionally excited thermoelastic instability. *Wear* **2009**, *266*, 139–146. [CrossRef]
11. Mao, J.-J.; Ke, L.-L.; Yang, J.; Kitipornchai, S.; Wang, Y.-S. Thermoelastic instability of functionally graded materials with interaction of frictional heat and contact resistance. *Mech. Based Des. Struct. Mach.* **2017**, *46*, 139–156. [CrossRef]
12. Mao, J.-J.; Ke, L.-L.; Yang, J.; Kitipornchai, S.; Wang, Y.-S. The coupled thermoelastic instability of FGM coatings with arbitrarily varying properties: In-plane sliding. *Acta Mech.* **2018**, *229*, 2979–2995. [CrossRef]
13. Kulchytsky-Zhyhailo, R.; Bajkowski, A.S. Axisymmetrical problem of thermoelasticity for halfspace with gradient coating. *Int. J. Mech. Sci.* **2016**, *106*, 62–71. [CrossRef]
14. Martínez-Pañeda, E. On the Finite Element Implementation of Functionally Graded Materials. *Materials* **2019**, *12*, 287. [CrossRef]
15. Mao, J.-J.; Ke, L.-L.; Yang, J.; Kitipornchai, S.; Wang, Y.-S. Thermoelastic instability of functionally graded coating with arbitrarily varying properties considering contact resistance and frictional heat. *Appl. Math. Model.* **2017**, *43*, 521–537. [CrossRef]
16. Polajnar, M.; Kalin, M.; Thorbjornsson, I.; Thorgrimsson, J.; Valle, N.; Botor-Probierz, A. Friction and wear performance of functionally graded ductile iron for brake pads. *Wear* **2017**, *382-383*, 85–94. [CrossRef]
17. Yevtushenko, A.; Topczewska, K.; Zamojski, P. The Effect of Functionally Graded Materials on Temperature during Frictional Heating: Under Uniform Sliding. *Materials* **2021**, *14*, 4285. [CrossRef] [PubMed]
18. Chichinadze, A.V. *Polymers in Friction Assembles of Machines and Devices: A Handbook*; Allerton Press Inc: New York, NY, USA, 1984.
19. Topczewska, K. Influence of the Time of Increase in Contact Pressure in the Course of Braking on the Temperature of a Pad–Disc Tribosystem. *Mater. Sci.* **2018**, *54*, 250–259. [CrossRef]
20. Yevtushenko, A.; Kuciej, M.; Topczewska, K. Some theoretical model for determining the temperature field of a multi-disk brake. *Adv. Mech. Eng.* **2020**, *12*. [CrossRef]
21. Abramowitz, M.; Stegun, I. *Handbook of Mathematical Functions with Formulas, Graphs, and Mathematical Tables*; United States Department of Commerce, National Bureau of Standards (NBS): Washington, WA, USA, 1964.
22. Özişik, N.M. *Heat conduction*; John Wiley: New York, NY, USA, 1993.
23. Chichinadze, A.V.; Braun, E.D.; Ginzburg, A.G.; Ignat'eva, E.V. *Calculation, Testing and Selection of Friction Couples*; Nauka: Moscow, Russia, 1979. (In Russian)
24. Yevtushenko, A.; Kuciej, M. Frictional heating during braking in a three-element tribosystem. *Int. J. Heat Mass Transf.* **2009**, *52*, 2942–2948. [CrossRef]
25. Tian, J.; Jiang, K. Heat conduction investigation of the functionally graded materials plates with variable gradient parameters under exponential heat source load. *Int. J. Heat Mass Transf.* **2018**, *122*, 22–30. [CrossRef]
26. Chichinadze, A.V.; Kozhemyakina, V.D.; Suvorov, A.V. Method of temperature-field calculation in model ring specimens during bilateral friction in multidisc aircraft brakes with the IM-58-T2 new multipurpose friction machine. *J. Frict. Wear* **2010**, *31*, 23–32. [CrossRef]
27. Evtushenko, O.; Kuciej, M.; Topczewska, K. Determination of the Maximal Temperature of a Pad–Disk Tribosystem during One-Time Braking. *Mater. Sci.* **2020**, *56*, 152–159. [CrossRef]
28. Yevtushenko, A.; Topczewska, K.; Kuciej, M. Analytical Determination of the Brake Temperature Mode during Repetitive Short-Term Braking. *Materials* **2021**, *14*, 1912. [CrossRef] [PubMed]

materials

MDPI

Article

Influence of Thermal Sensitivity of Functionally Graded Materials on Temperature during Braking

Aleksander Yevtushenko, Katarzyna Topczewska * and Przemysław Zamojski

Faculty of Mechanical Engineering, Bialystok University of Technology (BUT), 45C Wiejska Street, 15-351 Bialystok, Poland; a.yevtushenko@pb.edu.pl (A.Y.); p.zamojski@pb.edu.pl (P.Z.)
* Correspondence: k.topczewska@pb.edu.pl

Abstract: The model of the frictional heating process during single braking to determine the temperature of the functionally graded friction elements with an account of the thermal sensitivity of materials was proposed. The basis of this model is the exact solution of the one-dimensional thermal problem of friction during braking with constant deceleration. The formulas approximating the experimental data of the temperature dependencies of properties of the functionally graded materials (FGMs) were involved in the model to improve the accuracy of the achieved results. A comparative analysis was performed for data obtained for temperature-dependent FGMs and the corresponding data, calculated without consideration of thermal sensitivity. The results revealed that the assumption of thermal stability of FGMs during braking may cause a significant overestimation of temperature of the friction pair elements.

Keywords: thermal sensitivity; functionally graded materials; temperature; friction; braking

Citation: Yevtushenko, A.; Topczewska, K.; Zamojski, P. Influence of Thermal Sensitivity of Functionally Graded Materials on Temperature during Braking. *Materials* **2022**, *15*, 963. https://doi.org/10.3390/ma15030963

Academic Editors: Joe Sakai, Liaoliang Ke and Xiaoyan Li

Received: 23 December 2021
Accepted: 24 January 2022
Published: 26 January 2022

1. Introduction

During intensive braking, the volume temperature of the disc braking system may be higher than 450 °C [1] and the maximum temperature on the friction surfaces of the pad and the disc during single braking may even reach a level above 1000 °C [2]. In such severe conditions, the thermal and mechanical properties of materials may highly differ from the initial, reached at the ambient temperature. Therefore, in order to improve the theoretical analysis of the thermoelastic behavior of the braking systems, it is necessary to develop mathematical models taking into consideration the thermal sensitivity of friction materials. However, the introduction of the temperature-dependent properties in formulation of the thermal problems of friction leads to nonlinearity, so most of the published analyses have been performed using numerical methods, especially the finite element method [3,4]. One of the alternative techniques used to develop such nonlinear models of frictional heating is linearization by means of the Kirchhoff substitution [5]. This method relies on the reduction of the originally nonlinear heat conduction equation to the linear one. However, it works this way only for materials with simple nonlinearity, which means that their thermal conductivity and specific heat capacity are temperature-dependent, but the thermal diffusivity remains constant [6]. For materials with arbitrary nonlinearity, only the partial linearization by the Kirchhoff substitution of such a problem is possible; as a result, another nonlinear problem is obtained for which the method of solving is known [7]. The Kirchhoff transform has a similar effect in the heat conduction problems formulated for solids with simple nonlinear thermosensitivity under complex heat exchange. Some analytic–numerical methods for the solution of such problems have been proposed in the study [8]. Another technique to take into consideration the thermal sensitivity of materials is the method of successive approximations (iterations), in which the solution of the corresponding linear problem is adopted as the initial approximation, and then the solution found in the previous step is corrected. An iteration algorithm to solve the one-dimensional problem of heat conduction at braking has been proposed in the article [9]. Most models of the frictional heating process taking into account of the thermal sensitivity effect have been developed only for homogeneous materials. Modern friction materials often have a non-uniform, complex internal structure with a changing composition, microstructure, or porosity across the volume of material, such as functionally graded materials (FGMs), which are characterized by smooth variations of properties as a function of position along certain direction. In the case of devices operating at elevated temperatures, including braking systems, FGMs are primarily used in order to obtain high temperature resistance on the friction surface by dissipating heat from it to the inside of the element while maintaining low wear. FGMs of this type are usually two-component. Their friction surface is usually made of metal-ceramic, and the metal opposite surface (core) should have high thermal conductivity. The change of properties in the direction perpendicular to the friction surface is described by continuous functions, usually power or exponential. In the case of the latter, the material gradient parameters are responsible for the speed of transition from one material to another.

The problem of wear of an FGM strip with an account of the heating on sliding contact from friction has been considered in the study [10]. The exact solution of the problem was obtained with the help of the integral Laplace transform technique. It was assumed that the shear modulus is described by means of the function of the vertical coordinate. A comprehensive review of the literature concerning the thermal contact problems of frictional heating for functionally graded materials was provided in our previous article [11]. So far, investigations of the transient heat conduction in FGMs are limited, and most of them have ignored the temperature dependence of the material properties. Therefore, in general, those models are adequate only for relatively low temperatures in an FGM or the materials with insignificant thermal sensitivity. To accurately describe the thermomechanical behavior of FGMs, the temperature dependence of the material properties should be considered. The heat conduction problems formulated for FGMs with non-uniform spatially distributed and temperature-dependent properties are highly nonlinear. Nevertheless, several studies concerning such problems taking into consideration the thermal sensitivity of FGMs can be found, but most of them are solved by means of numerical or semi-analytical methods. The finite element method has been adopted in the paper [12], to perform the nonlinear transient thermal stress analysis of a thick-walled FGM cylinder with temperature-dependent material properties. Another nonlinear transient heat transfer and thermoelastic stress in thermosensitive functionally graded cylinder have been investigated using the Hermitian transfinite element method in the study [13]. The results showed that the effect of thermal sensitivity of materials has a significant influence on the thermal behavior of friction systems.

An analytical approach to solve the one-dimensional transient heat conduction problem for functionally graded materials with temperature-dependent properties has been presented in the article [14]. As for the analytical treatment, the temperature and thermal stress solutions have been obtained in approximate forms for a simplified, homogeneous, multi-layered model of materials. They concluded that the temperature dependence of the material properties is one of the most important factors in the accurate evaluation of temperature and stress distributions [14]. A similar multi-layered model was used to formulate another thermal problem of friction for a thermally sensitive FGM plate in the paper [15]. The authors made an attempt to optimize the functionally graded structure in order to enhance their thermal performance. The proper manufacturing process allows the design of an FGM according to the engineering demands by intentionally setting a specific distribution of the properties. A hybrid genetic algorithm has been developed for the optimization of the FGM composition with temperature-dependent material properties, in order to minimize the thermal stresses under steady-state thermal loads [15]. The optimum composition profile of the functionally graded materials for wide temperature ranges was also studied in the article [16]. The thermoelastic problem for functionally graded material with temperature-dependent properties was considered by means of the perturbation method. Additionally, the crack propagation path was predicted by introducing the fracture mechanics analysis. It was concluded that the proper selection of an FGM gradient can lead to a significant decrease in thermal stresses [16]. A transient thermoelastic behavior of the functionally graded plate with temperature-dependent properties due to a thermal shock was considered in the paper [17]. The temperature and thermal stress distributions in the Cu-W functionally graded composite were found by means of the semi-analytical micromechanical model.

The aim of this study was to investigate the influence of FGMs thermal sensitivity on the distribution of temperature in a disc brake system. This study is a continuation of our previous articles [11,18], which concern the transient thermal problem of friction under uniform sliding and during single braking with an exponential increase in the contact pressure. Due to the appearance of a high temperature level, there is a demand to improve the results by involving the variations of material properties dependent on the actual temperature, since the thermal sensitivity effect is particularly manifested in a high temperature range. In this article, the braking with constant deceleration is considered, when the nominal pressure is reached immediately at the beginning of the process, since the increase in the time of contact pressure growth causes a drop in the achieved temperature.

2. Statement to the Problem

To develop an analytical model of frictional heating process in the braking system, the following assumptions were taken into account:

1. The braking process with constant deceleration is considered;
2. At the initial time moment, the temperature of a brake is equal to the ambient temperature T_a;
3. In the heat conduction equation, only the change in temperature gradient in the perpendicular direction to the disc-pad contact surfaces is taken into consideration;
4. The thermal contact on the friction surfaces is perfect, i.e., the temperatures of its contact surfaces are equal, and the sum of frictional heat fluxes intensities, acting along the normal direction to the contact surface to the insides of the elements equal to the specific friction power;
5. Due to the symmetry of the system with respect to the mid plane of the disc, when determining the brake temperature, the contact of one pad and a disc with half of its thickness is considered;
6. The pads and the disc are made of two-component thermally sensitive functionally graded materials, in such a way that their friction surfaces are materials with low thermal conductivity (i.e., cermet), while the core materials are characterized by higher thermal conductivity (titanium alloys, aluminum, etc.);
7. The thermal conductivity of the disc and pads materials increases exponentially with the distance from the contact surface;
8. The whole initial kinetic energy of the vehicle is transformed into heat during braking, neglecting the small part of energy associated with wear on the contact surfaces of the disc and pads;

Based on the assumptions (1)–(5), in order to determine the temperature of the disc-pad system, the scheme of sliding with linearly decreasing velocity of two semi-spaces $z \geq 0$ (disc) and $z \leq 0$ (pad) has been adopted. Initiated by the frictional heating temperature field of such a system at a given time instant $t \geq 0$ depends only on the distance from the friction surface in a perpendicular direction—independent variable z: $T = T(z,t)$.

According to the assumption (6), the thermophysical properties of a friction pair are functions of temperature T:

$$K_{l,m} = K_{l,m}(T), c_{l,m} = c_{l,m}(T), \rho_{l,m} = \rho_{l,m}(T), \tag{1}$$

where $K_{l,m}$, $c_{l,m}$ and $\rho_{l,m}$—thermal conductivity, specific heat capacity and density of the first ($m = 1$) and second ($m = 2$) component of the materials of the disc ($l = 1$) and pad

$(l = 2)$, respectively. Corresponding values at the initial system temperature $T = T_0$ are marked as follows:

$$K_{l,m}^{(0)} \equiv K_{l,m}(T_0), c_{l,m}^{(0)} \equiv c_{l,m}(T_0), \rho_{l,m}^{(0)} \equiv \rho_{l,m}(T_0). \quad (2)$$

According to the mixture law, the effective specific heat capacities and densities were also determined:

$$c_l^{(0)} = c_{l,2}^{(0)} V_c + (1 - V_c) c_{l,1}^{(0)},\ \rho_l^{(0)} = \rho_{l,2}^{(0)} V_\rho + (1 - V_\rho) \rho_{l,1}^{(0)}, \quad (3)$$

where V_c, V_ρ—volume fractions of the phases $c_{l,m}^{(0)}$ and $\rho_{l,m}^{(0)}$, $l = 1,2$, $m = 1,2$, respectively.

Based on the assumption (7), the effective thermal conductivities K_l, $l = 1,2$ of the disc and pad materials were established from the equations:

$$K_1(z) = K_{1,1} e^{\gamma_1 z},\ 0 \le z \le a,\ K_2(z) = K_{2,1} e^{-\gamma_2 z},\ -a \le z \le 0, \quad (4)$$

where

$$\gamma_l = \frac{\gamma_l^*}{a},\ \gamma_l^* = \ln\left(\frac{K_{l,2}^{(0)}}{K_{l,1}^{(0)}}\right), \quad (5)$$

$$a = \max\{a_1, a_2\}, a_l = \sqrt{3 k_l^{(0)} t_s}, \quad (6)$$

$$k_l^{(0)} = \frac{K_{l,1}^{(0)}}{c_l^{(0)} \rho_l^{(0)}}, \quad (7)$$

and t_s—stop time, and parameters a_l, $l = 1,2$ (6) are the thicknesses of the subsurface layers actively participating in heat absorption in the disc and pads, respectively (the so-called effective depth of heat transfer [19]). During braking with constant deceleration, the specific friction power decreases linearly from the nominal value q_0 to zero [20]:

$$q(t) = q_0 q^*(t), q_0 = f p_0 V_0, q^*(t) = 1 - t\, t_s^{-1},\ 0 \le t \le t_s, \quad (8)$$

$$t_s = W_0\, Q_0^{-1},\ Q = q_0 A_a, A_a = 0.5\beta(R_e^2 - R_i^2), \quad (9)$$

where A_a—nominal area of the contact between the pad and the disc; f—friction coefficient; p_0—nominal pressure; Q_0—nominal friction power; $0 \le \beta \le 2\pi$—nominal friction power; —cover angle of the pad; R_i and R_e—respectively, the internal and external radii of the pads; V_0, W_0—the initial velocity and kinetic energy of the system, respectively. The latter, according to assumption (8), is equal to the total work of friction.

In order to solve the above-formulated nonlinear problem, we will use the idea of adapting an appropriate solution of the linear problem of thermal friction. This approach in the case of homogeneous materials was used in the studies [9,21].

3. Solution with Temperature-Independent FGMs Properties

The key element of the proposed approach is the precise solution of the linear thermal problem of friction during braking with constant deceleration. In the case of FGMs, such a solution for the above-adopted scheme of two sliding semi-spaces for the specific friction power $q(t)$ (8) and (9) can be written in the form [18]:

$$T(z,t) = T_0 + \Theta(z,t),\ 0 \le t \le t_s, \quad (10)$$

$$\Theta(z,t) = \Lambda e^{-\gamma_1 z/2}\left[\frac{e^{-\gamma_1 z/2}}{(1+\gamma_\varepsilon K_\varepsilon)} q^*(t) + \frac{4}{\gamma_\varepsilon}\sum_{n=1}^{\infty}\frac{\varphi_1(z,\mu_n)}{\Psi(\mu_n)} G_n(t)\right],\ z \ge 0, \quad (11)$$

$$\Theta(z,t) = \Lambda e^{\gamma_1 z/2}\left[\frac{e^{\gamma_2 z/2}}{(1+\gamma_\varepsilon K_\varepsilon)}q^*(t) + \frac{4}{\gamma_\varepsilon}\sum_{n=1}^{\infty}\frac{\varphi_2(z,\mu_n)}{\Psi(\mu_n)}G_n(t)\right],\ z \le 0, \tag{12}$$

$$G_n(t) = e^{-p_n t} - \frac{(1-e^{-p_n t})}{p_n t_s}, p_n = \frac{1}{4}k_1(\gamma_1\mu_n)^2, \tag{13}$$

$$\varphi_1(z,\mu_n) = J_1(\gamma_\varepsilon\mu_n)J_1(\mu_n e^{-\gamma_1 z/2}),\ \varphi_2(z,\mu_n) = J_1(\mu_n)J_1(\gamma_\varepsilon\mu_n e^{\gamma_2 z/2}), \tag{14}$$

$$\Psi(\mu_n) = \mu_n^2[(1+\gamma_\varepsilon K_\varepsilon)J_0(\mu_n)J_0(\gamma_\varepsilon\mu_n) - (\gamma_\varepsilon + K_\varepsilon)J_1(\mu_n)J_1(\gamma_\varepsilon\mu_n)], \tag{15}$$

$$K_\varepsilon = K^*(k^*)^{-1/2},\ \gamma_\varepsilon = \gamma^*(k^*)^{1/2}, \tag{16}$$

$$\Lambda = \frac{q_0}{\gamma_2 K_{2,1}^{(0)}},\ K^* = \frac{K_{1,1}^{(0)}}{K_{2,1}^{(0)}},\ k^* = \frac{k_1^{(0)}}{k_2^{(0)}},\ \gamma^* = \frac{\gamma_1}{\gamma_2}, \tag{17}$$

where $\mu_n > 0$, $n = 1,2,3,\ldots$, are the real roots of the functional equation:

$$J_0(\gamma_\varepsilon\mu_n)J_1(\mu_n) + K_\varepsilon J_0(\mu_n)J_1(\gamma_\varepsilon\mu_n) = 0. \tag{18}$$

$J_k(x)$—are the Bessel functions of the first kind of the kth order [22].

The temperature of the friction surfaces of both elements, in accordance with the assumption (4) of their perfect thermal contact of friction, should be the same. Substituting $z = 0$ in Equations (10)–(12) and (14), the following were obtained:

$$T(t) \equiv T(0^\pm,t) = T_0 + \Theta(t),\ 0 \le t \le t_s, \tag{19}$$

$$\Theta(t) \equiv \Theta(0^\pm,t) = \Lambda\left[\frac{q^*(t)}{(1+\gamma_\varepsilon K_\varepsilon)} + \frac{4}{\gamma_\varepsilon}\sum_{n=1}^{\infty}\frac{\hat{\varphi}(\mu_n)}{\Psi(\mu_n)}G_n(t)\right],\ 0 \le t \le t_s, \tag{20}$$

where

$$\hat{\varphi}(\mu_n) = J_1(\gamma_\varepsilon\mu_n)J_1(\mu_n). \tag{21}$$

Dimensionless variables and parameters were introduced:

$$\zeta = \frac{z}{a},\ \tau = \frac{k_1 t}{a^2},\ \tau_s = \frac{k_1 t_s}{a^2},\ \Theta_0 = \frac{q_0 a}{K_{1,1}^{(0)}},\ \Theta^* = \frac{\Theta}{\Theta_0}, \tag{22}$$

where parameters a and q_0 were determined accordingly from Formulas (6) and (8). Taking into account the indications (22) in Formulas (11)–(14), the dimensionless temperature rise of the friction pair elements can be presented in the form:

$$\Theta^*(\zeta,\tau) = \frac{K_0^*}{\gamma_2^*}e^{-\gamma_1^*\zeta/2}\left[\frac{e^{-\gamma_1^*\zeta/2}}{(1+\gamma_\varepsilon K_\varepsilon)}q^*(\tau) + \frac{4}{\gamma_\varepsilon}\sum_{n=1}^{\infty}\frac{\varphi_1^*(\zeta,\mu_n)}{\Psi(\mu_n)}G_n(\tau)\right], \zeta \ge 0,\ 0 \le \tau \le \tau_s, \tag{23}$$

$$\Theta^*(\zeta,\tau) = \frac{K_0^*}{\gamma_2^*}e^{\gamma_2^*\zeta/2}\left[\frac{e^{\gamma_2^*\zeta/2}}{(1+\gamma_\varepsilon K_\varepsilon)}q^*(\tau) + \frac{4}{\gamma_\varepsilon}\sum_{n=1}^{\infty}\frac{\varphi_2^*(\zeta,\mu_n)}{\Psi(\mu_n)}G_n(\tau)\right],\ \zeta \ge 0,\ 0 \le \tau \le \tau_s, \tag{24}$$

where:

$$\varphi_1^*(\zeta,\mu_n) = J_1(\gamma_\varepsilon\mu_n)J_1(\mu_n e^{-\gamma_1^*\zeta/2}),\ \varphi_2(\zeta,\mu_n) = J_1(\mu_n)J_1(\gamma_\varepsilon\mu_n e^{\gamma_2^*\zeta/2}), \tag{25}$$

$$G_n(\tau) = e^{-\lambda_n\tau} - \frac{1}{\lambda_n\tau_s}(1-e^{-\lambda_n\tau}),\ \lambda_n = \frac{1}{4}(\gamma_1^*\mu_n)^2, \tag{26}$$

and the remaining functions as well as parameters are given by Formulas (15)–(18). Substituting $\zeta = 0$ in Formulas (23)–(25), the dimensionless rise of the temperature on the friction surfaces was obtained:

$$\Theta^*(\tau) \equiv \Theta^*(0^{\pm}, \tau) = \frac{K_0^*}{\gamma_2^*}\left[\frac{q^*(\tau)}{(1+\gamma_\varepsilon K_\varepsilon)} + \frac{4}{\gamma_\varepsilon}\sum_{n=1}^{\infty}\frac{\hat{\varphi}(\mu_n)}{\Psi(\mu_n)}G_n(\tau)\right],\ 0 \le \tau \le \tau_s, \tag{27}$$

Based on Fourier's law, the intensity of heat fluxes directed along the normal to the contact surface $z = 0$ towards the insides of the friction pair elements were defined:

$$q_l(t) = (-1)^l K_{l,1}^{(0)} \frac{\partial T(z,t)}{\partial z}\bigg|_{z=0^{\pm}},\ 0 \le t \le t_s,\ l = 1,2. \tag{28}$$

Taking into account the indications (22) dimensionless intensities of heat fluxes $q_l^* = q_l q_0^{-1}, l = 1,2$ were written as:

$$q_1^*(\tau) = -\frac{\partial\Theta^*(\zeta,\tau)}{\partial\zeta}\bigg|_{\zeta=0^+},\ q_2^*(\tau) = \frac{\partial\Theta^*(\zeta,\tau)}{K_0^*\partial\zeta}\bigg|_{\zeta=0^-},\ 0 \le \tau \le \tau_s. \tag{29}$$

After differentiating the solution (23)–(26) with respect to the variable ζ and subsequent substitution of the found derivatives to the right side of Formula (30), the following was found:

$$q_1^*(\tau) = \frac{\gamma_\varepsilon K_\varepsilon}{(1+\gamma_\varepsilon K_\varepsilon)}q^*(\tau) + 2\gamma_\varepsilon K_\varepsilon \sum_{n=1}^{\infty}\frac{\tilde{\varphi}(\mu_n)}{\tilde{\Psi}(\mu_n)}G_n(\tau),\ 0 \le \tau \le \tau_s, \tag{30}$$

$$q_2^*(\tau) = \frac{1}{(1+\gamma_\varepsilon K_\varepsilon)}q^*(\tau) + 2\sum_{n=1}^{\infty}\frac{\tilde{\varphi}(\mu_n)}{\tilde{\Psi}(\mu_n)}G_n(\tau),\ 0 \le \tau \le \tau_s, \tag{31}$$

where:

$$\tilde{\varphi}(\mu_n) = J_0(\gamma_\varepsilon\mu_n)J_1(\mu_n),\ \tilde{\Psi}(\mu_n) = \mu_n^{-1}\Psi(\mu_n), \tag{32}$$

and functions $\Psi(\mu_n)$ and $G_n(\tau)$ can be determined from Equations (15) and (26), respectively.

It should be noted that in the case of homogeneous materials ($\gamma_i \to 0,\ i = 1,2$) of the disc and pads, the dimensionless temperature rise during braking with a constant deceleration has the form [23]:

$$\Theta^*(\zeta,\tau) = \frac{2K^*\sqrt{\tau}}{(1+K_\varepsilon)}\left\{\mathrm{ierfc}\left(\frac{\zeta}{2\sqrt{\tau}}\right) - \frac{\tau}{\tau_s}\left[\left(1+\frac{\zeta^2}{6\tau}\right)\mathrm{ierfc}\left(\frac{\zeta}{2\sqrt{\tau}}\right) - \frac{e^{-\frac{\zeta^2}{4\tau}}}{3\sqrt{\pi}}\right]\right\},\quad \zeta \ge 0,\ 0 \le \tau \le \tau_s, \tag{33}$$

$$\Theta^*(\zeta,\tau) = \frac{2K^*\sqrt{\tau}}{(1+K_\varepsilon)}\left\{\mathrm{ierfc}\left(-\frac{\zeta}{2}\sqrt{\frac{k_0^*}{\tau}}\right) - \frac{\tau}{\tau_s}\left[\left(1+\frac{\zeta^2 k_0^*}{6\tau}\right)\mathrm{ierfc}\left(-\frac{\zeta}{2}\sqrt{\frac{k_0^*}{\tau}}\right) - \frac{e^{-\frac{\zeta^2 k_0^*}{4\tau}}}{3\sqrt{\pi}}\right]\right\},\quad \zeta \le 0,\ 0 \le \tau \le \tau_s, \tag{34}$$

where $\mathrm{ierfc}(x) = \pi^{-1/2}e^{-x^2} - x\,\mathrm{erfc}(x)$, $\mathrm{erfc}(x) = 1 - \mathrm{erf}(x)$, $\mathrm{erf}(x)$—Gaussian error function. For $\zeta = 0$ from Equations (33) and (34), the known solution of Fazekas was obtained [24]:

$$\Theta^*(\tau) = \frac{2K^*}{(1+K_\varepsilon)}\sqrt{\frac{\tau}{\pi}}\left(1 - \frac{2\tau}{3\tau_s}\right),\ 0 \le \tau \le \tau_s. \tag{35}$$

4. Volume Temperature

With the given input parameters, solutions (19)–(27) make it possible to find the space–time distribution of the temperature inside and its evolution on the friction surfaces of the pad and disc, made of thermally insensitive FGMs. In order to take into account the thermal sensitivity of materials determining the temperature of the braking system using the above-mentioned solutions, the thermal conductivities $K_{l,m}^{(0)}$, specific heat capacities $c_{l,m}^{(0)}$ and densities $\rho_{l,m}^{(0)}$ at the initial temperature T_0 were replaced with corresponding values $K_{l,m}^{(\vartheta_l)}$, $c_{l,m}^{(\vartheta_l)}$ and $\rho_{l,m}^{(\vartheta_l)}$ found from Equations (1) and (2) for the volume temperature of the pad and the disc during braking [2,9]:

$$\vartheta_l = T_0 + \hat{\vartheta}_l,\ l = 1,2, \tag{36}$$

where:

$$\hat{\vartheta}_l = \frac{2\alpha_l W_0}{3G_l c_l^{(0)}}, \tag{37}$$

$$G_l = A_a a_l \rho_l^{(0)}, \tag{38}$$

a_l—the effective depths of heat penetration (6), $\alpha_1 = \alpha$, $\alpha_2 = 1-\alpha$, $0 \leq \alpha \leq 1$—heat partition ratio. Based on Formulas (30)–(32), the heat partition ratio was calculated from the formula:

$$\alpha \equiv \frac{q_1(t)}{q(t)} \approx \frac{\gamma_\varepsilon K_\varepsilon}{1+\gamma_\varepsilon K_\varepsilon}. \tag{39}$$

5. Numerical Analysis

The calculations were performed for the friction pair, one element of which was made of aluminium oxide Al_2O_3 (friction surface) and cooper Cu (core) [25]. The friction surface and core of the second element are manufactured of zirconium dioxide ZrO_2 and titanium alloy Ti-6Al-4V [14]. The temperature-dependent properties of these materials are as follows:

Al_2O_3 [26–28]

$$K_{1,1}(T) = 39.717 - 0.130T + 4.463 \cdot 10^{-4}T^2 - 2.836 \cdot 10^{-7}T^3 + 1.941 \cdot 10^{-10}T^4, \tag{40}$$

$$c_{1,1}(T) = 680.72 + 2.432T - 0.53 \cdot 10^{-2}T^2 + 0.6 \cdot 10^{-5}T^3 - 0.4 \cdot 10^{-8}T^4 + 10^{-12}T^5, \tag{41}$$

$$\rho_{1,1}(T) = 3992.2 - 0.062T - 0.6 \cdot 10^{-4}T^2 + 0.4 \cdot 10^{-7}T^3 - 0.9 \cdot 10^{-11}T^4, \tag{42}$$

Cu [17,29]

$$K_{1,2}(T) = 31.985 + 0.0099\,T - 0.1 \cdot 10^{-5}T^2, \tag{43}$$

$$c_{1,2}(T) = 523.3 + 1.4726\,T - 0.0024T^2 + 0.2 \cdot 10^{-5}T^3 - 0.5 \cdot 10^{-9}T^4, \tag{44}$$

$$\rho_{1,2}(T) = 492.45 - 0.01\,T - 0.1 \cdot 10^{-5}T^2, \tag{45}$$

ZrO_2 [27,30,31]

$$K_{2,1}(T) = 1.9365 + 0.7 \cdot 10^{-4}T + 0.5 \cdot 10^{-6}\,T^2 - 0.2 \cdot 10^{-9}T^3, \tag{46}$$

$$c_{2,1}(T) = 437.96 + 0.7767T - 0.17 \cdot 10^{-2}T^2, \tag{47}$$

$$\rho_{2,1}(T) = 6104.6 - 0.1212T - 0.4 \cdot 10^{-4}T^2 + 0.3 \cdot 10^{-7}T^3 - 0.1 \cdot 10^{-10}T^4, \tag{48}$$

Ti-6Al-4V [32,33]

$$K_{2,2}(T) = 6.6926 + 8.9177 \cdot 10^{-3}\,T + 6.8432 \cdot 10^{-6}T^2, \tag{49}$$

$$c_{2,2}(T) = 529.9316 + 0.4154T - 4.01646 \cdot 10^{-4}T^2 + 1.6364 \cdot 10^{-7}T^3, \tag{50}$$

$$\rho_{2,2}(T) = 4434 - 0.1088T - 0.8 \cdot 10^{-4}T^2 + 10^{-7}T^3 - 0.6 \cdot 10^{-10}T^4. \tag{51}$$

Graphs of dimensionless functions $K_{l,m}^* = K_{l,m}(T)/K_{l,m}^{(0)}$, $c_{l,m}^* = c_{l,m}(T)/c_{l,m}^{(0)}$ and $\rho_{l,m}^* = \rho_{l,m}(T)/\rho_{l,m}^{(0)}$ are illustrated in the Figures 1–3.

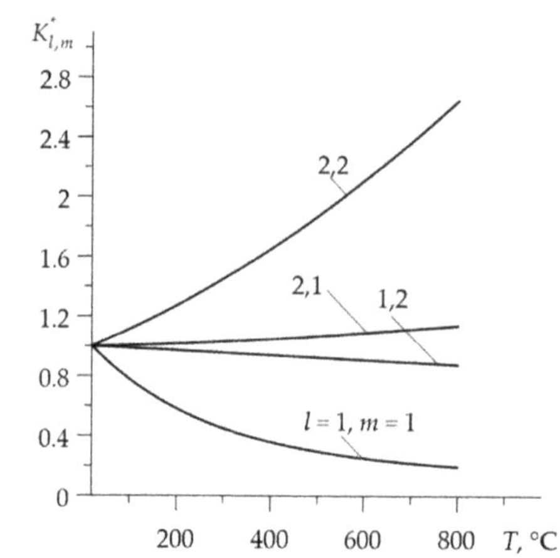

Figure 1. Dependencies of the dimensionless thermal conductivities $K^*_{l,m}$ on temperature T.

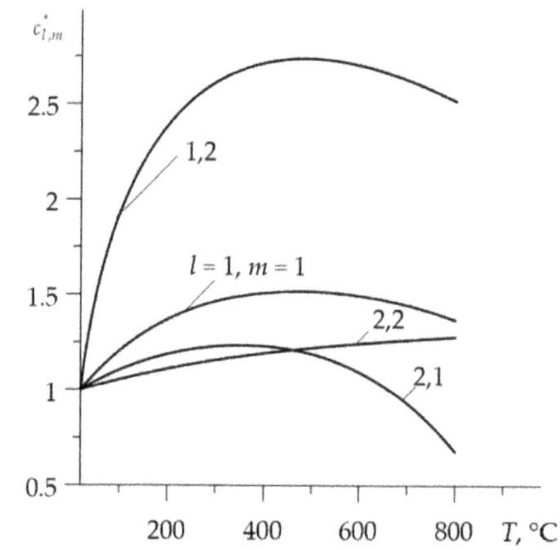

Figure 2. Dependencies of the dimensionless specific heat capacities $c^*_{l,m}$ on temperature T.

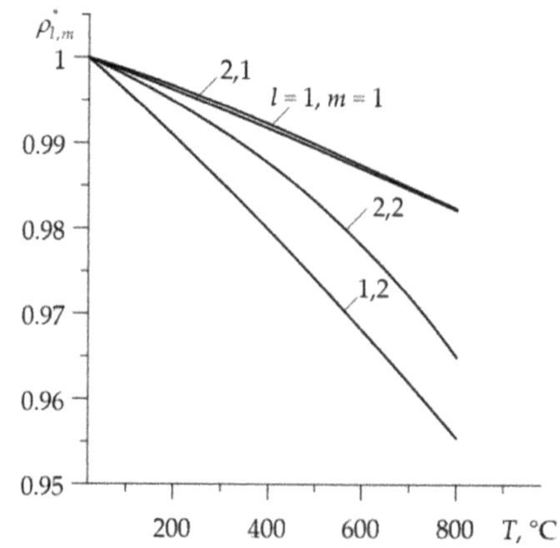

Figure 3. Dependencies of the dimensionless densities $\rho^*_{l,m}$ on temperature T.

The calculations were performed according to the following scheme:

(1) the values of the input parameters were given (Table 1), and then from Equations (8) and (9) the area of the nominal contact was calculated $A_a = 0.0022\,\mathrm{m}^2$, specific friction power $q_0 = 3.87\,\mathrm{MW\,m}^{-2}$, friction power $Q_0 = 8510$ W and stop time $t_s = 12.1$ s;

Table 1. Input parameters.

Friction Coefficient f	Nominal Pressure p_0, MPa	Initial Sliding Speed V_0, ms^{-1}	Initial Kinetic Energy W_0, kJ	Outer Radius R_e, mm	Inner Radius R_i, mm	Initial Temperature T_0, °C
0.27	0.602	23.8	103.54	37.5	26.5	20

(2) using the dependencies (40)–(51) the materials properties $K_{l,m}^{(0)}$, $c_{l,m}^{(0)}$ and $\rho_{l,m}^{(0)}$, $l, m = 1,2$ at the initial temperature $T_0 = 20\,°\mathrm{C}$ were established (Table 2);

Table 2. Material properties at the initial temperature T_0.

Element Index	Material Index	Material	Thermal Conductivity $K_{l,m}^{(0)}$, Wm^{-1}K^{-1}	Specific Heat Capacity $c_{l,m}^{(0)}$, J kg^{-1}K^{-1}	Density $\rho_{l,m}^{(0)}$, kgm^{-3}
$l = 1$	$m = 1$	Al_2O_3	37.24	727.29	3990.92
	$m = 2$	Cu	402.65	147.35	8947.92
$l = 2$	$m = 1$	ZrO_2	1.94	452.83	6102.16
	$m = 2$	Ti-6Al-4V	6.87	538.08	4431.79

(3) the effective values of: the specific heat $c_l^{(0)}$, density $\rho_l^{(0)}$, thermal diffusivity $k_l^{(0)}$, the effective depths of heat penetration a_l and the dimensionless gradient parameters of materials γ_l^*, $l = 1,2$ were found from Equations (3) and (5)–(7). Then, the dimensionless parameters K_ε and γ_ε were determined from the Formulas (16) and (17), and also the weight G_l and heat partition ratios α_l, $l = 1,2$ were calculated from the Equations (38) and (39) (Table 3);

Table 3. Calculated parameters at the initial temperature T_0.

Element Index	$l = 1$	$l = 2$
$c_l^{(0)}$, J kg^{-1}K^{-1}	437.3	495.5
$\rho_l^{(0)}$, kgm^{-3}	6469.4	5267
$k_l^{(0)} \times 10^6$, m^2 s^{-1}	13.2	0.743
γ_l^*	2.381	1.266
a_l, mm	21.854	5.193
G_l, kg	0.3127	0.0605
α_l	0.896	0.104

(4) the volume temperature values $\vartheta_1^{(0)} = 471.97$ of the disc and $\vartheta_2^{(0)} = 260.92$ the pad were obtained from the Equations (36) and (37);

(5) the values of materials properties $K_{l,m}^{(\vartheta_l^{(0)})}$, $c_{l,m}^{(\vartheta_l^{(0)})}$, $\rho_{l,m}^{(\vartheta_l^{(0)})}$, $l, m = 1,2$, corresponding to the volume temperature $\vartheta_l^{(0)}$ were determined from the Formulas (40)–(51);

(6) the steps (3)–(5) were repeated resulting in the corrected values for the volume temperature $\vartheta_l^{(1)} = 624.93$, and $\vartheta_2^{(1)} = 292.98$;

(7) by means of the formula $\vartheta_l = 0.5(\vartheta_l^{(0)} + \vartheta_l^{(1)})$, $l = 1,2$ final values of the volume temperature $\vartheta_1 = 548.45\,°\mathrm{C}$, and $\vartheta_2 = 267.95\,°\mathrm{C}$ were found;

(8) based on the dependencies (40)–(51) the values of materials properties $K_{l,m}^{(\vartheta_l)}$, $c_{l,m}^{(\vartheta_l)}$, $\rho_{l,m}^{(\vartheta_l)}$, $l,m = 1,2$ corresponding to the volume temperature ϑ_l were established (Table 4) and other parameters necessary to perform the calculations (Table 5);

Table 4. Material properties at volume temperature ϑ_l, $l = 1,2$.

Element Index	Material Index	Material	Thermal Conductivity $K_{l,m}^{(\vartheta_l)}$, $\mathrm{Wm^{-1}K^{-1}}$	Specific Heat Capacity $c_{l,m}^{(\vartheta_l)}$, $\mathrm{J\,kg^{-1}K^{-1}}$	Density $\rho_{l,m}^{(\vartheta_l)}$, $\mathrm{kgm^{-3}}$
$l = 1$	$m = 1$	Al_2O_3	10.19	1097.93	3945.59
	$m = 2$	Cu	367.15	401.89	8690.20
$l = 2$	$m = 1$	ZrO_2	1.99	552.67	6069.84
	$m = 2$	Ti-6Al-4V	9.57	615.44	4399.06

Table 5. Calculated parameters at volume temperature ϑ_l, $l = 1,2$.

Element Index	$l = 1$	$l = 2$
$c_l^{(\vartheta_l)}$, $\mathrm{J\,kg^{-1}K^{-1}}$	749.7	584.9
$\rho_l^{(\vartheta_l)}$, $\mathrm{kgm^{-3}}$	6317.2	5233.8
$k_l^{(\vartheta_l)} \times 10^6$, $\mathrm{m^2\,s^{-1}}$	2.15	0.65
γ_l^*	3.585	1.583
a_l, mm	8.834	4.854
G_l, kg	0.1234	0.0562
α_l	0.863	0.137

(9) the temperature field $\Theta^*(\zeta,\tau)$ (23)–(26), the temperature evolution $\Theta^*(\tau)$ (27), and temporal profiles of heat fluxes intensities $q_l^*(\tau)$, $l = 1,2$ (30)–(32) were determined.

In order to calculate the values of Bessel functions $J_k(x)$, $k = 0, 1$ the programs BESSJ0 and BESSJ1 from the Numerical Recipes package [34] were used. The roots of the characteristic Equation (18) were searched for by the bisection method with the RTBIS program from this package. In summation of the series in solutions (23), (24), and (30), (31) was performed with an accuracy of 5×10^{-5}. For this accuracy, the minimum number of components was equal to 70.

Changes in the dimensionless temperature rise $\Theta^*(\zeta,\tau)$ during braking, at few selected distances from the contact surface are presented in Figure 4. The temperature calculated with an account of the thermal sensitivity of the materials (solid lines) is significantly lower in both friction elements compared to the results achieved without taking into account the temperature dependencies of FGMs properties (dashed lines). The maximum dimensionless temperature on the contact surface $\zeta = 0$ without and taking into account the thermal sensitivity of the materials are 0.816 and 0.277, respectively (reduction of about 2.94 times) and are reached at the time moments $\tau_{\max} = 0.37$ and $\tau_{\max} = 0.29$ (reduction of 21.6%).

Increasing the distance from the contact surface $\zeta = 0$, the temperature level of both elements drops (Figure 5). The temperature of components made of thermally sensitive materials is lower than their temperature, found for the constant material properties. The greatest difference between these results is on the contact surface.

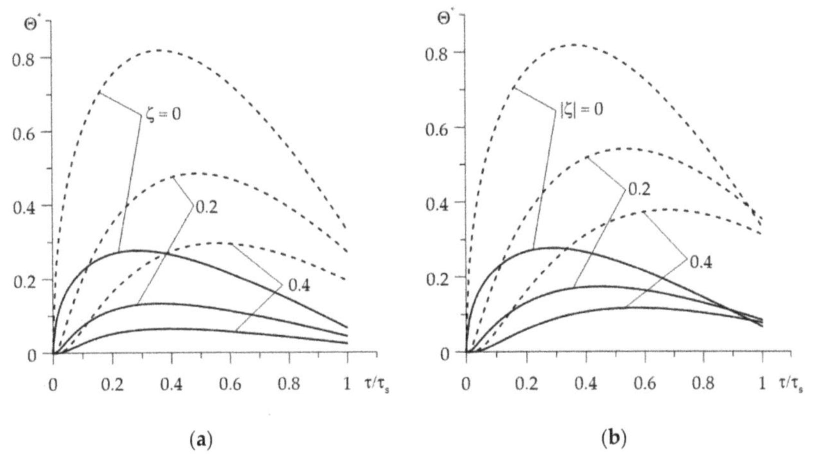

Figure 4. Evolutions of the dimensionless temperature $\Theta^*(\zeta, \tau)$ during braking at different distances ζ from the surface of friction with (solid lines) and without (dashed lines) taking into account the thermal sensitivity of the materials: (**a**) Al_2O_3—Cu; (**b**) ZrO_2—Ti-6Al-4V.

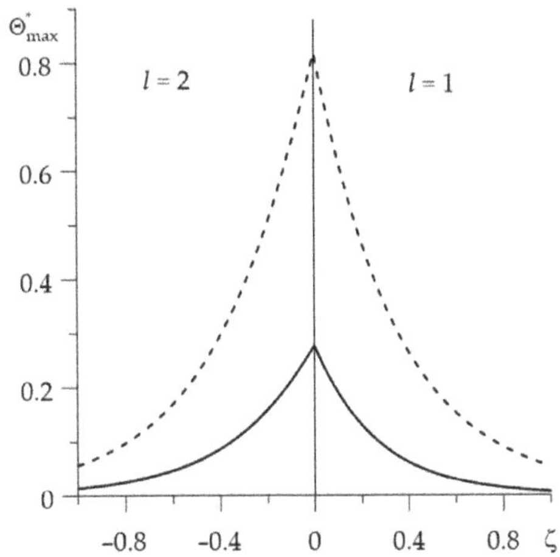

Figure 5. Distribution of the dimensionless temperature $\Theta^*_{max}(\zeta) = \Theta^*(\zeta, \tau_{max})$ reached at the time moment $\tau = \tau_{max}$ along the distance ζ from the surface of friction with (solid lines) and without (dashed lines) taking into account the thermal sensitivity of the materials.

The conclusions established on the basis of Figures 4 and 5 confirm the results of the calculations, presented in Figure 6. It shows the dimensionless temperature isotherms $\Theta^*(\zeta, \tau)$. It can be seen that the effective depth of heat transfer is much greater in the case that material properties remain unchanged under the influence of temperature, than in the case of considering the thermally sensitive FGMs. This effect is most noticeable for the first one ($l = 1$), the Al_2O_3-Cu element. This result is also confirmed by the parameter values a_l, $l = 1, 2$ presented in Tables 3 and 5.

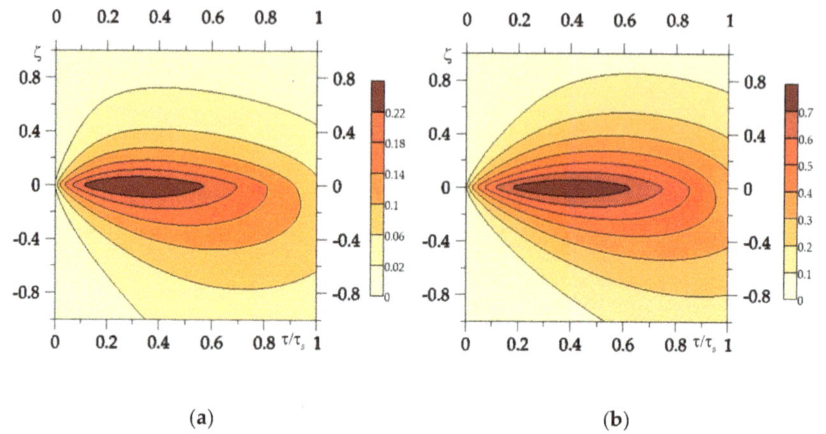

Figure 6. Isotherms of the dimensionless temperature $\Theta^*(\zeta,\tau)$ for: (a) thermally sensitivity materials; (b) materials with properties at the initial temperature.

The time profiles of the dimensionless intensities of heat fluxes $q_l^*(\tau)$, $l = 1,2$ are shown in Figure 7. They decrease linearly during the braking process from the maximum value at the initial moment to zero at the stop. Most of the frictional heat generated is absorbed by the first element ($l = 1$) Al_2O_3-Cu. The linear change in $q_l^*(\tau)$ is the result of the specific friction power $q^*(\tau)$ (8), which decreases linearly during braking with a constant deceleration, and the requirement to meet the boundary condition $q_1^*(\tau) + q_2^*(\tau) = q^*(\tau)$, $0 \leq \tau \leq \tau_s$. The influence of thermal sensitivity on the intensity of heat fluxes is much smaller than on the temperature. For thermally sensitive materials, the maximum values of the intensity of heat fluxes are $q_{1,\max}^* = 0.864$ and $q_{2,\max}^* = 0.136$, and for constant properties of the materials, we have $q_{1,\max}^* = 0.895$ and $q_{2,\max}^* = 0.105$.

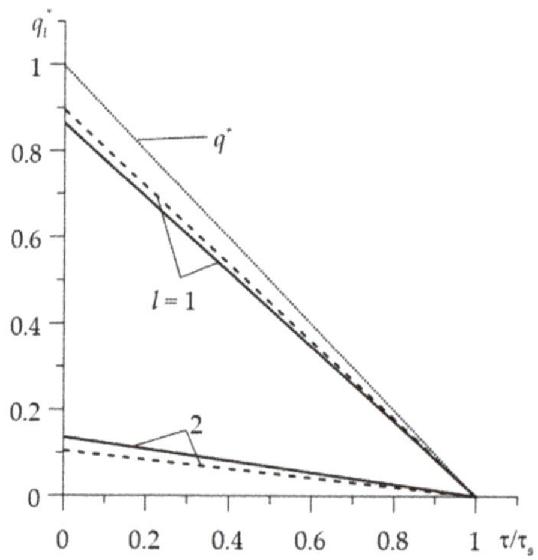

Figure 7. Temporal profiles of the dimensionless heat fluxes $q_l^*(\tau)$, $l = 1,2$ during braking with (continuous lines) and without (dashed lines) taking into account the thermal sensitivity of the materials. Dotted lines represent the dimensionless specific power of friction q^*.

6. Conclusions

A calculation scheme was proposed to determine the temperature field of the friction elements of a disc brake, taking into account the changes in the FGMs properties depending on the actual temperature. The main part of the scheme was the adaptation of a linear solution (with temperature-independent material properties) to the thermal problem of friction during braking to thermally sensitive FGMs. A numerical analysis was performed in the case of braking with constant deceleration of elements made of two-component functionally graded materials with exponential variations in thermal conductivities in the axial direction, across the volume of the materials. It was found that:

- the influence of thermal sensitivity on the temperature of FGMs may be more significant than in the case of homogeneous materials;
- for the selected friction pair, taking into account the thermal sensitivity caused an almost threefold reduction in the maximum temperature in comparison to the appropriate temperature values, found with the same properties of the materials;
- the influence of thermal sensitivity on the intensity of heat fluxes directed from the friction surface to the interior of the friction pair elements is insignificant. This means that to estimate the amount of heat absorbed by the individual elements of the friction pair, appropriate solutions to linear problems can be used.

A verification of the developed theoretical model based on empirical results would be advisable. However, no information on this kind of experimental data has been found in the literature. In particular, it concerns the frictional heating of braking systems with friction elements made of thermally sensitive FGMs. Therefore, the verification of the exact solution was obtained carried out by determining from it, in cases of limit parameters, known solutions of other authors for homogeneous materials, which were verified with appropriate experimental data. A new element, significantly differentiating the results of a given article from those published earlier by us, is the incorporation in the model of the possibility of changing the frictional properties of FGMs under temperature influence. This model includes many new elements, such as determining the intensity of heat fluxes to obtain the form of the heat partition ratio, finding the volume temperature of FGMs, developing a calculation algorithm that takes into account the thermal sensitivity of all materials components, etc. We have shown that taking into consideration the thermal sensitivity of materials can significantly reduce the surface temperature contact of the pad and disc. We proposed a theoretical computational model. We hope that it will be verified with the data obtained from other authors' research positions. An indirect confirmation of the correctness of our model is also the time profiles of temperature and heat fluxes obtained on its basis, characteristic for braking with a constant deceleration.

It should be noted that all three of our papers constitute a monothematic cycle of interrelated research. We also want to develop a suitable model for braking systems operating in a short-term, repetitive mode. The problem of lowering the temperature level in such systems is up to date.

Author Contributions: Conceptualization and methodology, A.Y; software, P.Z.; validation, K.T., formal analysis, A.Y.; investigation, A.Y., K.T. and P.Z.; writing—original draft preparation K.T. and A.Y.; writing—review and editing, A.Y. and K.T.; visualization and figures preparation, P.Z.; supervision, A.Y.; project administration, K.T. All authors have read and agreed to the published version of the manuscript.

Funding: This study was performed within the framework of research project No. 2017/27/B/ST8/01249, funded by the National Science Centre, Poland and project financing through the program of the Minister of Education and Science of Poland named "Regional Initiative of Excellence" in 2019–2022, project No. 011/RID/2018/19 (amount of financing 12,000,000 PLN).

Institutional Review Board Statement: Not applicable.

Informed Consent Statement: Not applicable.

Data Availability Statement: No new data were created or analyzed in this study. Data sharing is not applicable to this article.

Conflicts of Interest: The authors declare no conflict of interest.

Nomenclature

a_l	Effective depth of heat penetration (m)
A_a	Area of the nominal contact region (m^2)
$c_{l,m}$	Specific heat capacity ($J\,kg^{-1}K^{-1}$)
f	Coefficient of friction (dimensionless)
G_l	Weight of the friction elements (kg)
$J_k(\cdot)$	The Bessel functions of the first kind of the kth order
$k_{l,m}$	Thermal diffusivity (m^2s^{-1})
$K_{l,m}$	Thermal conductivity ($W\,m^{-1}K^{-1}$)
p	Contact pressure (Pa)
p_0	Nominal value of the contact pressure (Pa)
R_e	External radius of the pads (m)
R_i	Internal radius of the pads (m)
q	Specific power of friction ($W\,m^{-2}$)
q_0	Nominal value of the specific power of friction ($W\,m^{-2}$)
Q_0	Nominal friction power (W)
t	Time (s)
t_s	Stop time (s)
T	Temperature (°C)
T_0	Initial temperature (°C)
V	Velocity ($m\,s^{-1}$)
V_c, V_ρ	Volume fractions of the material phases
V_0	Initial velocity ($m\,s^{-1}$)
W_0	Initial kinetic energy of the system (J)
z	Spatial coordinate in axial direction (m)
lower l	Number of the main ($l = 1$) and frictional ($l = 2$) elements of the friction pair
lower m	Number of the component material $m = 1, 2$ of selected friction element
α_l	Heat partition ratio (dimensionless)
β	Cover angle of the pads (rad)
γ_l	Parameter of material gradient (m^{-1})
γ_l^*	Parameter of material gradient (dimensionless)
Θ_l	Temperature rise (°C)
Θ_l^*	Temperature rise (dimensionless)
Θ_0	Temperature scaling factor (°C)
$\rho_{l,m}$	Density ($kg\,m^{-3}$)
τ	Time (dimensionless)
τ_s	Time of braking (dimensionless)
ζ	Spatial coordinate in axial direction (dimensionless)
ϑ_l	Volume temperature (°C)

References

1. Balakin, V.; Sergienko, V. *Thermal Calculation of Brakes and Frictional Assemblies*; Belyi Metal-Polymer Research Institute of the National Academy of Sciences of Belarus: Gomel, Belarus, 1999. (In Russian)
2. Chichinadze, A.V.; Braun, E.D.; Ginzburg, A.G.; Ignat'eva, E.V. *Calculation, Testing and Selection of Friction Couples*; Nauka: Moscow, Russia, 1979. (In Russian)
3. Adamowicz, A.; Grześ, P. Three-dimensional FE model for calculation of temperature of a thermosensitive disc. *Appl. Therm. Eng.* **2013**, *50*, 572–581. [CrossRef]
4. Yevtushenko, A.A.; Grześ, P.; Adamowicz, A. The temperature mode of the carbon-carbon multi-disc brake in the view of the interrelations of its operating characteristics. *Materials* **2020**, *13*, 1878. [CrossRef] [PubMed]
5. Kushnir, R.M.; Popovych, V.S. *Heat Conduction Problems of Thermosensitive Solids under Complex Heat Exchange*; Vyacheslav, V., Ed.; Heat Conduction-Basic Research: Rijeka, Croatia, 2011; ISBN 978-953-307-404-7.
6. Yevtushenko, A.A.; Kuciej, M.; Och, E. Some Methods for Calculating Temperature during the Friction of Thermosensitive Materials. *Numer. Heat Transf. Part A Appl.* **2015**, *67*, 696–718. [CrossRef]

7. Yevtushenko, A.; Kuciej, M.; Och, E.; Yevtushenko, O. Effect of the thermal sensitivity in modeling of the frictional heating during braking. *Adv. Mech. Eng.* **2016**, *8*, 1–10. [CrossRef]
8. Popovich, V.S.; Garmatii, G.Y. Solution of nonstationary heat conduction problems for thermosensitive bodies under convective heat exchange. *J. Math. Sci.* **1998**, *90*, 2037–2041. [CrossRef]
9. Evtushenko, O.; Kuciej, M.; Topczewska, K. Determination of the maximal temperature of a pad-disk tribosystem during one-time braking. *Mater. Sci.* **2020**, *56*, 152–159. [CrossRef]
10. Zelentsov, V.B.; Lapina, P.A.; Mitrin, B.I. Wear of Functionally Graded Coatings under Frictional Heating Conditions. *Nanomaterials* **2022**, *12*, 142. [CrossRef]
11. Yevtushenko, A.; Topczewska, K.; Zamojski, P. The Effect of Functionally Graded Materials on Temperature during Frictional Heating: Under Uniform Sliding. *Materials* **2021**, *14*, 4285. [CrossRef]
12. Abbas, I.A. Nonlinear transient thermal stress analysis of thick-walled FGM cylinder with temperature-dependent material properties. *Meccanica* **2014**, *49*, 1697–1708. [CrossRef]
13. Azadi, M.; Azadi, M. Nonlinear transient heat transfer and thermoelastic analysis of thick-walled FGM cylinder with temperature-dependent material properties using Hermitian transfinite element. *J. Mech. Sci. Technol.* **2009**, *23*, 2635. [CrossRef]
14. Tanigawa, Y.; Akai, T.; Kawamura, R.; Oka, N. Transient heat conduction and thermal stress problems of a nonhomogeneous plate with temperature-dependent material properties. *J. Therm. Stresses* **1996**, *19*, 77–102. [CrossRef]
15. Ding, S.; Wu, C.-P. Optimization of material composition to minimize the thermal stresses induced in FGM plates with temperature-dependent material properties. *Int. J. Mech. Mater. Des.* **2018**, *14*, 527–549. [CrossRef]
16. Noda, N. Thermal stresses in functionally graded materials. *J. Therm. Stresses* **1999**, *22*, 477–512. [CrossRef]
17. Ueda, S. Thermoelastic analysis of W-Cu functionally graded materials subjected to a thermal shock using a micromechanical model. *J. Therm. Stresses* **2001**, *24*, 19–46. [CrossRef]
18. Yevtushenko, A.; Topczewska, K.; Zamojski, P. The Effect of Functionally Graded Materials on Temperature during Frictional Heating at Single Braking. *Materials* **2021**, *14*, 6241. [CrossRef]
19. Chichinadze, A.V. *Polymers in Friction Assembles of Machines and Devices: A Handbook*; Allerton Press Inc.: New York, NY, USA, 1984.
20. Yevtushenko, A.; Kuciej, M.; Topczewska, K. Frictional Heating during Braking of the C/C Composite Disc. *Materials* **2020**, *13*, 2691. [CrossRef]
21. Yevtushenko, A.; Topczewska, K.; Kuciej, M. Analytical Determination of the Brake Temperature Mode during Repetitive Short-Term Braking. *Materials* **2021**, *14*, 1912. [CrossRef]
22. Watson, G.N. *A Treatise of the Theory of Bessel Functions*, 2nd ed.; Cambridge University Press: Cambridge, UK, 1995.
23. Yevtushenko, A.A.; Kuciej, M. Two heat conduction problems with frictional heating during braking. *J. Theor. Appl. Mech.* **2010**, *48*, 367–380.
24. Fazekas, G.A.G. Temperature gradients and heat stresses in brake drums. *SAE Trans.* **1953**, *61*, 279–284.
25. Strojny-Nędza, A.; Pietrzak, K.; Gili, F.; Chmielewski, M. FGM based on copper–alumina composites for brake disc applications. *Arch. Civ. Mech. Eng.* **2020**, *20*, 83. [CrossRef]
26. Burghartz, S.; Schulz, B. Thermophysical properties of sapphire, AlN and MgAl2O4 down to 70 K. *J. Nucl. Mater.* **1994**, *212–215 Pt B*, 1065–1068. [CrossRef]
27. Kingery, W.D.; Francl, J.; Coble, R.L.; Vasilos, T. Thermal Conductivity: X, Data for Several Pure Oxide Materials Corrected to Zero Porosity. *J. Am. Ceram. Soc.* **1954**, *37*, 107–110. [CrossRef]
28. Archer, D.A. Thermodynamic Properties of Synthetic Sapphire (α-Al_2O_3), Standard Reference Material 720 and the Effect of Temperature-Scale Differences on Thermodynamic Properties. *J. Phys. Chem. Ref. Data* **1993**, *22*, 1441–1453. [CrossRef]
29. Simon, N.J.; Drexler, E.S.; Reed, R.P. *Properties of Copper and Copper Alloys at Cryogenic Temperatures*; Monograph 177; National Institute of Standards and Technology: New York, NY, USA, 1992.
30. Pankratz, L.B. *Thermodynamic Properties of Elements and Oxides*; Bulletin 672; U.S. Bureau of Mines: Washington, DC, USA, 1982.
31. Taylor, D. Thermal expansion data. II: Binary oxides with the fluorite and rutile structures, MO_2, and the antifluorite structure, M_2O. *Trans. J. Br. Ceram. Soc.* **1984**, *83*, 32–37.
32. Deem, H.W.; Wood, W.D.; Lucks, C.F. The relationship between Electrical and Thermal Conductivities of Titanium Alloys. *Trans. Metall. Soc. AIME* **1958**, *212*, 520–523.
33. Cezairrliyan, A.; McClure, J.L.; Taylor, R.J. Thermophysical Measurements on 90Ti-6Al-4V Alloy Above 1450 K Using a Transient (Subsecond) Technique. *J. Res. Nat. Bur. Stand.-A Phys. Chem.* **1977**, *81*, 251–256. [CrossRef]
34. Press, W.H.; Teukolsky, S.A.; Flannery, B.P.; Vetterling, W.T. *Numerical Recipes in Fortran 77: Volume 1, Volume 1 of Fortran Numerical Recipes: The Art of Scientific Computing*; Cambridge University Press: Cambridge, UK, 1992.

materials

MDPI

Article

Temperature in the Friction Couple Consisting of Functionally Graded and Homogeneous Materials

Aleksander Yevtushenko, Michał Kuciej, Katarzyna Topczewska * and Przemysław Zamojski

Faculty of Mechanical Engineering, Bialystok University of Technology (BUT), 45C Wiejska Street, 15-351 Bialystok, Poland; a.yevtushenko@pb.edu.pl (A.Y.); m.kuciej@pb.edu.pl (M.K.); p.zamojski@pb.edu.pl (P.Z.)

* Correspondence: k.topczewska@pb.edu.pl

Abstract: An analytical model was developed to determine the temperature of friction coupling, in which one element was made of a functionally graded material (FGM) and the other was homogeneous. First, for such a system, the boundary–value problem of heat conduction was formulated with consideration of the heat generation due to friction. Then, using the Laplace integral transform, an exact solution to this problem was obtained for uniform sliding, and braking with constant deceleration. A numerical analysis was performed for the selected friction pair consisting of the FGM (zircon dioxide + titanium alloy) and cast iron. It was established that the use of elements made of a FGM consisting of ZrO_2 and Ti-6Al-4V can significantly reduce the maximum temperature achieved in the friction system.

Keywords: frictional heating; functionally graded materials; temperature; braking

Citation: Yevtushenko, A.; Kuciej, M.; Topczewska, K.; Zamojski, P. Temperature in the Friction Couple Consisting of Functionally Graded and Homogeneous Materials. *Materials* **2022**, *15*, 3600. https://doi.org/10.3390/ma15103600

Academic Editor: Hansang Kwon

Received: 6 April 2022
Accepted: 16 May 2022
Published: 18 May 2022

1. Introduction

Reviews of investigations on methods for establishing the temperature of systems containing friction elements made of functionally gradient materials (FGMs) can be found in previous articles [1–3]. In these studies, the methodology of determining the temperature in such friction couples under uniform sliding [1], during braking with time-dependent contact pressure [2], and considering the thermal sensitivity of component materials of FGMs was investigated [3]. The main factor in this methodology is an exact solution to the boundary–value heat conduction problem, taking into account the frictional heating of two semi-infinite bodies made of FGMs. It should be noted, however, that the obtained solutions did not allow determining automatically, with the help of limit transformations, solutions to the problems in the case when one of the friction pair elements is made of FGM and the other is homogeneous. Moreover, this type of friction pair is one of the most common [4]. Therefore, in this study, an attempt was made to develop a mathematical model for determining the temperature of a friction pair consisting of a body made of a two-component FGM, sliding on the surface of a homogeneous body. An exponential change in the thermal conductivity of the FGM with distance from the friction surface was assumed. Two modes of changing the sliding velocity over time were considered: uniform and linearly decreasing.

2. Statement of the Problem

The object of study is the transient temperature field, initiated in the process of frictional heating of the friction pair elements of a braking system, corresponding to the brake pad and disc. Taking into account the fact that the heat generated as a result of friction during braking is mainly directed along the normal from the friction surface to the inside of both elements [5,6], for the description of the heating process of the system, a contact scheme of two semi-infinite bodies was adopted, related to the Cartesian coordinate system (Figure 1).

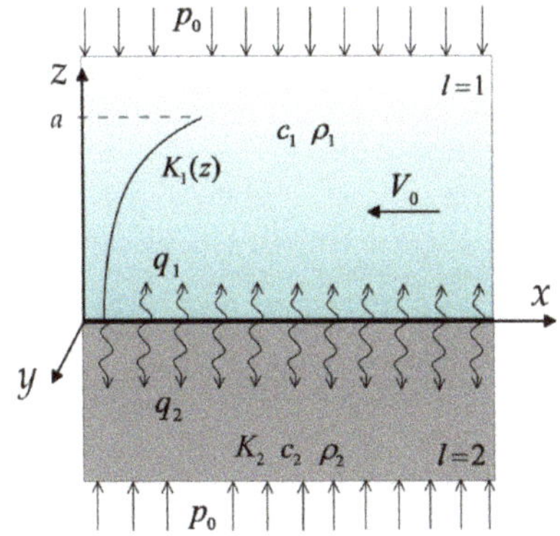

Figure 1. Scheme of the problem.

The pad (body 1) is made of a two-component functionally graded material (FGM), in such a way that the friction surface is a material of low thermal conductivity and high wear resistance (ceramics etc.), while the core material has high thermal conductivity (metal alloys, copper, iron etc.). The increase in thermal conductivity of the pad material in the distance from the friction surface is exponential. On the other hand, the disc (body 2) is made of a homogeneous material (cast iron etc.). A more detailed description of the adopted model assumptions is presented in our previous articles [1,2].

The analytical model presented in the manuscript concerns the frictional system of two semi-infinite bodies, in which it is not possible to take into consideration the heat exchange between the heated elements and the surrounding environment. It is known, however, that consideration of convection cooling, would lead to a lower maximum temperature; the most important parameters in the design process of frictional systems. It should be ensured that the theoretical value of the permissible temperature for a given material (i.e., the melting point) is not exceeded. For this reason, at the design stage, calculations should be performed for the maximum temperature achieved for adiabatic conditions on the free surfaces of the friction system.

The braking process with constant deceleration was considered when the contact pressure achieved its nominal value p_0 immediately at the beginning of the braking, with simultaneously reduction of velocity from the initial value V_0 to zero at the stopping moment $t = t_s$. For such braking, the specific friction power was written in the form:

$$q(t) = q_0 q^*(t), q_0 = f_0 p_0 V_0, q^*(t) = 1 - \frac{t}{t_s},\ 0 \le t \le t_s,\ t_s = \frac{W_0}{q_0 A_a}, \tag{1}$$

where f_0—friction coefficient, W_0—initial kinetic energy of the system, and A_a— area of nominal contact between one brake pad and the disc.

The temperature field $T(z,t)$ in the system consisting of two sliding semi-spaces was sought based on the solution to the following thermal problem of friction:

$$\frac{\partial}{\partial z}\left[K_1(z)\frac{\partial T(z,t)}{\partial z}\right] = c_1\rho_1\frac{\partial T(z,t)}{\partial t},\ z > 0,\ 0 < t \le t_s, \tag{2}$$

$$K_2\frac{\partial^2 T(z,t)}{\partial z^2} = c_2\rho_2\frac{\partial T(z,t)}{\partial t},\ z < 0,\ 0 < t \le t_s, \tag{3}$$

$$T(0^+,t) = T(0^-,t) \equiv T(t),\ 0 < t \le t_s, \tag{4}$$

$$K_2 \frac{\partial T(z,t)}{\partial z}\bigg|_{z=0^-} - K_1(z)\frac{\partial T(z,t)}{\partial z}\bigg|_{z=0^+} = q(t),\ 0 < t \le t_s, \tag{5}$$

$$T(z,t) \to T_0\,,\ |z| \to \infty,\ 0 < t \le t_s, \tag{6}$$

$$T(z,0) = T_0,\ |z| < \infty. \tag{7}$$

where

$$K_1(z) = K_{1,1}e^{\gamma z},\ z \ge 0,\ \gamma \ge 0, \tag{8}$$

$$c_1 = c_{1,1}(1-v) + c_{1,2}v,\ \rho_1 = \rho_{1,1}(1-v) + \rho_{1,2}v,\ 0 \le v \le 1, \tag{9}$$

temporal profile of specific friction power $q(t)$ was determined from Equation (1), $K_{1,m}$, $c_{1,m}$, and $\rho_{1,m}$—thermal conductivity, specific heat, and density of the first ($m = 1$) and the second ($m = 2$) component of pad material, respectively, and parameters K_2, c_2, and ρ_2 —correspond to the disc material, v—the relative volumetric fraction of the first component of the pad material, and T_0—temperature of the system at the initial time moment $t = 0$.

The dimensionless variables and parameters were introduced:

$$\zeta = \frac{z}{a},\ \tau = \frac{k_1 t}{a^2},\ \tau_s = \frac{k_1 t_s}{a^2},\ K^* = \frac{K_2}{K_{1,1}},\ k^* = \frac{k_2}{k_1},\ \Theta^* = \frac{T - T_0}{\Theta_0},\ \Theta_0 = \frac{q_0 a}{K_{1,1}^{(0)}}, \tag{10}$$

where

$$a = \sqrt{3k_1 t_s}, \tag{11}$$

$$k_1 = \frac{K_{1,1}}{c_1\rho_1},\ k_2 = \frac{K_2}{c_2\rho_2}. \tag{12}$$

Taking into account the designations (10)–(12), the problem (2)–(9) was written in the form:

$$\frac{\partial^2\Theta^*(\zeta,\tau)}{\partial\zeta^2} + \gamma^*\frac{\partial\Theta^*(\zeta,\tau)}{\partial\zeta} - e^{-\gamma^*\zeta}\frac{\partial\Theta^*(\zeta,\tau)}{\partial\tau} = 0,\ \zeta > 0,\ 0 < \tau \le \tau_s, \tag{13}$$

$$\frac{\partial^2\Theta^*(\zeta,\tau)}{\partial\zeta^2} - \frac{1}{k^*}\frac{\partial\Theta^*(\zeta,\tau)}{\partial\tau} = 0,\ \zeta < 0,\ 0 < \tau \le \tau_s, \tag{14}$$

$$\Theta^*(0^+,\tau) = \Theta^*(0^-,\tau) \equiv \Theta^*(\tau),\ 0 < \tau \le \tau_s, \tag{15}$$

$$K^*\frac{\partial\Theta^*(\zeta,\tau)}{\partial\zeta}\bigg|_{\zeta=0^-} - \frac{\partial\Theta^*(\zeta,\tau)}{\partial\zeta}\bigg|_{\zeta=0^+} = q^*(\tau),\ 0 < \tau \le \tau_s, \tag{16}$$

$$\Theta^*(\zeta,\tau) \to 0\,,\ |\zeta| \to \infty,\ 0 < \tau \le \tau_s, \tag{17}$$

$$\Theta^*(\zeta,0) = 0,\ |\zeta| < \infty, \tag{18}$$

where

$$q^*(\tau) = 1 - \frac{\tau}{\tau_s},\ 0 < \tau \le \tau_s, \tag{19}$$

$$\gamma^* \equiv a\gamma = \ln\left(\frac{K_{1,2}}{K_{1,1}}\right). \tag{20}$$

3. Frictional Heating under Uniform Sliding

First, the case of frictional heating process during sliding of the pad on the disc surface with constant velocity V_0 was considered. Then for $\tau_s \to \infty$ from the Equation (19), it follows that $q^*(\tau) = 1$. For the boundary–value heat conduction problem (13)–(20) with a constant temporal profile of specific friction power $q^*(\tau) = 1$, the integral Laplace transform was applied [7]:

$$\overline{\Theta}^*(\zeta, p) \equiv L[\Theta^*(\zeta, \tau); p] = \int_0^\infty \Theta^*(\zeta, \tau) e^{-p\tau} d\tau, \tag{21}$$

it was obtained:

$$\frac{d^2\overline{\Theta}^*(\zeta, p)}{d\zeta^2} + \gamma^* \frac{d\overline{\Theta}^*(\zeta, p)}{d\zeta} - pe^{-\gamma^*\zeta}\overline{\Theta}^*(\zeta, p) = 0,\ \zeta > 0, \tag{22}$$

$$\frac{d^2\overline{\Theta}^*(\zeta, p)}{d\zeta^2} - \frac{p}{k^*}\overline{\Theta}^*(\zeta, p) = 0,\ \zeta < 0, \tag{23}$$

$$\overline{\Theta}^*(0^+, p) = \overline{\Theta}^*(0^-, p) \equiv \overline{\Theta}^*(p), \tag{24}$$

$$K^* \frac{d\overline{\Theta}^*(\zeta, p)}{d\zeta}\bigg|_{\zeta=0^-} - \frac{d\overline{\Theta}^*(\zeta, p)}{d\zeta}\bigg|_{\zeta=0^+} = \frac{1}{p}, \tag{25}$$

$$\overline{\Theta}^*(\zeta, p) \to 0\,,\ |\zeta| \to \infty. \tag{26}$$

An exact solution to the ordinary differential Equations (22) and (23), which meet the boundary conditions (24)–(26) has the form:

$$\overline{\Theta}^*(\zeta, p) = \frac{\Delta_1(\zeta, p)}{p\sqrt{p}\Delta(p)},\ \zeta \geq 0,\ \overline{\Theta}^*(\zeta, p) = \frac{\Delta_2(\zeta, p)}{p\sqrt{p}\Delta(p)},\ \zeta \leq 0, \tag{27}$$

where

$$\Delta_1(\zeta, p) = e^{-0.5\gamma^*\zeta} I_1\left(\frac{2}{\gamma^*}\sqrt{p}e^{-0.5\gamma^*\zeta}\right),\ \Delta_2(\zeta, p) = e^{\sqrt{\frac{p}{k^*}}\zeta} I_1\left(\frac{2}{\gamma^*}\sqrt{p}\right), \tag{28}$$

$$\Delta(p) = I_0\left(\frac{2}{\gamma^*}\sqrt{p}\right) + K_\varepsilon I_1\left(\frac{2}{\gamma^*}\sqrt{p}\right), \tag{29}$$

$I_k(x)$—modified Bessel functions of the first kind of the kth order $k = 0, 1$ [8].

Using the inverse Laplace transform to the solution (27)–(29), the dimensional temperature rise was found in the form:

$$\Theta^*(\zeta, \tau) \equiv L^{-1}[\overline{\Theta}^*(\zeta, p); \tau] = \frac{1}{2\pi i} \int_{\omega - i\infty}^{\omega + i\infty} \overline{\Theta}^*(\zeta, p) e^{p\tau} dp,\ \tau \geq 0,\ \omega \equiv \mathrm{Re}p > 0,\ i \equiv \sqrt{-1}. \tag{30}$$

The presence of $\sqrt{p}$, as well as the lack of the roots of function $\Delta(p)$, testifies that the solution (36)–(39) has a branch point for $p = 0$. Therefore, to perform the integration on the complex plane (Rep, Im p), the closed contour Γ was chosen, as demonstrated in Figure 2. The contour Γ consists of the straight line Γ_ω Re $p = \omega$, the circles Γ_R and Γ_δ with the radii R and δ, respectively, with the center $p = 0$, and a cut of a complex p–plane along negative real axis Re $p < 0$ and two boundaries $\Gamma_\pm$. Within the contour Γ, the integral function $\overline{\Theta}^*(\zeta, p)$ in the Equation (30) is unambiguous and analytical.

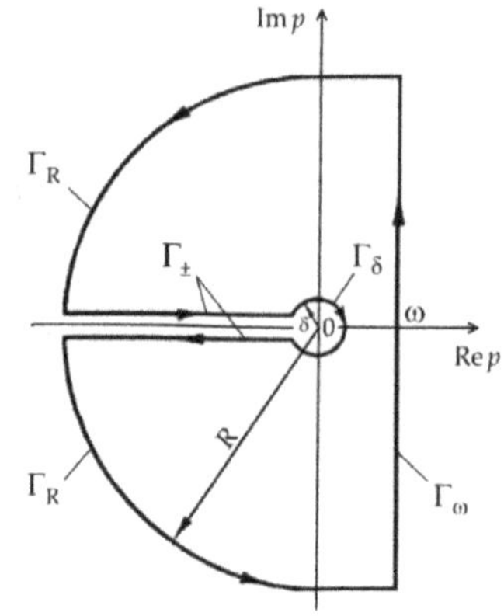

Figure 2. Integration contour Γ.

Then, based on Cauchy's theorem we obtained [9]:

$$\frac{1}{2\pi i}\oint_{\Gamma} \overline{\Theta}^*(\zeta, p)e^{p\tau}dp = 0. \tag{31}$$

Since the transform $\overline{\Theta}^*(\zeta,\tau)$ carries out the conditions of Jordan's lemma [7]:

$$\left|\frac{\Delta_l(\zeta,p)}{p\sqrt{p}\Delta(p)}\right| \le \frac{const.}{p\sqrt{p}},\ l = 1,2, \tag{32}$$

integrands on arcs Γ_R in the Equation (31) tend to zero for $R \to \infty$; therefore, on the basis of the relations (30) and (31), the dimensional temperature rise was written in the form:

$$\Theta^*(\zeta,\tau) + \Theta^*_+(\zeta,\tau) + \Theta^*_-(\zeta,\tau) + \Theta^*_\delta(\zeta,\tau) = 0,\ |\zeta| < \infty,\ \tau \ge 0, \tag{33}$$

where

$$\Theta^*_\pm(\zeta,\tau) = \frac{1}{2\pi i}\int_{\Gamma_\pm} \overline{\Theta}^*(\zeta,p)e^{p\tau}dp,\ \Theta^*_\delta(\zeta,\tau) = \frac{1}{2\pi i}\int_{\Gamma_\delta} \overline{\Theta}^*(\zeta,p)e^{p\tau}dp. \tag{34}$$

In the polar coordinate system (r,φ) with center in the point $p = 0$, parameter of the Laplace transform $p = re^{i\varphi}$, $r \ge 0$, and $|\varphi| \le \pi$. Then on the boundary Γ_+ we obtained $p = re^{i\pi} = -r$, $\sqrt{p} = i\sqrt{r}$, and on the edge Γ_-, respectively, $p = re^{-i\pi} = -r$, $\sqrt{p} = -i\sqrt{r}$ and the first two integrals (34) took the form:

$$\Theta^*_\pm(\zeta,\tau) = \pm\frac{1}{2\pi i}\int_0^\infty \overline{\Theta}^*_\pm(\zeta,r)e^{-r\tau}dr,\ |\zeta| < \infty,\ \tau \ge 0, \tag{35}$$

where $\overline{\Theta}^*_\pm(\zeta,r) \equiv \overline{\Theta}^*(\zeta,re^{\pm i\pi})$.

Taking into account the dependencies [8]:

$$I_0(x) = J_0(ix),\ I_1(x) = -iJ_1(ix), \tag{36}$$

(where $J_k(x)$ are the Bessel functions of the first kind of the kth order $k = 0,1$), from Equations (27)–(29) was obtained:

$$\overline{\Theta}^*_\pm(\zeta,r) = \frac{\Delta_1^\mp(\zeta,r)}{r\sqrt{r}\Delta^\mp(r)},\ \zeta \ge 0,\ \overline{\Theta}^*_\pm(\zeta,r) = \frac{\Delta_2^\mp(\zeta,r)}{r\sqrt{r}\Delta^\mp(r)},\ \zeta \le 0, \tag{37}$$

where:

$$\Delta_1^{\pm}(\zeta,r) = \pm ie^{-0.5\gamma^*\zeta} J_1\left(\frac{2}{\gamma^*}\sqrt{r}e^{-0.5\gamma^*\zeta}\right),\ \Delta_2^{\pm}(\zeta,r) = \pm ie^{\pm i\sqrt{\frac{r}{k^*}}\zeta} J_1\left(\frac{2}{\gamma^*}\sqrt{r}\right), \quad (38)$$

$$\Delta^{\pm}(\zeta,r) = K_\varepsilon J_1\left(\frac{2}{\gamma^*}\sqrt{r}\right) \pm iJ_0\left(\frac{2}{\gamma^*}\sqrt{r}\right). \quad (39)$$

On the circle Γ_δ it is $p = \delta e^{i\varphi}$, $\sqrt{p} = \sqrt{\delta}e^{0.5i\varphi}$, $|\varphi| \le \pi$. Approaching the limit $\delta \to 0$ with consideration of the solutions forms (27)–(29), the third integral (34) was written as:

$$\Theta_\delta^*(\zeta,\tau) = \lim_{\delta\to 0}\left(-\frac{1}{2\pi i}\int_{-\pi}^{\pi} \overline{\Theta}_\delta^*(\zeta,\delta e^{i\varphi}) e^{\delta e^{i\varphi}\tau} i\delta e^{i\varphi} d\varphi\right),\ \tau \ge 0, \quad (40)$$

where

$$\overline{\Theta}_\delta^*(\zeta,\delta e^{i\varphi}) = \frac{\Delta_1(\zeta,\delta e^{i\varphi})}{\delta\sqrt{\delta}e^{1.5i\varphi}\Delta(\delta e^{i\varphi})}, \zeta \ge 0,\ \overline{\Theta}_\delta^*(\zeta,\delta e^{i\varphi}) = \frac{\Delta_2(\zeta,\delta e^{i\varphi})}{\delta\sqrt{\delta}e^{1.5i\varphi}\Delta(\delta e^{i\varphi})}, \zeta \le 0, \quad (41)$$

$$\Delta_1(\zeta,\delta e^{i\varphi}) = e^{-0.5\gamma^*\zeta} I_1\left(\frac{2}{\gamma^*}\sqrt{\delta}e^{0.5i\varphi}e^{-0.5\gamma^*\zeta}\right), \quad (42)$$

$$\Delta_2(\zeta,\delta e^{i\varphi}) = e^{\sqrt{\frac{\delta}{k^*}}\zeta e^{0.5i\varphi}} I_1\left(\frac{2}{\gamma^*}\sqrt{\delta}e^{0.5i\varphi}\right), \quad (43)$$

$$\Delta^{\pm}(\zeta,r) = K_\varepsilon J_1\left(\frac{2}{\gamma^*}\sqrt{r}\right) \pm iJ_0\left(\frac{2}{\gamma^*}\sqrt{r}\right). \quad (44)$$

Substituting the functions (41)–(44) into Equation (40), it was found:

$$\Theta_\delta^*(\zeta,\tau) = \lim_{\delta\to 0}\left(-\frac{1}{2\pi}\int_{-\pi}^{\pi} \frac{\Delta_1(\zeta,\delta e^{i\varphi})}{\sqrt{\delta}e^{0.5i\varphi}\Delta(\delta e^{i\varphi})} e^{\delta e^{i\varphi}\tau} d\varphi\right),\ \zeta \ge 0,\ \tau \ge 0, \quad (45)$$

$$\Theta_\delta^*(\zeta,\tau) = \lim_{\delta\to 0}\left(-\frac{1}{2\pi}\int_{-\pi}^{\pi} \frac{\Delta_2(\zeta,\delta e^{i\varphi})}{\sqrt{\delta}e^{0.5i\varphi}\Delta(\delta e^{i\varphi})} e^{\delta e^{i\varphi}\tau} d\varphi\right),\ \zeta \le 0,\ \tau \ge 0. \quad (46)$$

For small values of the argument [8]:

$$I_0(x) \cong 1,\ I_1(x) \cong 0.5x, \quad (47)$$

from Equations (45) and (46), the following was obtained:

$$\Theta_\delta^*(\zeta,\tau) = -\frac{1}{\gamma^*}e^{-0.5\gamma^*\zeta}, \zeta \ge 0,\ \Theta_\delta^*(\zeta,\tau) = -\frac{1}{\gamma^*}, \zeta \le 0,\ \tau \ge 0. \quad (48)$$

Applying the function $\Theta_\pm^*(\zeta,\tau)$ (35), (37)–(39), and $\Theta_\delta^*(\zeta,\tau)$ (48) into the Equation (33) and introducing the notation: $\sqrt{r} = x$, $r = x^2$, the dimensional rise of temperature was found in the form:

$$\Theta^*(\zeta,\tau) = \frac{1}{\gamma^*}\left[e^{-0.5\gamma^*\zeta} - \frac{4}{\pi}\int_0^\infty F(x)G_1(\zeta,x)e^{-(0.5\gamma^*x)^2\tau}dx\right],\ \zeta \ge 0,\ \tau \ge 0, \quad (49)$$

$$\Theta^*(\zeta,\tau) = \frac{1}{\gamma^*}\left[1 - \frac{4}{\pi}\int_0^\infty F(x)G_2(\zeta,x)e^{-(0.5\gamma^*x)^2\tau}dx\right],\ \zeta \le 0,\ \tau \ge 0, \quad (50)$$

where

$$F(x) = \frac{J_1(x)}{x^2\left\{[J_0(x)]^2 + [K_\varepsilon J_1(x)]^2\right\}}, \tag{51}$$

$$G_1(\zeta, x) = K_\varepsilon e^{-0.5\gamma^*\zeta} J_1(xe^{-0.5\gamma^*\zeta}), \tag{52}$$

$$G_2(\zeta, x) = K_\varepsilon J_1(x) \cos\left(\frac{\gamma^*\zeta}{2\sqrt{k^*}}x\right) - J_0(x) \sin\left(\frac{\gamma^*\zeta}{2\sqrt{k^*}}x\right). \tag{53}$$

Substituting $\zeta = 0$ into Equations (49)–(53) it was established that the temperature rise on the contact surface included in the boundary condition (24) has the form:

$$\Theta^*(\tau) = \frac{1}{\gamma^*}\left[1 - \frac{4}{\pi}\int_0^\infty G(x)e^{-(0.5\gamma^* x)^2\tau}dx\right], \ \tau \geq 0, \tag{54}$$

where

$$G(x) = \frac{K_\varepsilon [J_1(x)]^2}{x^2\left\{[J_0(x)]^2 + [K_\varepsilon J_1(x)]^2\right\}}. \tag{55}$$

On the basis of the Fourier's law, the intensities of heat fluxes directed from the contact surface towards the inside of the friction pair elements were defined:

$$q_1(t) = -K_{1,1}\frac{\partial\Theta(z,t)}{\partial z}\bigg|_{z=0^+}, \ q_2(t) = K_2\frac{\partial\Theta(z,t)}{\partial z}\bigg|_{z=0^-}, \ t \geq 0, \tag{56}$$

The dimensionless form of dependencies (56) can be found as:

$$q_l^*(\tau) = \frac{q_l(t)}{q_0}, \ l = 1,2, \tag{57}$$

and taking account of (8) and (18), it was obtained:

$$q_1^*(\tau) = -\frac{\partial\Theta^*(\zeta,\tau)}{\partial\zeta}\bigg|_{\zeta=0^+}, \ q_2^*(\tau) = K^*\frac{\partial\Theta^*(\zeta,\tau)}{\partial\zeta}\bigg|_{\zeta=0^-}, \ \tau \geq 0. \tag{58}$$

Substituting the dimensionless temperature rise (49)–(53) into Equation (58) and differentiating, it was found:

$$q_1^*(\tau) = 1 + \frac{2}{\pi}\int_0^\infty Q(x)e^{-(0.5\gamma^* x)^2\tau}dx, \ q_2^*(\tau) = -\frac{2}{\pi}\int_0^\infty Q(x)e^{-(0.5\gamma^* x)^2\tau}dx, \ \tau \geq 0, \tag{59}$$

where

$$Q(x) = \frac{K_\varepsilon J_0(x) J_1(x)}{x\left\{[J_0(x)]^2 + [K_\varepsilon J_1(x)]^2\right\}}. \tag{60}$$

From Equations (59) and (60) it follows that $q_1^*(\tau) + q_2^*(\tau) = 1$, which confirms the fulfillment of the boundary condition (16) for $q^*(\tau) = 1$, $\tau \geq 0$.

4. Asymptotic Solutions

It should be noted that solutions (49)–(55) have the form of a quadrature; thus, using them, numerical integration should be performed each time on the range of bounded fields. However, in the case of small and large values of dimensionless time τ (Fourier number), the corresponding asymptotic solution will be obtained in the analytical form, not requiring numerical integration.

Small values of the Fourier number $0 \leq \tau << 1$ (large values of the parameter p of the Laplace integral transform (30)). At large values of arguments, the modified Bessel functions behave as follows [8]:

$$I_0(x) \cong \frac{e^x}{\sqrt{2\pi x}}\left(1+\frac{1}{8x}+\frac{9}{128x^2}+\ldots\right),\ I_1(x) \cong \frac{e^x}{\sqrt{2\pi x}}\left(1-\frac{3}{8x}-\frac{15}{128x^2}-\ldots\right). \quad (61)$$

Limiting only to the first two components in the formula (61), the transforms of the dimensionless temperature rise (27)–(29) were written in the form:

$$\overline{\Theta}^*(\zeta,p) \cong \frac{e^{-0.25\gamma^*\zeta-\alpha\sqrt{p}}}{(1+K_\varepsilon)p\sqrt{p}}\left(1-\frac{3\gamma^* e^{0.5\gamma^*\zeta}}{16\sqrt{p}}\right)\left[1+\frac{\gamma^*(1-3K_\varepsilon)}{16(1+K_\varepsilon)\sqrt{p}}\right]^{-1},\ \zeta \geq 0, \quad (62)$$

$$\overline{\Theta}^*(\zeta,p) \cong \frac{e^{\sqrt{\frac{p}{k^*}}\zeta}}{(1+K_\varepsilon)p\sqrt{p}}\left(1-\frac{3\gamma^*}{16\sqrt{p}}\right)\left[1+\frac{\gamma^*(1-3K_\varepsilon}{16(1+K_\varepsilon)\sqrt{p}}\right]^{-1},\ \zeta \leq 0, \quad (63)$$

where

$$\alpha = \frac{2}{\gamma^*}(1-e^{-0.5\gamma^*\zeta}),\ \zeta \geq 0. \quad (64)$$

Taking into consideration that:

$$\left(1-\frac{3\gamma^* e^{0.5\gamma^*\zeta}}{16\sqrt{p}}\right)\left[1+\frac{\gamma^*(1-3K_\varepsilon}{16(1+K_\varepsilon)\sqrt{p}}\right]^{-1} \approx 1-\frac{\gamma^*}{16\sqrt{p}}\left(3e^{0.5\gamma^*\zeta}+\frac{1-3K_\varepsilon}{1+K_\varepsilon}\right), \quad (65)$$

$$\left(1-\frac{3\gamma^*}{16\sqrt{p}}\right)\left[1+\frac{\gamma^*(1-3K_\varepsilon}{16(1+K_\varepsilon)\sqrt{p}}\right]^{-1} \approx 1-\frac{\gamma^*}{4(1+K_\varepsilon)\sqrt{p}}, \quad (66)$$

the transforms (62)–(64) were obtained in the form:

$$\overline{\Theta}^*(\zeta,p) \cong \frac{e^{-0.25\gamma^*\zeta-\alpha\sqrt{p}}}{(1+K_\varepsilon)p\sqrt{p}}\left[1-\frac{\gamma^*}{16\sqrt{p}}\left(3e^{0.5\gamma^*\zeta}+\frac{1-3K_\varepsilon}{1+K_\varepsilon}\right)\right],\ \zeta \geq 0, \quad (67)$$

$$\overline{\Theta}^*(\zeta,p) \cong \frac{e^{\sqrt{\frac{p}{k^*}}\zeta}}{(1+K_\varepsilon)p\sqrt{p}}\left(1-\frac{\gamma^*}{4(1+K_\varepsilon)\sqrt{p}}\right),\ \zeta \leq 0. \quad (68)$$

Taking account of the relations [10]:

$$L^{-1}\left[\frac{e^{-a\sqrt{p}}}{p\sqrt{p^n}};\tau\right] = (4\tau)^{\frac{n}{2}}\mathrm{i}^n\mathrm{erfc}\left(\frac{a}{2\sqrt{\tau}}\right),\ n=1,2,\ a \geq 0, \quad (69)$$

from the transforms of solutions (67) and (68), the dimensionless temperature rises were found:

$$\Theta^*(\zeta,\tau) \cong \frac{2e^{-0.25\gamma^*\zeta}\sqrt{\tau}}{(1+K_\varepsilon)}\left[\mathrm{ierfc}\left(\frac{\alpha}{2\sqrt{\tau}}\right)-\frac{\gamma^*\sqrt{\tau}}{8}\left(3e^{0.5\gamma^*\zeta}+\frac{1-3K_\varepsilon}{1+K_\varepsilon}\right)\mathrm{i}^2\mathrm{erfc}\left(\frac{\alpha}{2\sqrt{\tau}}\right)\right], \quad \zeta \geq 0, \quad (70)$$

$$\Theta^*(\zeta,\tau) \cong \frac{2\sqrt{\tau}}{(1+K_\varepsilon)}\left[\mathrm{ierfc}\left(\frac{|\zeta|}{2\sqrt{k^*\tau}}\right)-\frac{\gamma^*\sqrt{\tau}}{2(1+K_\varepsilon)}\mathrm{i}^2\mathrm{erfc}\left(\frac{|\zeta|}{2\sqrt{k^*\tau}}\right)\right],\ \zeta \leq 0, \quad 0 \leq \tau << 1, \quad (71)$$

where

$$\mathrm{i}^2\mathrm{erfc}(x) = 0.25[\mathrm{erfc}(x)-2x\,\mathrm{ierfc}(x)],\ \mathrm{ierfc}(x) = \pi^{-0.5}e^{-x^2}-x\,\mathrm{erfc}(x), \quad \mathrm{erfc}(x) = 1-\mathrm{erf}(x), \quad (72)$$

$\mathrm{erf}(x)$—Gauss error function [8]. On the contact surface $\zeta = 0$ from Equations (70) and (71) it was obtained:

$$\Theta^*(\tau) \cong \frac{2\sqrt{\tau}}{(1+K_\varepsilon)}\left[\frac{1}{\sqrt{\pi}} - \frac{\gamma^*\sqrt{\tau}}{8(1+K_\varepsilon)}\right],\ 0 \le \tau << 1. \tag{73}$$

Approaching in Equations (70)–(73) the limit $\gamma^* \to 0$ $(\alpha \to \zeta)$, the solution for homogeneous materials was obtained [11]:

$$\Theta^*(\zeta,\tau) \cong \frac{2\sqrt{\tau}}{(1+K_\varepsilon)}\mathrm{ierfc}\left(\frac{\zeta}{2\sqrt{\tau}}\right),\ \zeta \ge 0, \tag{74}$$

$$\Theta^*(\zeta,\tau) \cong \frac{2\sqrt{\tau}}{(1+K_\varepsilon)}\mathrm{ierfc}\left(\frac{|\zeta|}{2\sqrt{k^*\tau}}\right),\ \zeta \le 0, \tag{75}$$

$$\Theta^*(\tau) \cong \frac{2}{(1+K_\varepsilon)}\sqrt{\frac{\tau}{\pi}},\ 0 \le \tau << 1. \tag{76}$$

Large values of Fourier number $\tau >> 1$(small values of the parameter p of the Laplace integral transform (30)). Distributions of the modified Bessel functions for small values of argument in the power series have the form [8]:

$$I_0(x) \cong 1 + \frac{x^2}{4} + \ldots,\ I_1(x) \cong \frac{x}{2}\left(1 + \frac{x^2}{8} + \ldots\right). \tag{77}$$

Taking into account the relations (77), the Laplace transforms of dimensionless temperature rise (27)–(29) were written as:

$$\overline{\Theta}^*(\zeta,p) \cong \frac{e^{-\gamma^*\zeta}}{\gamma^*}\left[\frac{\beta}{p(\beta+\sqrt{p})} + \frac{e^{-\gamma^*\zeta}}{2K_\varepsilon\gamma^*(\beta+\sqrt{p}}\right],\ \zeta \ge 0, \tag{78}$$

$$\overline{\Theta}^*(\zeta,p) \cong \frac{e^{\sqrt{\frac{p}{k^*}}\zeta}}{\gamma^*}\left[\frac{\beta}{p(\beta+\sqrt{p})} + \frac{1}{2K_\varepsilon\gamma^*(\beta+\sqrt{p}}\right],\ \zeta \le 0, \tag{79}$$

where

$$\beta = \frac{\gamma^*}{K_\varepsilon}. \tag{80}$$

Using the dependencies [10]:

$$L^{-1}\left[\frac{e^{-a\sqrt{p}}}{(\beta+\sqrt{p})};\tau\right] = \frac{e^{-\frac{a^2}{4\tau}}}{\sqrt{\pi\tau}} - \beta\, e^{a\beta+\beta^2\tau}\mathrm{erfc}\left(\frac{a}{2\sqrt{\tau}} + \beta\sqrt{\tau}\right), \tag{81}$$

$$L^{-1}\left[\frac{\beta\, e^{-a\sqrt{p}}}{p(\beta+\sqrt{p})};\tau\right] = \mathrm{erfc}\left(\frac{a}{2\sqrt{\tau}}\right) - e^{a\beta+\beta^2\tau}\mathrm{erfc}\left(\frac{a}{2\sqrt{\tau}} + \beta\sqrt{\tau}\right),\ a \ge 0, \tag{82}$$

from the transform solutions (78) and (79), the dimensionless temperature rises were obtained in the form:

$$\Theta^*(\zeta,\tau) \cong \frac{e^{-\gamma^*\zeta}}{\gamma^*}\left\{1 - e^{\beta^2\tau}\mathrm{erfc}(\beta\sqrt{\tau}) + \frac{e^{-\gamma^*\zeta}}{2K_\varepsilon\gamma^*}\left[\frac{1}{\sqrt{\pi\tau}} - \beta\, e^{\beta^2\tau}\mathrm{erfc}(\beta\sqrt{\tau})\right]\right\},\ \zeta \ge 0,\ \tau >> 1, \tag{83}$$

$$\begin{gathered}\Theta^*(\zeta,\tau) \cong \tfrac{1}{\gamma^*}\Big\{\mathrm{erfc}\Big(\tfrac{|\zeta|}{2\sqrt{k^*\tau}}\Big) - e^{\frac{\beta|\zeta|}{\sqrt{k^*}}+\beta^2\tau}\mathrm{erfc}\Big(\tfrac{|\zeta|}{2\sqrt{k^*\tau}} + \beta\sqrt{\tau}\Big) + \\ + \tfrac{1}{2K_\varepsilon\gamma^*}\left[\frac{e^{-\frac{\zeta^2}{4k^*\tau}}}{\sqrt{\pi\tau}} - \beta e^{\frac{\beta|\zeta|}{\sqrt{k^*}}+\beta^2\tau}\mathrm{erfc}\Big(\tfrac{|\zeta|}{2\sqrt{k^*\tau}} + \beta\sqrt{\tau}\Big)\right]\Big\},\ \zeta \le 0,\ \tau >> 1.\end{gathered} \tag{84}$$

Substituting $\zeta = 0$ into Equations (83) and (84), it was found:

$$\Theta^*(\tau) \cong \frac{1}{\gamma^*}\left\{1 - e^{\beta^2\tau}\mathrm{erfc}(\beta\sqrt{\tau}) + \frac{1}{2K_\varepsilon\gamma^*}\left[\frac{1}{\sqrt{\pi\tau}} - \beta e^{\beta^2\tau}\mathrm{erfc}(\beta\sqrt{\tau})\right]\right\},\ \tau >> 1. \quad (85)$$

5. Temperature Field during Braking with Constant Deceleration

Based on Duhamel's theorem [12], the dimensionless temperature rise during braking with constant deceleration was sought in the form:

$$\hat{\Theta}^*(\zeta,\tau) = \frac{\partial}{\partial\tau}\int_0^\tau q^*(\tau - s)\Theta^*(\zeta,s)ds,\ |\zeta| < \infty,\ 0 \le \tau \le \tau_s, \quad (86)$$

where the temporal profiles of the specific friction power $q^*(\tau)$ and function $\Theta^*(\zeta,\tau)$ were determined from Equations (19), (49), and (50), respectively. Performing the integration first, and then differentiating, from the Equation (86) we obtained:

$$\hat{\Theta}^*(\zeta,\tau) = \frac{1}{\gamma^*}\left[e^{-0.5\gamma^*\zeta}q^*(\tau) - \frac{4}{\pi}\int_0^\infty F(x)G_1(\zeta,x)P(\tau,x)dx\right],\ \zeta \ge 0,\ 0 \le \tau \le \tau_s, \quad (87)$$

$$\hat{\Theta}^*(\zeta,\tau) = \frac{1}{\gamma^*}\left[1 - \frac{4}{\pi}\int_0^\infty F(x)G_2(\zeta,x)P(\tau,x)dx\right],\ \zeta \le 0,\ 0 \le \tau \le \tau_s, \quad (88)$$

where

$$P(\tau,x) = e^{-(0.5\gamma^*x)^2\tau} - \frac{(1 - e^{-(0.5\gamma^*x)^2\tau})}{(0.5\gamma^*x)^2\tau_s}, \quad (89)$$

and functions $F(x)$, $G_1(\zeta,x)$, and $G_2(\zeta,x)$ can be found from the Formulas (51)–(53).

The temperature change on the friction surface was found, substituting $\zeta = 0$ into the Equations (87) and (88), in the form:

$$\hat{\Theta}^*(\tau) = \frac{1}{\gamma^*}\left[1 - \frac{4}{\pi}\int_0^\infty G(x)P(\tau,x)dx\right],\ \tau \ge 0, \quad (90)$$

where functions $G(x)$ and $P(\tau,x)$ were determined from relations (55) and (89), respectively.

Knowing the dimensionless temperature rise (87), (88), from Formulas (58) the dimensionless intensities of frictional heat fluxes were found:

$$\hat{q}_1^*(\tau) = q^*(\tau) + \frac{2}{\pi}\int_0^\infty Q(x)P(\tau,x)dx,\ \hat{q}_2^*(\tau) = -\frac{2}{\pi}\int_0^\infty Q(x)P(\tau,x)dx,\ 0 \le \tau \le \tau_s, \quad (91)$$

where functions $Q(x)$ and $P(\tau,x)$ have the forms (60) and (89), respectively. From Equation (91) it follows that $\hat{q}_1^*(\tau) + \hat{q}_2^*(\tau) = q^*(\tau)$, which confirms the fulfillment of the boundary condition (16) with the dimensionless specific friction power $q^*(\tau)$ in the form (19).

6. Numerical Analysis

Calculations were performed for a friction pair, where the first element (pad) is made of two-component FGM: zircon dioxide ZrO_2 (friction surface) and titanium alloy Ti − 6Al − 4V(core). While the second material (brake disc) is homogeneous: cast iron ChNMKh. The properties of the materials are included in Table 1.

Table 1. Material properties at the initial temperature T_0 [3,13].

Material	Thermal Conductivity $Wm^{-1}K^{-1}$	Specific Heat Capacity $J\,kg^{-1}K^{-1}$	Density $kg\,m^{-3}$
ZrO_2	1.94	452.83	6102.16
Ti-6Al-4V	6.87	538.08	4431.79
ChNMKh	52.17	444.6	7100

The values of the remaining input parameters used to perform the calculations are listed in Table 2.

Table 2. Input parameters [14].

Friction Coefficient f_0	Nominal Pressure p_0, MPa	Initial Sliding Speed V_0, ms^{-1}	Initial Kinetic Energy W_0, kJ	Nominal Contact Area A_a, m^2	Initial Temperature T_0, °C
0.27	0.602	23.8	103.54	0.00221	20

Then, from formulas (1) and (19), the nominal value of specific friction power $q_0 = 3.87\ \mathrm{MW\,m^{-2}}$, braking time $t_s = 12.1\,\mathrm{s}$, and gradient parameter $\gamma^* = 1.26$ were determined. Based on Equation (9), for an equal volumetric fraction of FGM component ($v = 0.5$), the effective values of specific heat capacity and density of the pad material were obtained, $c_1 = 495.45\,\mathrm{J\,kg^{-1}K^{-1}}$, $\rho_1 = 5266.97\,\mathrm{kg\,m^{-3}}$, respectively. Thereafter, the following parameters were calculated sequentially: thermal diffusivity $k_1 = 0.743 \cdot 10^{-6}\,\mathrm{m^2s^{-1}}$ and $k_2 = 1.65 \cdot 10^{-5}\,\mathrm{m^2s^{-1}}$, the effective depth of heat penetration of the pad $a = 5.2\,\mathrm{mm}$, the dimensionless braking time $\tau_s = 0.33$, and the temperature scaling factor $\Theta_0 = 10{,}373\,°\mathrm{C}$, based on Equations (10)–(12).

The integrals in the obtained solutions were calculated numerically using the QAGI procedure of the QUADPACK package [15]. Changes of the dimensionless temperature rise and intensities of heat fluxes during sliding with a constant velocity are presented in Figures 3–6.

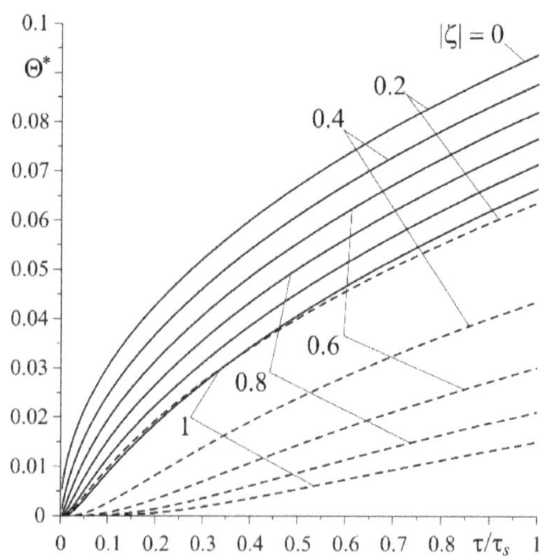

Figure 3. Evolutions of dimensionless temperature rise $\Theta^*(\zeta, \tau)$ on the established distances $|\zeta|$ from the friction surface during sliding with a constant velocity: cast iron—solid lines; FGM—dashed lines.

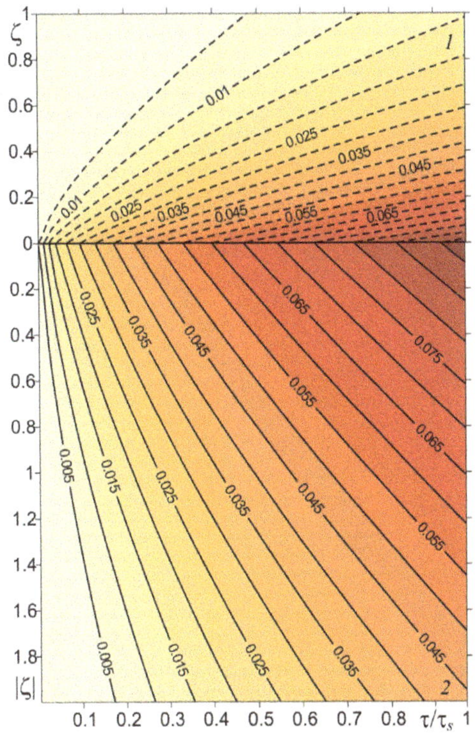

Figure 4. Isotherms of dimensionless temperature rise $\Theta^*(\zeta, \tau)$ during sliding with constant velocity: cast iron (*1*)—solid lines; FGM (*2*)—dashed lines.

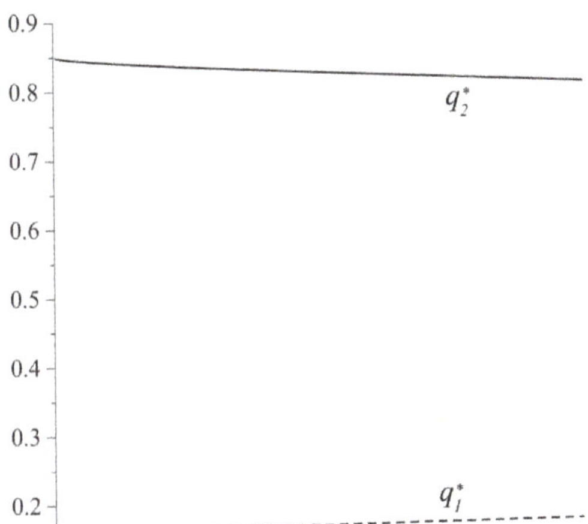

Figure 5. Evolutions of dimensionless intensities of heat fluxes q_l^*, $l = 1, 2$, directed along the normal from the friction surface to the insides of the elements made of cast iron (solid line) and FGM (dashed line) under uniform sliding.

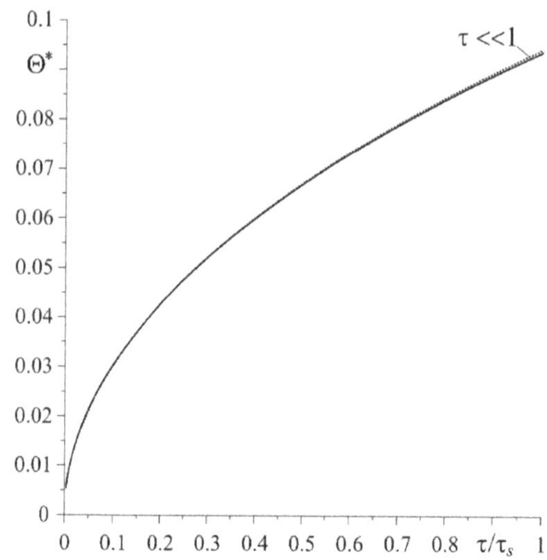

Figure 6. Change with time of the dimensionless temperature rise $\Theta^*(\tau)$ on the friction surface $\zeta = 0$ during sliding with constant velocity: exact solution—solid line; asymptotic solution—dashed line.

Temporal profiles of the dimensionless temperature rise $\Theta^*(\zeta, \tau)$ (49)–(53) at a few distances from the friction surface are shown in Figure 3. The temperature of both elements increased monotonically over time. The highest temperature was reached on the friction surface, and decreasing moving away from it. For a given distance from this surface, the temperature of the homogeneous cast iron element was always higher than the temperature of the functionally graded element. Having a much greater thermal conductivity, the cast iron was heated to a much deeper extent than the FGM (Figure 4).

Temporal profiles of dimensionless heat flux intensities $q_l^*(\tau)$, $l = 1, 2$ (59), (60) are demonstrated in Figure 5. It was found that the main element that absorbs frictional heat is the cast-iron disc, especially at the initial stage of the heating process. The amount of heat directed from the friction surface towards the inside of the pad increases with time, and towards the inside of the disc it decreases. A comparison of dimensionless temperature values $\Theta^*(\tau)$ of the friction surface, found by means of the exact (54), (55) and asymptotic solutions (74), (75) are shown in Figure 6. In the considered range of Fourier number $0 \le \tau \le \tau_s$, the respective temperature values were almost the same.

Relevant results, obtained in the case of a linearly decreasing velocity (so-called braking with a constant deceleration), are presented in Figures 7–10. The temporal profile of the dimensionless temperature rise $\hat{\Theta}^*(\zeta, \tau)$ (87)–(89) during the braking process was different than during uniform sliding (Figure 7). The dimensionless time to reach the maximum temperature on the friction surface was $\tau_{max} \approx 0.5\tau_s$ and became higher when increasing the distance from it. After reaching the maximum value, the temperature dropped. More vividly, such a concentration of high temperature near the friction surface is shown in the distribution of isotherms, as illustrated in Figure 8. Apparently, as in the case of uniform sliding, the greater part of the frictionally-generated heat is absorbed by the cast iron disc ($\approx 85\%$) (Figure 9). The intensities of heat fluxes $\hat{q}_l^*(\tau)$, $l = 1, 2$ (91) are almost unchanged during the entire braking process.

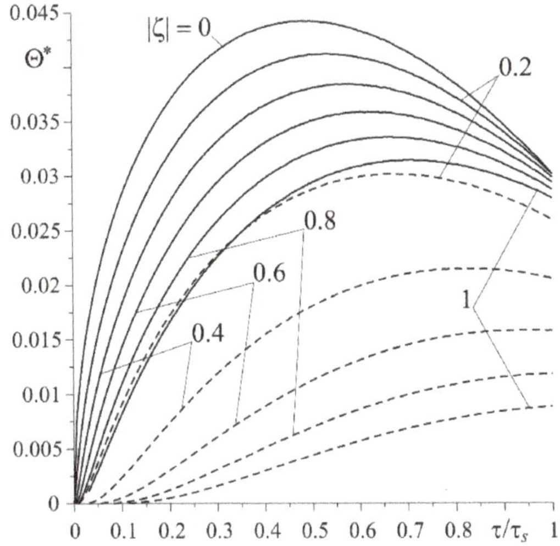

Figure 7. Evolutions of dimensionless temperature rise $\hat{\Theta}^*(\zeta,\tau)$ on the established distances $|\zeta|$ from the friction surface during braking with constant deceleration: cast iron—solid lines; FGM—dashed lines.

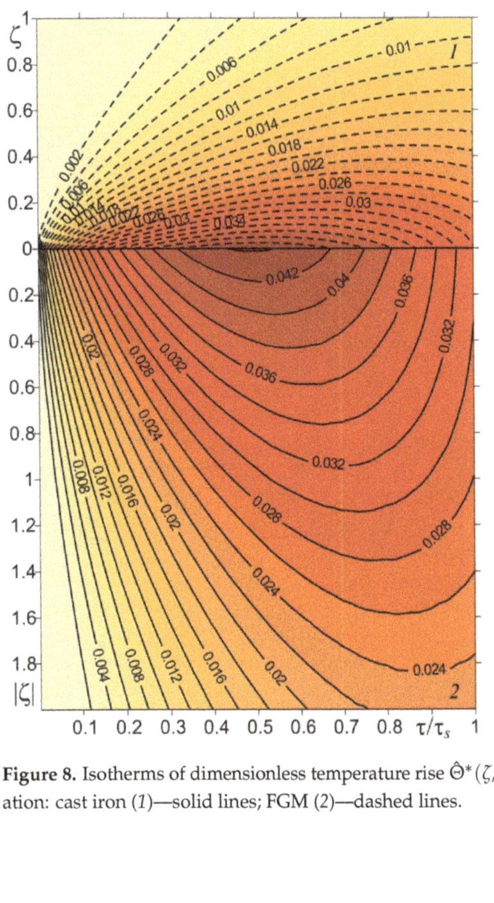

Figure 8. Isotherms of dimensionless temperature rise $\hat{\Theta}^*(\zeta,\tau)$ during braking with constant deceleration: cast iron (*1*)—solid lines; FGM (*2*)—dashed lines.

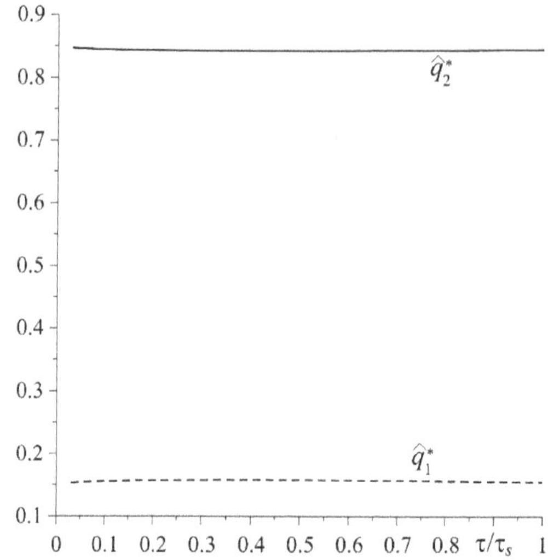

Figure 9. Evolutions of dimensionless intensities of heat fluxes $\hat{q}_l^*$, $l = 1, 2$ directed along the normal from the friction surface to the insides of the elements made of cast iron (solid line) and FGM (dashed line) during braking with constant deceleration.

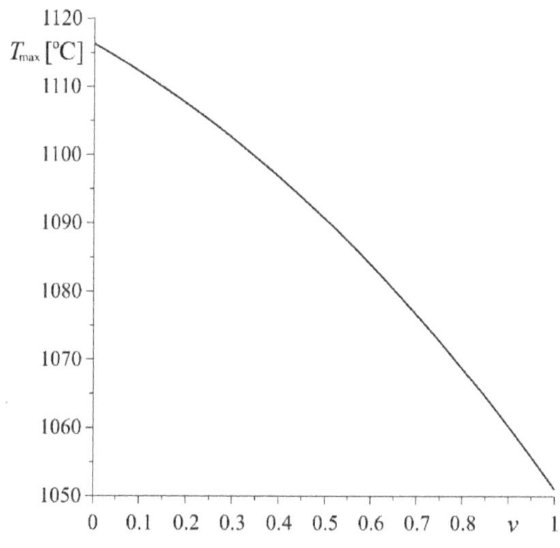

Figure 10. Dependency of maximum temperature T_{max} during braking with constant deceleration on the volumetric fraction v.

The results presented in Figures 3–9 were obtained with the same ($v = 0.5$) volumetric components fractions of ZrO_2 and Ti-6Al-4V, determining the effective specific heat capacity and density using formula (9). On the other hand, the change of the maximum temperature $T_{max} \equiv T(0, t_{max})$ with the increase of the parameter v is presented in Figure 10. The highest value $T_{max} = 1117\,^\circ C$ was achieved in the case of the pad made of pure zirconium dioxide, and the lowest $T_{max} = 1052\,^\circ C$, when it was made of the titanium alloy.

7. Conclusions

A mathematical model was proposed to determine the transient temperature field in a friction pair, in which one element is made of a functionally graded material and the other is made of a homogeneous material. It was assumed that the thermal conductivity of a FGM increases exponentially with the distance from the contact surface. An exact solution of the appropriate boundary–value problem of heat conduction was formulated and then solved, with consideration of frictional heat generation. Two cases of the friction power temporal profiles were analyzed in detail: constant (uniform sliding), and linearly decreasing in time (braking with constant deceleration). A numerical analysis was performed for a two-component FGM (ZrO_2 + Ti-6Al-4V) sliding on the cast-iron disc. It was found that the greater part of heat generated due to friction was absorbed by the cast iron (about 85%), which resulted in a greater depth of effective heat penetration in this element, due to the high thermal conductivity of cast iron. At a fixed distance from the friction surface, the temperature of the cast iron element is higher than that of the FGM element, in both the considered cases: uniform sliding, and during braking. Thus, in order to protect systems against such undesirable phenomena as overheating and thermal cracking etc., the use of FGM on the friction elements may be justified. It is also worth emphasizing that in the analyzed range of the Fourier number change $0 \leq \tau \leq 0.33$, the appropriate asymptotic solution can be effectively used, giving a high accuracy of calculations, without the inconveniences related to numerical integration in an exact solution.

It should be noted that the shape of the friction pair elements, as well as their positional relationship, can be considered in some spatial problems of friction solved by numerical methods, in particular the finite element method (FEM). The temperature evolution obtained by them oscillates, as a result of the heating area moving on the surface of the brake disc. The model proposed in this paper is one-dimensional, based on a physically-justified assumption that heat, generated by friction of two elements, propagates in the direction perpendicular to the contact surface. This allows determining the mean temperature (from the above-mentioned oscillations) on the friction surfaces of both elements.

According to the current state of knowledge [16,17], the temperature of the friction surface is the sum of the volume temperature (average temperature in volume), the mean temperature, and the flash temperature. The flash temperature is the component that takes into consideration the texture of the friction surfaces. The flash temperature calculation models need appropriate experimental data as input parameters. In the case of homogeneous materials, such data can be found in the article in ref. [13]. However, we have not found such data for the considered friction pair. The development of models for determining the flash temperature of such couples is a potential direction for our research. In the future, we intend to expand the proposed mathematical model with the possibility of testing the temperature of friction systems of this type (functionally graded and homogeneous materials) made of thermally sensitive materials and the temperature-dependent friction coefficient.

Author Contributions: Conceptualization and methodology, A.Y. and P.Z.; software, M.K. and P.Z.; validation, P.Z. and K.T.; formal analysis, P.Z. and K.T.; investigation, A.Y., M.K., K.T. and P.Z.; writing—original draft preparation A.Y., P.Z. and K.T.; writing—review and editing, K.T. and P.Z.; visualization and figures preparation, P.Z. and M.K.; supervision, A.Y.; project administration, K.T. All authors have read and agreed to the published version of the manuscript.

Funding: This investigation was performed within the framework of research project No. 2017/27/B/ST8/01249, funded by the National Science Centre, Poland, and with project financing through the program of the Minister of Education and Science of Poland named "Regional Initiative of Excellence" in 2019–2022, project No. 011/RID/2018/19, amount financing 12,000,000 PLN.

Institutional Review Board Statement: Not applicable.

Informed Consent Statement: Not applicable.

Data Availability Statement: No new data were created or analyzed in this study. Data sharing is not applicable to this article.

Conflicts of Interest: The authors declare no conflict of interest.

Nomenclature

a	Effective depth of heat penetration (m)
A_a	Area of the nominal contact region (m^2)
c	Specific heat capacity (J kg^{-1} K^{-1})
f_0	Coefficient of friction (dimensionless)
$I_k(\cdot)$	Modified Bessel functions of the first kind of the kth order
$J_k(\cdot)$	Bessel functions of the first kind of the kth order
k	Thermal diffusivity (m^2 s^{-1})
K	Thermal conductivity (W m^{-1} K^{-1})
p	Dimensionless parameter of the Laplace integral transform
p_0	Nominal value of the contact pressure (Pa)
r	Radial coordinate in the polar system (m)
R	Radius of integration contour (m)
q	Specific power of friction (W m^{-2})
q_0	Nominal value of the specific power of friction (W m^{-2})
t	Time (s)
t_s	Stop time (s)
T	Temperature (°C)
T_0	Initial temperature (°C)
v	Volume fraction of the material phases (dimensionless)
V_0	Initial velocity (m s^{-1})
W_0	Initial kinetic energy of the system (J)
z	Spatial coordinate in axial direction (m)

Greek Symbols

γ	Parameter of material gradient (m^{-1})
γ^*	Dimensionless parameter of material gradient
Γ	Integration contour
δ	Radius of integration contour (m)
Θ	Temperature rise (°C)
Θ^*	Dimensionless temperature rise
Θ_0	Temperature rise scaling factor (°C)
$\overline{\Theta}^*$	Dimensionless transform of temperature rise
ρ	Density (kg m^{-3})
τ	Dimensionless time
τ_s	Dimensionless time of braking
ζ	Dimensionless spatial coordinate in axial direction
φ	Angular coordinate in the polar system (rad)

References

1. Yevtushenko, A.; Topczewska, K.; Zamojski, P. The Effect of Functionally Graded Materials on Temperature during Frictional Heating: Under Uniform Sliding. *Materials* **2021**, *14*, 4285. [CrossRef] [PubMed]
2. Yevtushenko, A.; Topczewska, K.; Zamojski, P. The Effect of Functionally Graded Materials on Temperature during Frictional Heating at Single Braking. *Materials* **2021**, *14*, 6241. [CrossRef] [PubMed]
3. Yevtushenko, A.; Topczewska, K.; Zamojski, P. Influence of Thermal Sensitivity of Functionally Graded Materials on Temperature during Braking. *Materials* **2022**, *15*, 963. [CrossRef] [PubMed]
4. Zelentsov, V.B.; Lapina, P.A.; Mitrin, B.I. Wear of Functionally Graded Coatings under Frictional Heating Conditions. *Nanomaterials* **2022**, *2*, 142. [CrossRef] [PubMed]
5. Day, A.J.; Newcomb, T.P. The dissipation of frictional energy from the interface of an annular disc brake. *Proc. Inst. Mech. Sci. Part D J. Automob. Eng.* **1984**, *198*, 201–209. [CrossRef]
6. Qi, H.S.; Day, A.J. Investigation of disc/pad interface temperatures in friction braking. *Wear* **2007**, *262*, 505–513. [CrossRef]
7. Sneddon, I.N. *The Use of Integral Transforms*; McGraw-Hill: New York, NY, USA, 1972.
8. Abramowitz, M.; Stegun, I. *Handbook of Mathematical Functions with Formulas, Graphs, and Mathematical Tables*; United States Department of Commerce: Washington, DC, USA; National Bureau of Standards (NBS): Washington, DC, USA, 1964.

9. Wong, M.W. *Complex Analysis*; York University: Toronto, ON, Canada, 2008. [CrossRef]
10. Bateman, H.; Erdélyi, A. *Tables of Integral Transforms*; McGraw-Hill: New York, NY, USA, 1954; Volume 1.
11. Carslaw, H.S.; Jaeger, J.C. *Conduction of Heat in Solids*; Clarendon Press: Oxford, UK, 1959.
12. Özisik, N.M. *Heat Conduction*; John Wiley: New York, NY, USA, 1993.
13. Yevtushenko, A.; Topczewska, K.; Kuciej, M. Analytical Determination of the Brake Temperature Mode during Repetitive Short-Term Braking. *Materials* **2021**, *14*, 1912. [CrossRef] [PubMed]
14. Chichinadze, A.V.; Kozhemyakina, V.D.; Suvorov, A.V. Method of Temperature-Field Calculation in Model Ring Specimens during Bilateral Friction in Multidisc Aircraft Brakes with the IM-58-T2 New Multipurpose Friction Machine. *J. Frict. Wear* **2010**, *31*, 23–32. [CrossRef]
15. Piessens, R.; De Doneker-Kapenga, E.; Überhuber, C.W.; Kahaner, D.K. *Quadpack: A Subroutine Package for Automatic Integration*; Springer Science & Business Media: Berlin/Heidelberg, Germany, 2012; Volume 1.
16. Kennedy, F.E.; Tian, X. Modeling sliding contact temperatures including effects of the surface roughness and convection. *J. Tribol.* **2016**, *138*, 042101. [CrossRef]
17. Choudhry, J.; Almqvist, A.; Larsson, R. Validation of a multi-scale contact temperature model for dry sliding rough surfaces. *Lubricants* **2022**, *10*, 41. [CrossRef]

materials

MDPI

Article

The Heat Partition Ratio during Braking in a Functionally Graded Friction Couple

Aleksander Yevtushenko , Katarzyna Topczewska * and Przemysław Zamojski

Faculty of Mechanical Engineering, Bialystok University of Technology (BUT), 45C Wiejska Street, 15-351 Bialystok, Poland; a.yevtushenko@pb.edu.pl (A.Y.); p.zamojski@pb.edu.pl (P.Z.)
* Correspondence: k.topczewska@pb.edu.pl

Abstract: The theoretical scheme for determining the heat partition ratio (HPR) in a friction couple made of functionally graded materials (FGMs) was proposed. As a result, the formula for the calculation of the HPR was found, which depends on the thermal properties and the parameters of the material's gradient. In specific cases of these parameters, the known formulas for estimating the HPR for homogeneous materials were obtained. Calculations were carried out for the friction couple consisting of the following two-component FGMs: Al_2O_3–Cu (first body) and ZrO_2–Ti–6Al–4V (second body), under the conditions corresponding to a single braking with a constant deceleration. It was established that the vast majority (almost 90%) of heat that was generated by friction was absorbed by the first body in the selected couple. The possibilities of using the obtained results were discussed herein.

Keywords: heat partition ratio; functionally graded materials; frictional heating; temperature; braking

Citation: Yevtushenko, A.; Topczewska, K.; Zamojski, P. The Heat Partition Ratio during Braking in a Functionally Graded Friction Couple. *Materials* **2022**, *15*, 4623. https://doi.org/10.3390/ma15134623

Academic Editor: Pawel Pawlus

Received: 1 June 2022
Accepted: 28 June 2022
Published: 30 June 2022

1. Introduction

Functionally graded materials (FGMs) are materials in which, along at least one specific direction, a continuous change in the functional or structural properties has been obtained in a selected technological process. The possibilities of using graded materials in technology seem to be almost unlimited. Proposals of possible FGM applications appear for many industries, such as the following: optical, energy [1,2], aviation [3,4], medical [5], etc. Moreover, they have been used in friction elements, including braking systems [6,7].

The significant influence of temperature on the friction and wear characteristics of friction materials is well known and has been studied by many researchers who are involved in determining the temperature mode of disc brakes [8–10]. The basis for establishing the temperature are the solutions (analytical, numerical, or analytical–numerical) to the thermal problems of friction, i.e., the boundary-value problems of heat conduction with two specific boundary conditions on the contact surface of the pad and the disc [11,12]. One of them, the energetic solution, determines the equality of the specific friction power q and the sum of the heat flux intensities $q_l, l = 1, 2$ that are directed along the normal surface to the contact surface towards the insides of the friction elements $q_1 + q_2 = q$. Whereas the second solution concerns the type of thermal contact—with (imperfect) or without (perfect) consideration of the thermal resistance on the friction surfaces. In the latter case, the surface temperatures of the friction elements are the same. The coupling of the temperature fields of both of the friction elements through the above-mentioned boundary conditions means that obtaining the analytical [13–16] or the numerical [17–19] solutions of the thermal problems of friction requires the performance of complex mathematical transformations or long-term calculations.

In order to reduce these difficulties, there is also another approach to estimate the temperature of the braking systems, on the basis of solutions to the thermal problems of friction. It is based on a virtual separation of the friction pair elements and the subsequent replacement of the frictional interaction on their working surfaces with the heat fluxes of the following intensities: $q_l = \alpha_l q$, $l = 1, 2$, $\alpha_1 = \alpha$, and $\alpha_2 = 1 - \alpha$, respectively, where α is the heat partition ratio [20,21]. Reviews of the experimental methods, as well as the theoretical methods, for determining the HPR in braking systems have been developed [22–25]. The theoretical methods primarily rely on the fact that the temperature field, which is first found by means of analytical solutions, contains *a priori* unknown coefficient α, which is then determined from the condition of the equality of the maximum or the mean temperature on the friction surfaces of the pad and the disc. As a result, the formulas for the calculation of α contain the thermo-physical properties of the friction materials and, in some cases, the operating parameters of the process. Substituting thus determines the value of α back into the analytical or the numerical solutions, and the temperature mode of the braking system is estimated.

It should be noted that the formulas that have been obtained so far for determining the HPR, based on the analytical solutions to the thermal problems of friction, concern homogeneous materials [26–30]. The proportion of the heat distribution by means of the HPR between polymer and steel elements was established experimentally [31]. The aim of this study is to obtain the theoretical formulas for estimating the HPR during the braking of friction pair elements that are made of functionally graded materials (FGMs). One of the proposed applications of the results could be the determination of the bulk temperature of friction elements that are made of thermally sensitive FGM, during a single braking process [32] or a repetitive short-term braking mode [33].

2. Heating of the FGM Semi-Space by the Heat Flux with Constant Intensity

Consider a temperature of T of a semi-limited body $z \geq 0$, which is uniform and equal to T_0 at the initial point in time of $t = 0$. The body is made of a functionally graded material (FGM) with the thermal conductivity of K, increasing exponentially in the positive direction of the axis z as follows:

$$K(z) = K_0 e^{\gamma z}, z \geq 0, K_0 \equiv K(0), \quad (1)$$

where $\gamma > 0$ is the gradient of the material. Next, the semi-space is heated on the outer surface as $z = 0$ by the heat flux with a constant intensity of $q = q_0$ in time as $t > 0$ (Figure 1).

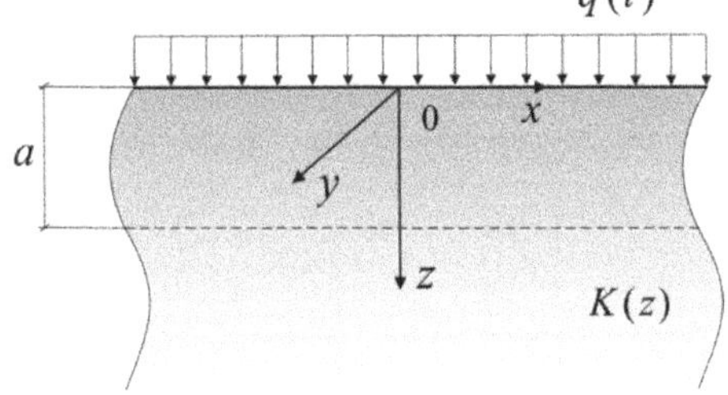

Figure 1. Scheme of the heating of the FGM semi-space.

The transient temperature field of the semi-space was searched in following the form:

$$T(z,t) = T_0 + \Theta(z,t),\ z \geq 0,\ t \geq 0, \quad (2)$$

where the temperature rise of $\Theta(z,t)$ was determined from the solution to the following boundary-value heat conduction problem:

$$\frac{\partial}{\partial z}\left[K(z)\frac{\partial \Theta(z,t)}{\partial z}\right] = \rho c \frac{\partial \Theta(z,t)}{\partial t},\ z > 0,\ t > 0, \quad (3)$$

$$K_0 \frac{\partial \Theta(z,t)}{\partial z}\bigg|_{z=0^+} = -q_0,\ t > 0, \tag{4}$$

$$\Theta(z,t) \to 0\ ,\ z \to \infty,\ t > 0, \tag{5}$$

$$\Theta(z,0) = 0,\ z \geq 0, \tag{6}$$

and ρ, c are the density and the specific heat capacity of the material, respectively.

The following dimensionless variables and parameters were introduced:

$$\zeta = \frac{z}{a},\ \gamma = \frac{\gamma^*}{a},\ \tau = \frac{kt}{a^2},\ k = \frac{K_0}{c\rho},\ q^* = \frac{q}{q_0},\ \Theta^* = \frac{\Theta}{\Theta_0},\ \Theta_0 = \frac{q_0 a}{K_0}, \tag{7}$$

where a is the effective heating depth, i.e., the distance from the heated surface, at which the temperature is equal to 5% of the maximum temperature on the heated surface [2]. Taking into account the designations (7), the problem (3)–(6) was written in following the form:

$$\frac{\partial^2 \Theta^*(\zeta,\tau)}{\partial \zeta^2} + \gamma^* \frac{\partial \Theta^*(\zeta,\tau)}{\partial \zeta} - e^{-\gamma^* \zeta} \frac{\partial \Theta^*(\zeta,\tau)}{\partial \tau} = 0,\ \zeta > 0,\ \tau > 0, \tag{8}$$

$$\frac{\partial \Theta^*(\zeta,\tau)}{\partial \zeta}\bigg|_{\zeta=0} = -1,\ \tau > 0, \tag{9}$$

$$\Theta^*(\zeta,\tau) \to 0\ ,\ \zeta \to \infty,\ \tau > 0, \tag{10}$$

$$\Theta^*(\zeta,0) = 0,\ \zeta \geq 0. \tag{11}$$

Applying the following Laplace integral transform [34]:

$$\overline{\Theta}^*(\zeta,p) \equiv L[\Theta^*(\zeta,\tau);\, p] = \int_0^\infty \Theta^*(\zeta,\tau) e^{-p\tau} d\tau,\ \mathrm{Re}p \geq 0, \tag{12}$$

to the boundary-value problem (8)–(11), the following were obtained:

$$\frac{d^2 \overline{\Theta}^*(\zeta,p)}{d\zeta^2} + \gamma^* \frac{d\overline{\Theta}^*(\zeta,p)}{d\zeta} - p e^{-\gamma^* \zeta} \overline{\Theta}^*(\zeta,p) = 0,\ \zeta > 0, \tag{13}$$

$$\frac{d\overline{\Theta}^*(\zeta,p)}{d\zeta}\bigg|_{\zeta=0^+} = -\frac{1}{p}, \tag{14}$$

$$\overline{\Theta}^*(\zeta,p) \to 0\ ,\ \zeta \to \infty. \tag{15}$$

The solution to the boundary problem (13)–(15) takes the following form:

$$\overline{\Theta}^*(\zeta,p) = e^{-\frac{1}{2}\gamma_1^* \zeta} \frac{\varphi(\zeta,p)}{\Phi(p)},\ \zeta \geq 0, \tag{16}$$

where

$$\varphi(\zeta,p) = I_1\left(\frac{2}{\gamma^*}\sqrt{p}\, e^{-\frac{1}{2}\gamma^* \zeta}\right), \tag{17}$$

$$\Phi(p) = p\sqrt{p}\, I_0\left(\frac{2}{\gamma^*}\sqrt{p}\right), \tag{18}$$

$I_k(x)$, $k = 0,1$ are the modified Bessel functions of the first kind [35]. Differentiating the function (18) with consideration of the relation of $I'_0(x) = I_1(x)$, the following was found:

$$\Phi'(p) = \frac{3}{2}\sqrt{p}\, I_0\left(\frac{2}{\gamma^*}\sqrt{p}\right) + \frac{p}{\gamma^*} I_1\left(\frac{2}{\gamma^*}\sqrt{p}\right). \tag{19}$$

The transition from the space of the transforms to the originals in the solution (16)–(19) was carried out on the basis of the Vashchenko-Zakharchenko theorem [36,37] as follows:

$$\Theta^*(\zeta,\tau) = e^{-\frac{1}{2}\gamma^*\zeta}\left[\lim_{p\to 0}\frac{\varphi(\zeta,p)p}{\Phi(p)} + \sum_{n=1}^{\infty}\frac{\varphi(\zeta,p_n)}{\Phi'(p_n)}e^{p_n\tau}\right],\ \zeta \geq 0,\ \tau \geq 0, \tag{20}$$

where

$$I_0\left(\frac{2}{\gamma_1^*}\sqrt{p_n}\right) \equiv 0,\ n = 1,2,\ldots \tag{21}$$

By using the expressions of the modified Bessel functions [35] as follows:

$$I_0(x) \cong 1 + \frac{x^2}{4} + \frac{x^4}{64} + \ldots,\ I_1(x) \cong \frac{x}{2} + \frac{x^3}{16} + \ldots \tag{22}$$

and by limiting them to the first components, the following representations of functions (17) and (18) for the small values of the parameter p were obtained:

$$\varphi(\zeta,p) \cong \frac{1}{\gamma^*}e^{-\frac{1}{2}\gamma^*\zeta}\sqrt{p},\ \Phi(p) \cong p\sqrt{p}. \tag{23}$$

Taking into account the Formulas (23) and relationships [35] as follows:

$$I_0(x) = J_0(ix),\ I_1(x) = -iJ_1(ix),\ i \equiv \sqrt{-1}, \tag{24}$$

where $J_k(x), k = 0,1$ are the Bessel functions of the first kind, solutions (20) and (21) were written in the following form:

$$\Theta^*(\zeta,\tau) = \frac{1}{\gamma^*}e^{-\frac{1}{2}\gamma^*\zeta}\left[e^{-\frac{1}{2}\gamma^*\zeta} - 4\sum_{n=1}^{\infty}\frac{J_1(\mu_n e^{-\frac{1}{2}\gamma^*\zeta})}{\mu_n^2 J_1(\mu_n)}e^{-\lambda_n\tau}\right],\ \zeta \geq 0,\ \tau \geq 0, \tag{25}$$

where

$$\lambda_n = \left(\frac{1}{2}\gamma^*\mu_n\right)^2, \tag{26}$$

$$J_0(\mu_n) \equiv 0,\ n = 1,2,\ldots \tag{27}$$

By adopting $\zeta = 0$ in the solution (25)–(27), the dimensionless temperature rise on the heated surface was found in the following form:

$$\Theta^*(\tau) \equiv \Theta^*(0,\tau) = \frac{1}{\gamma^*}\left(1 - 4\sum_{n=1}^{\infty}\frac{e^{-\lambda_n\tau}}{\mu_n{}^2}\right),\ \tau \geq 0. \tag{28}$$

Verification of the developed model was carried out by checking the boundaries (9), (10), and the initial (11) conditions. For this, by differentiating the solution (25) with respect to the spatial variable ζ, and taking into account the relationship [35] as follows:

$$J'_1(x) = J_0(x) - x^{-1}J_1(x) \tag{29}$$

the following was found:

$$\frac{\partial\Theta^*(\zeta,\tau)}{\partial\zeta} = -e^{-\gamma^*\zeta}\left[1 - 2\sum_{n=1}^{\infty}\frac{J_0(\mu_n e^{-\frac{1}{2}\gamma^*\zeta})}{\mu_n J_1(\mu_n)}e^{-\lambda_n\tau}\right],\ \zeta \geq 0,\ \tau \geq 0. \tag{30}$$

Approaching Formula (30) to the limit of $\zeta \to 0$, with consideration of the fact that μ_n are the roots of Equation (27), the following was achieved:

$$\left.\frac{\partial\Theta^*(\zeta,\tau)}{\partial\zeta}\right|_{\zeta=0^+} = -1, \tag{31}$$

which confirms that the boundary condition (9) was satisfied. However, by approaching to the limit $\zeta \to \infty$ in the solution (25), we have confirmed the fulfillment of the condition of the temperature fade (10). In addition, it was numerically established that the initial condition (11) was met. In particular, according to the Formula (25), on the heated surface $\zeta = 0$ at the initial point in time, the dimensionless temperature rise is zero if the following occurs:

$$\sum_{n=1}^{\infty} \frac{1}{\mu_n{}^2} = \frac{1}{4}. \tag{32}$$

Based on the calculations, it was found that the sum of the first 10^2 components of the series on the left-hand side of Equation (32) is equal to 0.248985.

In addition to the exact (25), the appropriate asymptotic solutions to the problem at small and large values of the Fourier number τ (dimensionless time) were also found.

Small values τ (large values of the parameter p). Taking into account that, in Formulas (17) and (18), the first components of the asymptotic of modified Bessel functions at large values of the argument [35] were as follows:

$$I_0(x) \cong \frac{e^x}{\sqrt{2\pi x}}\left(1+\frac{1}{8x}+\ldots\right),\ I_1(x) \cong \frac{e^x}{\sqrt{2\pi x}}\left(1+\frac{3}{8x}+\ldots\right), \tag{33}$$

the transformed solution (16) was written as follows:

$$\overline{\Theta}^*(\zeta,p) = e^{-\frac{1}{4}\gamma^*\zeta}\frac{e^{-b\sqrt{p}}}{p\sqrt{p}},\ b = \frac{2}{\gamma^*}\left(1-e^{-\frac{1}{2}\gamma^*\zeta}\right),\ \zeta \geq 0. \tag{34}$$

Using the relation [38] as follows:

$$L^{-1}\left[\frac{e^{-b\sqrt{p}}}{p\sqrt{p}};\tau\right] = 2\sqrt{\tau}\,\mathrm{ierfc}\left(\frac{b}{2\sqrt{\tau}}\right), \tag{35}$$

from the Formula (34), the following form of dimensionless temperature rise at the initial moments of heating was obtained:

$$\Theta^*(\zeta,\tau) \cong 2e^{-\frac{1}{4}\gamma^*\zeta}\sqrt{\tau}\,\mathrm{ierfc}\left(\frac{b}{2\sqrt{\tau}}\right),\ \zeta \geq 0,\ 0 \leq \tau << 1, \tag{36}$$

where $\mathrm{ierfc}(x) = \pi^{-\frac{1}{2}}e^{-x^2} - x\mathrm{erfc}(x)$, $\mathrm{erfc}(x) = 1 - \mathrm{erf}(x)$, $\mathrm{erf}(x)$ is the Gauss error function [35].

At $\zeta = 0$ from the solution (36), a known result for the evolution of the temperature on the heated surface of a homogeneous semi-space was obtained [18] as follows:

$$\Theta^*(\tau) \cong 2\sqrt{\frac{\tau}{\pi}},\ 0 \leq \tau << 1, \tag{37}$$

Large values τ (small values of the parameter p). By including the first two components in the distributions (22), from the transformed solution (16)–(18) it follows that:

$$\overline{\Theta}^*(\zeta,p) \cong \frac{e^{-\gamma^*\zeta}}{\gamma^*}\left[\frac{1}{p} - \left(1-\frac{1}{2}e^{-\gamma^*\zeta}\right)\frac{1}{(\gamma^{*2}+p)}\right],\ \zeta \geq 0. \tag{38}$$

Considering the relations [38] as follows:

$$L^{-1}[p^{-1};\tau]=1,\ L^{-1}[(\gamma^{*2}+p)^{-1};\tau]=e^{-\gamma^{*2}\tau}, \tag{39}$$

from the solution (38), the following asymptotic representation for the dimensionless temperature rise at large values of time was obtained:

$$\Theta^*(\zeta,\tau)\cong\frac{e^{-\gamma^*\zeta}}{\gamma^*}\left[1-\left(1-\frac{1}{2}e^{-\gamma^*\zeta}\right)e^{-\gamma^*\tau}\right],\ \zeta\geq 0,\ \tau>>1. \tag{40}$$

The temperature rise of the heated surface was found, substituting to the formula (40) $\zeta=0$ as follows:

$$\Theta^*(\tau)\cong\frac{1}{\gamma^*}\left(1-\frac{1}{2}e^{-\gamma^*\tau}\right),\ \tau>>1. \tag{41}$$

3. Heating of the FGM Semi-Space by Heat Flux with the Intensity Linearly Decreasing in Time

As presented above, the exact (25), asymptotic (36), and (40) solutions were obtained at a constant intensity of heat flux of $q=q_0$. This section concerns the heating of the surface of the semi-space by heat flux with the following time profile of intensity:

$$q(t)=q_0q^*(t),\ q^*(t)=1-\frac{t}{t_s},\ 0\leq t\leq t_s, \tag{42}$$

where t_s is the stop moment of heating. The evolution of the heat flux intensity (42) corresponds to the temporal profile of the specific friction power during braking with constant deceleration [39]. The corresponding dimensionless temperature rise of $\hat{\Theta}^*$ was searched based on Duhamel's theorem [40] as follows:

$$\hat{\Theta}^*(\zeta,\tau)=\frac{\partial}{\partial\tau}\int_0^\tau q^*(\tau-s)\Theta^*(\zeta,s)ds,\ \zeta\geq 0, 0\leq\tau\leq\tau_s, \tag{43}$$

where Θ^* is the dimensionless temperature rise (25)–(27), $q^*(\tau)$ is the function (42), and τ_s is the dimensionless stop time as follows:

$$\tau_s=\frac{kt_s}{a^2}. \tag{44}$$

After the integration with the next differentiation on the right-hand side of Formula (42), the following was obtained:

$$\hat{\Theta}^*(\zeta,\tau)=\frac{1}{\gamma^*}e^{-\frac{1}{2}\gamma^*\zeta}\left[q^*(\tau)e^{-\frac{1}{2}\gamma^*\zeta}-4\sum_{n=1}^{\infty}\frac{J_1(\mu_ne^{-\frac{1}{2}\gamma^*\zeta})}{\mu_n^2J_1(\mu_n)}G_n(\tau)\right],\zeta\geq 0, 0\leq\tau\leq\tau_s, \tag{45}$$

where

$$G_n(\tau)=e^{-\lambda_n\tau}-\frac{(1-e^{-\lambda_n\tau})}{\lambda_n\tau_s}, \tag{46}$$

the coefficients λ_n were determined from Formula (26), and the numbers μ_n, $n=1,2,\ldots$, are the single positive roots of Equation (27).

By putting $\zeta=0$ in the solution (45), the following formula was used to determine the evolution of the dimensionless temperature rise of the heated surface that was obtained:

$$\hat{\Theta}^*(\tau)=\frac{1}{\gamma^*}\left[q^*(\tau)-4\sum_{n=1}^{\infty}\frac{G_n(\tau)}{\mu_n^2}\right],\ 0\leq\tau\leq\tau_s. \tag{47}$$

4. The Heat Partition Ratio

We will generalize the results that were obtained above to the case of two ($l = 1, 2$) FGM semi-spaces that exponentially increased with the distance from the surface $z = 0$ of thermal conductivity as follows:

$$K_l(z) = K_{l,0}e^{\gamma_l z},\ z \geq 0,\ K_{l,0} \equiv K_l(0),\ \gamma_l \geq 0,\ l = 1, 2. \tag{48}$$

The surfaces of $z = 0$ of each semi-space were heated with the heat flux at the following intensities:

$$q_l(t) = \alpha_l q(t),\ 0 \leq t \leq t_s,\ l = 1, 2,\ \alpha_1 = \alpha,\ \alpha_2 = 1 - \alpha, \tag{49}$$

where $q(t)$ is the function (42) and α is the unknown heat partition ratio (HPR) (Figure 2).

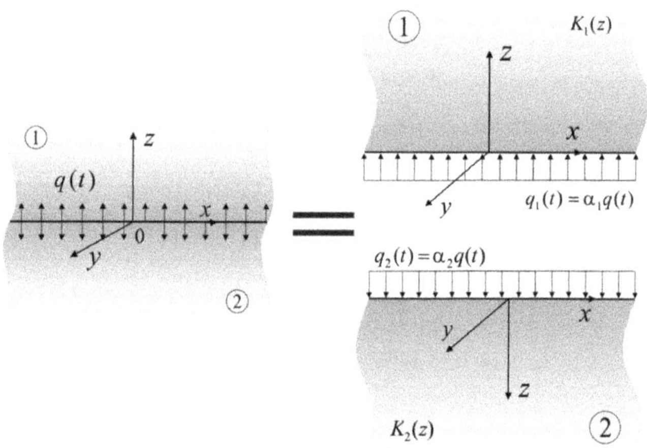

Figure 2. Scheme of separation of the friction pair elements.

Taking into account the form of the solutions (45) and (46), the temperature field in each semi-space heated by heat fluxes of intensity (49) respectively, was written in the following form:

$$T_l(z,t) = T_0 + \Theta_{l,0}\hat{\Theta}_l^*(\zeta_l, \tau^*),\ z \geq 0,\ 0 \leq t \leq t_s,\ l = 1, 2, \tag{50}$$

where

$$\hat{\Theta}_l^*(\zeta_l, \tau^*) = \frac{1}{\gamma_l^*}e^{-\frac{1}{2}\gamma_l^*\zeta_l}\left[(1-\tau^*)\,e^{-\frac{1}{2}\gamma_l^*\zeta_l} - 4\sum_{n=1}^{\infty}\frac{J_1(\mu_n e^{-\frac{1}{2}\gamma_l^*\zeta_l})}{\mu_n^2 J_1(\mu_n)}G_{l,n}(\tau^*)\right],\ \zeta_l \geq 0,\ 0 \leq \tau^* \leq 1, \tag{51}$$

$$G_{l,n}(t^*) = e^{-\lambda_{l,n}\tau^*} - \frac{(1 - e^{-\lambda_{l,n}\tau^*})}{\lambda_{l,n}\tau_{l,s}}, \tag{52}$$

$$\lambda_{l,n} = \left(\frac{1}{2}\gamma_l^*\mu_n\right)^2 \tau_{l,s}, \tag{53}$$

$$\tau^* = \frac{t}{t_s},\ \zeta_l = \frac{z}{a_l},\ \tau_{l,s} = \frac{k_l t_s}{a_l^2},\ \gamma_l = \frac{\gamma_l^*}{a_l},\ k_l = \frac{K_{l,0}}{c_l\rho_l},\ \hat{\Theta}_l^* = \frac{\hat{\Theta}_l}{\Theta_{l,0}},\ \Theta_{l,0} = \frac{\alpha_l q_0 a_l}{K_{l,0}}, l = 1, 2, \tag{54}$$

c_l, ρ_l are the specific heat capacity and the density of the materials, respectively, $\mu_n > 0$, $n = 1, 2, \ldots$ are the single roots of Equation (27).

Substituting $z = 0$ ($\zeta_l = 0$) to the solution (51)–(54), the change in the temperature on the heated surface was obtained in the following form:

$$T_l(t) = T_0 + \Theta_{l,0}\hat{\Theta}_l^*(\tau^*),\ 0 \le t \le t_s,\ l = 1,2, \tag{55}$$

where

$$\hat{\Theta}_l^*(\tau^*) = \frac{1}{\gamma_l^*}\left[1 - \tau^* - 4\sum_{n=1}^{\infty}\frac{G_{l,n}(\tau^*)}{\mu_n^2}\right],\ 0 \le \tau^* \le 1. \tag{56}$$

The coefficients of $\Theta_{l,0}$ (54) in Formulas (50) and (55) contain an unknown heat partition ratio of α. In order to determine it, the equality condition of the mean temperature in time on the heated surfaces was used as follows:

$$\widetilde{T}_1 = \widetilde{T}_2, \tag{57}$$

where

$$\widetilde{T}_l = \frac{1}{t_s}\int_0^{t_s} T_l(t)dt,\ l = 1,2, \tag{58}$$

Considering the temperature on the heated surfaces (55) and (56) in the Formula (58), after integration was found as follows:

$$\widetilde{T}_l = T_0 + \Theta_{l,0}\widetilde{\Theta}_l^*,\ l = 1,2, \tag{59}$$

where

$$\widetilde{\Theta}_l^* = \frac{1}{2\gamma_l^*}\left(1 - 8\sum_{n=1}^{\infty}\frac{\widetilde{G}_{l,n}}{\mu_n^2}\right), \tag{60}$$

$$\widetilde{G}_{l,n} = \frac{(1 - e^{-\lambda_{l,n}})}{\lambda_{l,n}}\left(1 + \frac{1}{\lambda_{l,n}\tau_{l,s}}\right) - \frac{1}{\lambda_{l,n}\tau_{l,s}},\ l = 1,2, \tag{61}$$

and the coefficients of $\lambda_{l,n}$ were determined from the formula (53).

Substituting the mean temperature (59)–(61) to Equation (57), the HPR was obtained in the following form:

$$\alpha = \frac{K^*}{a^*\widetilde{\Theta}^* + K^*}, \tag{62}$$

where

$$a^* = \frac{a_1}{a_2},\ K^* = \frac{K_{1,0}}{K_{2,0}},\ \widetilde{\Theta}^* = \frac{\widetilde{\Theta}_1^*}{\widetilde{\Theta}_2^*}. \tag{63}$$

If the effective heating depths are determined from the Formula [9] as follows:

$$a_l = \sqrt{3k_l t_s},\ l = 1,2, \tag{64}$$

then from Formula (54) it follows that the dimensionless heating time of each semi-space is equal to $\tau_{l,s} = 3^{-1} \cong 0.333$, and from Formulas (62) and (63) it follows that:

$$\alpha = \frac{K_\varepsilon}{\widetilde{\Theta}^* + K_\varepsilon}, \tag{65}$$

where

$$K_\varepsilon = \frac{K^*}{\sqrt{k^*}},\ k^* = \frac{k_1}{k_2}. \tag{66}$$

Omitting the component containing a series, on the right-hand side of Formula (60), i.e., assuming the following:

$$\widetilde{\Theta}_l^* \cong \frac{1}{2\gamma_l^*},\ l = 1,2, \tag{67}$$

where

$$\alpha \cong \frac{K^*\gamma^*}{a^* + K^*\gamma^*}, \tag{68}$$

and Formula (65) is as follows:

$$\alpha \cong \frac{K_\varepsilon\gamma^*}{1 + K_\varepsilon\gamma^*}, \tag{69}$$

where

$$\gamma^* = \frac{\gamma_1^*}{\gamma_2^*}. \tag{70}$$

With the same effective heating depths ($a^* = 1$) and dimensionless gradients of materials ($\gamma^* = 1$) from Formula (68), the known Block' result [20] as follows:

$$\alpha \cong \frac{K^*}{1 + K^*}, \tag{71}$$

and from Formula (69), the classic Charron's Formula [41] was achieved as follows:

$$\alpha \cong \frac{K_\varepsilon}{1 + K_\varepsilon}, \tag{72}$$

which is often used in analytical and numerical modeling of the frictional heating of homogeneous materials during braking [42–44]. For the same materials of both semi-spaces ($K^* = k^* = \gamma^* = a^* = 1$), all of the above obtained formulas give the value of the heat partition ratio equal to $\alpha = 0.5$.

It should be noted that the received formulas for the determination of HPR need to be verified in the future with appropriate experimental data. Due to the lack of such opportunity at that moment, the authors would be grateful for the provision of such data or for carrying out the cooperative research. This would also allow us to establish the limits of the applicability of the formulas. However, already at this stage of research, we can assume the possibility of the practical use of the obtained formulas to determine HPR, due to the usage of the classic methodology that has been approved for homogeneous materials. Furthermore, the experimentally confirmed formulas for homogeneous materials can be obtained from the proposed solution for FGMs, as a results of the proper limit approach.

An important element of the most general Formula (62) for determining the HPR are the dimensionless time-averaged temperature rises Formulas (60) and (61) found in the case of the heat flux intensity linearly decreasing in time. This case is often considered when calculating the temperature of the brake systems that are operating in the mode with a sudden increase in the contact pressure to the nominal value at the beginning of braking. A classification of the remaining heat flux intensity temporal profiles has been proposed for homogeneous materials, without and with consideration of the contact pressure rise [45,46]. Obtaining the appropriate solutions in the case of FGM is also one of the directions of our research in the future.

5. Example of Calculation of the Heat Partition Ratio for an FGM Couple

Calculations were performed for two semi-limited bodies made of two-component FGMs. The first ($l = 1$) element forms aluminum oxide Al_2O_3 (base, $m = 0$) and copper Cu (core, $m = 1$), and the other ($l = 2$) contains zircon dioxide ZrO_2 (base, $m = 0$) and titanium alloy Ti-6Al-4V (core, $m = 1$). The thermo-physical properties of these materials are demonstrated in Table 1.

Table 1. Thermo-physical properties of FGM components [32].

Element Subscript	Material Subscript, m	Material	Thermal Conductivity $K_{l,m}^{(0)}$, $\mathrm{Wm^{-1}K^{-1}}$	Specific Heat Capacity $c_{l,m}^{(0)}$, $\mathrm{J\,kg^{-1}K^{-1}}$	Density $\rho_{l,m}^{(0)}$, $\mathrm{kgm^{-3}}$
$l = 1$	*base*, 0	Al_2O_3	37.24	727.29	3990.92
	core, 1	Cu	402.65	147.35	8947.92
$l = 2$	*base*, 0	ZrO_2	1.94	452.83	6102.16
	core, 1	Ti-6Al-4V	6.87	538.08	4431.79

The other operating parameters are the nominal heat flux intensity of $q_0 = 3.78\ \mathrm{MW\,m^{-2}}$, the braking time of $t_s = 121$ s, and the temperature at the initial time moment of $T_0 = 20\ ^\circ\mathrm{C}$ [32]. The dimensionless material gradient parameters of each element were calculated from the relation [47] as follows:

$$\gamma_l^* = \ln\left(\frac{K_{l,1}}{K_{l,0}}\right),\ l = 1,2. \tag{73}$$

Obtaining the values of $\gamma_1^* = 2.38$, $\gamma_2^* = 1.26$. The specific heat capacity and density of materials of the heated elements were determined according to the following mixture law:

$$c_l = c_{l,1}v + (1-v)c_{l,0},\ \rho_l = \rho_{l,1}v + (1-v)\rho_{l,0},\ l = 1,2, \tag{74}$$

where v is the volume fractions of the base and core components. For their equal participation ($v = 0.5$), the following properties of FGM were found:

$$c_1 = 437.32\ \mathrm{J\,kg^{-1}K^{-1}},\ \rho_1 = 6469.42\ \mathrm{kgm^{-3}},\ k_1 = 1.32{\cdot}10^{-5}\ \mathrm{m^2s^{-1}}, \tag{75}$$

$$c_2 = 495.46\ \mathrm{J\,kg^{-1}K^{-1}},\ \rho_2 = 5266.96\ \mathrm{kgm^{-3}},\ k_2 = 7.43{\cdot}10^{-7}\ \mathrm{m^2s^{-1}}. \tag{76}$$

From Equation (64) the effective depths of the heat penetration were determined as $a_1 = 21.8$ mm, $a_2 = 5.2$ mm. Next, based on Formulas (63), (66), and (70), the values of the dimensionless parameters were calculated as follows: $a^* = 4.208$, $K^* = 19.196$, $k^* = 17.706$, $K_\varepsilon = 4.562$, $\gamma^* = 1.883$. This allowed for an estimation by means of Equations (60), (61) dimensionless, time-averaged temperature rises of $\widetilde{\Theta}_1^* = 0.329$, $\widetilde{\Theta}_2^* = 0.703$, and their ratio of $\widetilde{\Theta}^* = 0.468$ (63).

By substituting the found parameter values successively to the right-hand sides of Formulas (65) and (69), the proper values of the HPR were calculated as $\alpha = 0.907$ and $\alpha = 0.896$. A slight (1.2%) difference in the obtained results allowed us to analyze the influence of the dimensionless parameters of the thermal activity K_ε (66) and the relative gradient of the FGMs γ^* (70) on the HPR value (Figure 3), on the basis of Formula (69) only. For a fixed value of the thermal activity coefficient K_ε, a rise of the parameter γ^* causes an increase in the amount of heat that is directed to the first element of the heated couple. Conversely, by increasing the thermal activity of the friction pair at a predetermined value γ^*, the amount of heat that is directed to the first element increases.

For the HPR values $\alpha_1 = 0.896$, $\alpha_2 = 0.104$ (49), from Equation (54), the scaling factors of the temperature rises were determined to be $\Theta_{1,0} = 2034\ ^\circ\mathrm{C}$ and $\Theta_{2,0} = 1080\ ^\circ\mathrm{C}$. Then, based on the relations Formulas (55) and (56) the evolutions of the temperature on the heated surfaces of each element were found (Figure 4).

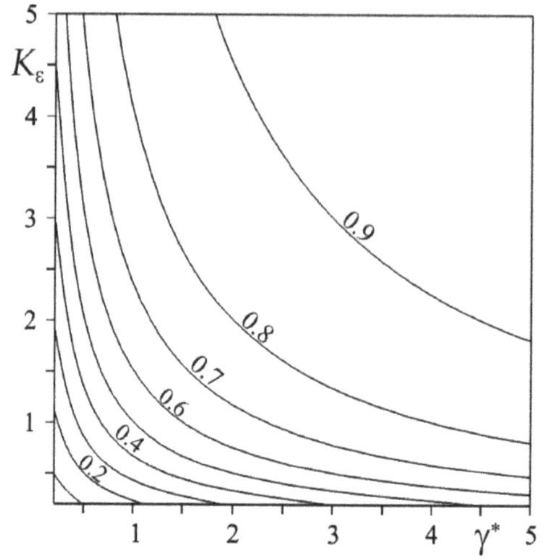

Figure 3. Isolines of the heat partition ratio α (69) in the coordinate system (γ^*, K_ε).

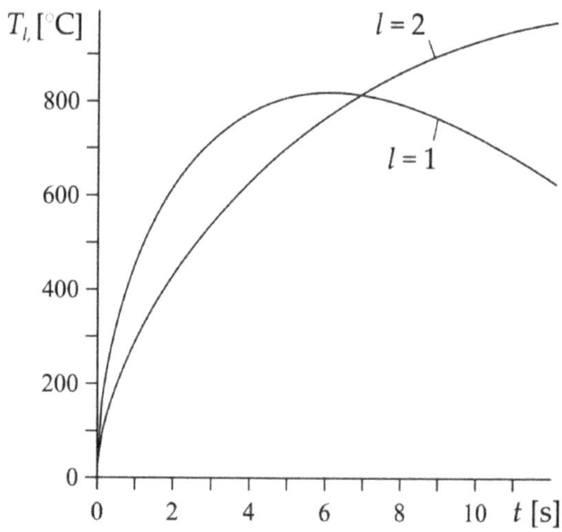

Figure 4. Evolutions of temperature T_l, $l = 1,2$ (55) and (56) on the heated surfaces $z = 0$ of the considered semi-spaces made of FGMs.

It has been established that the temporal profiles of the temperature of the elements are different, in particular in the final stage of the heating process. The time course of the temperature of the first element ($l = 1$, Al_2O_3–Cu) is typical for the evolution of the friction surface temperature during braking with a constant deceleration—a rapid increase in the temperature at the beginning of braking, reaching its maximum value in the middle of the process, followed by a temperature reduction until the standstill. However, in the second element ($l = 2$, ZrO_2–Ti–6Al–4V) a rise of temperature on the heated surface is monotonic during the whole process. Such temperature behavior is decisively influenced by the thermo-physical properties of the component materials of each element. In the functionally graded friction couple under consideration, the materials of both of the components of the first element have a significantly greater ability to dissipate the heat from the heated surface than the materials of the second element (Table 1). Moreover, this is confirmed, by the values of the coefficients $\alpha_1 = 0.896$, $\alpha_2 = 0.104$, which prove that the first element absorbs almost 90%, and the second only slightly more than 10% of the entire heat flux intensity q (42). Due to the low thermal conductivity of the component materials of the second element, in particular the zircon dioxide, the temperature of the heated surface of this element continues to rise during heating, even with a linearly decreasing intensity of the heat flux.

The results that are demonstrated in Figures 3 and 4 were obtained by using the HPR value $\alpha = 0.896$ that was found from the dependency (69). The approximated formula for determining the α (69), as well as the exact Equation (65), was obtained by assuming that the effective depths of the heating of the elements a_l, $l = 1,2$ were determined from the empirical Formula (64), so the corresponding values of the Fourier numbers $\tau_{l,s}$, $l = 1,2$ (54) were the same and equal to 1/3. The introduction of Formula (64) to the model is justified when the thickness of the heated element is greater than the appropriate value a_l, $l = 1, 2$ [2]. In general, for the determination of the temperature mode at the design stage of the brake, the Fourier numbers $\tau_{l,s}$, $l = 1,2$ are given, and the effective heating depths based on the relation (54) are calculated in the following form:

$$a_l = \sqrt{\frac{k_l t_s}{\tau_{l,s}}},\ l = 1,2. \tag{77}$$

With consideration of Equation (75) for the selected friction couple and the established time of heating (braking) $t_s = 12.1$ s the ratio of the depths of heating a^* (63) is determined from the following relation:

$$a^* = \sqrt{\frac{k^*}{\tau_s^*}},\ \tau_s^* = \frac{\tau_{1,s}}{\tau_{2,s}}, \tag{78}$$

where k^* is the ratio of the thermal diffusivities of the FGMs (66). By treating the Fourier numbers $\tau_{l,s}$, $l = 1,2$ (54) as independent variables in Formula (62), their influence on the HPR α was investigated with the previously determined values of the dimensionless parameters, $K^* = 19.196$, $k^* = 17.706$, $\gamma_1^* = 2.38$, and $\gamma_2^* = 1.26$ for a given pair of elements. The dimensionless parameter $\widetilde{\Theta}^*$ (63) included in Formula (62) was determined from Formulas (60) and (61), and the finding of a^* was made by using Formula (76). The results of the calculations are presented in Figure 5.

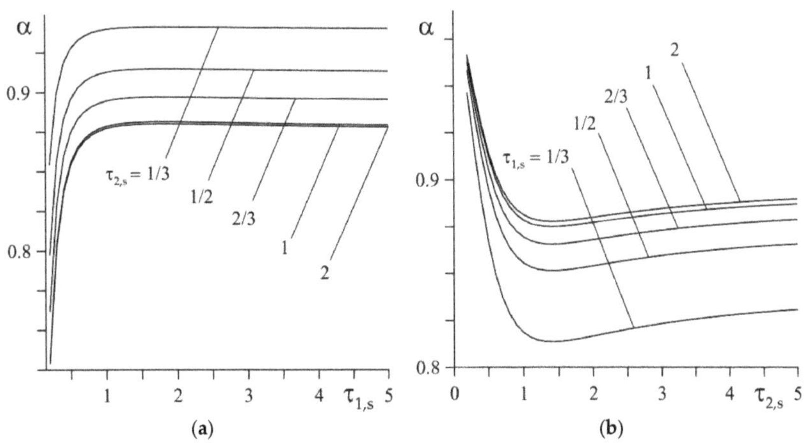

Figure 5. Dependences of the heat partition ratio α (62) on: (**a**) Fourier number $\tau_{1,s}$ for different values of $\tau_{2,s}$; (**b**) Fourier number $\tau_{2,s}$ for different values of $\tau_{1,s}$.

At a fixed value $\tau_{2,s}$, the HPR α quickly increases with increasing $\tau_{1,s}$, reaching its nominal value at $\tau_{1,s} \cong 1.5$ (Figure 5a). The largest nominal value α is achieved for $\tau_{1,s} = 1/3$, i.e., when determining the effective heating depth using Formula (64). The results obtained at $\tau_{2,s} = 1$ and $\tau_{2,s} = 2$ practically coincide. A different nature of the HPR α change occurs with a fixed value of the Fourier number $\tau_{1,s}$ and increasing the values of $\tau_{2,s}$ (Figure 5b). By increasing $\tau_{2,s}$, the amount of heat that is absorbed by the first element decreases, reaching a minimum value at $\tau_{2,s} \cong 1.5$. It should be noted that the Fourier numbers $\tau_{l,s}$, $l = 1,2$ (54) are dimensionless input parameters that play an important role in estimating the temperature mode of the brakes. The designers calculate their values after determining the geometric parameters a_l, $l = 1,2$ (63). In the quality of these parameters, they choose the thickness of the friction elements, or, if they are greater than the effective depth of the heat penetration (64), then the latter. Then, having previously found the maps of those, as shown in Figure 5, the designer can easily estimate the HPR of a given braking system.

6. Conclusions

A methodology was proposed in order to determine the heat partition ratio (HPR) in the braking systems with friction pair elements that were made of functionally graded materials (FGMs). The basis of this methodology is an exact solution to the initial-boundary problem of heat conduction that is formulated for the FGM semi-space that is heated on its surface by a heat flux with an intensity that decreases linearly with time and includes an unknown a priori HPR. For its determination, the condition of the equality of the time-averaged temperature on the heated surfaces of the two different semi-spaces was used.

As an example, the heat partition ratio was found for the FGM friction couple that consisted of Al_2O_3–Cu and ZrO_2–Ti–6Al–4V. It was established that element Al_2O_3–Cu absorbs most of the heat that is generated due to friction (almost 90%). The maximum temperature on the friction surface of the Al_2O_3–Cu element is about 820 °C and is achieved in the middle of the braking time. However, the highest temperature of the friction surface of the ZrO_2–Ti–6Al–4V element is achieved at the stop moment and amounts to 970 °C. Thus, the crucial influence on the evolution of the temperature in the functionally graded friction couple have the thermo-physical properties of the component materials.

Then, the simulation of the HPR dependency on the dimensionless parameters, such as the thermal activity of the friction couple, the gradient materials, and the Fourier numbers was carried out. It was found that increasing the values of the friction couple thermal activity or the gradient materials ratio causes an increase in the HPR. The scheme of the map development for a given functionally graded friction pair was proposed, which allows for a quick estimation of the HPR, depending on such input parameters as their Fourier numbers. The importance of having such maps for the designer is indicated by the results that have been obtained in the case of the homogeneous materials [9,28].

Author Contributions: Conceptualization and methodology, A.Y.; software, K.T. and P.Z.; validation, P.Z.; formal analysis, K.T. and P.Z.; investigation, A.Y., K.T. and P.Z.; writing—original draft preparation, K.T. and P.Z.; writing–review and editing, A.Y., K.T. and P.Z.; visualization and figures preparation, K.T.; supervision, A.Y.; project administration, K.T.; funding acquisition, K.T. All authors have read and agreed to the published version of the manuscript.

Funding: This investigation was performed within the framework of research project No. 2017/27/B/ST8/01249, funded by the National Science Centre, Poland, and project financing was obtained through the program of the Minister of Education and Science of Poland named "Regional Initiative of Excellence" in 2019–2022, project No. 011/RID/2018/19, amount financing 12,000,000 PLN.

Institutional Review Board Statement: Not applicable.

Informed Consent Statement: Not applicable.

Data Availability Statement: No new data were created or analyzed in this study. Data sharing is not applicable to this article.

Conflicts of Interest: The authors declare no conflict of interest.

Nomenclature

a	Effective depth of heat penetration (m)
c	Specific heat capacity (J kg^{-1}K^{-1})
f_0	Coefficient of friction
$I_k(\cdot)$	Modified Bessel functions of the first kind of the kth order
$J_k(\cdot)$	Bessel functions of the first kind of the kth order
k	Thermal diffusivity (m^{2}s^{-1})
K	Thermal conductivity (Wm^{-1}K^{-1})
K_ε	Dimensionless coefficient of thermal activity of friction couple
p	Dimensionless parameter of the Laplace integral transform
q	Intensity of heat flux (Wm^{-2})
q_0	Nominal intensity of the heat flux (Wm^{-2})
t	Time (s)
t_s	Braking time (s)
T	Temperature (°C)
T_0	Initial temperature (°C)
v	Volume fraction of the material phases (dimensionless)
z	Spatial coordinate in axial direction (m)
α	Heat partition ratio
γ	Parameter of material gradient (m^{-1})
γ^*	FGMs gradient ratio
Θ	Temperature rise (°C)
Θ^*	Dimensionless temperature rise
Θ_0	Scaling factor of temperature rise (°C)
ρ	Density (kgm^{-3})
τ^*	Dimensionless time
τ_s	Dimensionless braking time
ζ	Dimensionless spatial coordinate in axial direction

References

1. Padture, N.P.; Gell, M.; Jordan, E.H. Thermal Barrier Coatings for Gas-Turbine Engine Applications. *Science* **2002**, *296*, 280–284. [CrossRef] [PubMed]
2. Mahamood, R.M.; Akinlabi, E.T. *Functionally Graded Materials*; Springer: Cham, Switzerland, 2017.
3. Zhou, W.; Ai, S.; Chen, M.; Zhang, R.; He, R.; Pei, Y.; Fang, D. Preparation and thermodynamic analysis of the porous $ZrO_2/(ZrO_2$ + Ni) functionally graded bolted joint. *Compos. Part B Eng.* **2015**, *82*, 13–22. [CrossRef]
4. Zhou, W.; Zhang, R.; Ai, S.; He, R.; Pei, Y.; Fang, D. Load distribution in trends of porous metal-ceramic functionally graded composite joints subjected to thermomechanical loading. *Compos. Struct.* **2015**, *134*, 680–688. [CrossRef]
5. Mahmoud, D.; Elbestawi, M. Lattice Structures and Functionally Graded Materials. Applications in Additive Manufacturing of Orthopedic Implants: A Review. *J. Manuf. Mater. Process.* **2017**, *1*, 13. [CrossRef]
6. Hovorun, T.P.; Berladir, K.V.; Pererva, V.I.; Rudenko, S.G.; Martynov, A.I. Modern Materials for Automotive Industry. *J. Eng. Sci.* **2017**, *4*, 8–18. [CrossRef]
7. Strojny-Nędza, A.; Pietrzak, K.; Gili, F.; Chmielewski, M. FGM based on copper-alumina composites for brake disc applications. *Arch. Civ. Mech. Eng.* **2020**, *20*, 83. [CrossRef]
8. Newcomb, T.P. Temperatures reached in disc brakes. *J. Mech. Eng. Sci.* **1960**, *2*, 167–177. [CrossRef]
9. Chichinadze, A.V. *Polymers in Friction Assembles of Machines and Devices: A Handbook*; Allerton Press Inc.: New York, NY, USA, 1984.
10. Day, A.J. *Braking of Road Vehicles*; Butterworth-Heinemann: Oxford, UK, 2014.
11. Ling, F.F. *Surface Mechanics*; John Wiley & Sons: New York, NY, USA, 1973.
12. Barber, J.R.; Comninou, M. Thermoelastic contact problems. In *Thermal Stresses*, 3rd ed.; Hetnarsky, R.R., Ed.; Elsevier: Amsterdam, The Netherlands, 1989.
13. Awrejcewicz, J.; Pyryev, Y. *Nonsmooth Dynamics of Contacting Thermoelastic Bodies*; Springer: New York, NY, USA, 2009.
14. Yevtushenko, A.A.; Kuciej, M. One-dimensional thermal problem of friction during braking: The history of development and actual state. *Int. J. Heat Mass Transf.* **2012**, *55*, 4148–4153. [CrossRef]
15. Evtushenko, O.; Kuciej, M.; Topczewska, K. Determination of the maximal temperature of a pad-disk tribosystem during one-time braking. *Mater. Sci.* **2020**, *56*, 152–159. [CrossRef]
16. Pyryev, Y.; Yevtushenko, A. The influence of the brakes friction elements thickness on the contact temperature and wear. *Heat Mass Transf.* **2000**, *36*, 319–323. [CrossRef]

17. Yevtushenko, A.A.; Grześ, P. The FEM-modeling of the frictional heating phenomen in the pad/disc tribosystem (a review). *Num. Heat Trans. Part A-Appl.* **2010**, *58*, 207–226. [CrossRef]
18. Wasilewski, P. Frictional heating in railway brakes: A review of numerical models. *Arch. Computat. Methods Eng.* **2020**, *27*, 45–58. [CrossRef]
19. Deressa, K.T.; Ambie, D.A. Thermal load simulations in railway disc brake: A systematic review of modelling temperature, stress and fatigue. *Arch. Computat. Methods Eng.* **2022**, *29*, 2271–2283. [CrossRef]
20. Blok, H. Theoretical field study of temperature rise at surfaces of actual contact under oiliness lubricating conditions. *Proc. Instn. Mech. Engrs. (Gen. Discuss. Lubr. Lubr.)* **1937**, *45*, 222–235.
21. Day, A.J.; Newcomb, T.P. Dissipation of frictional energy from the interface of an annular disc brake. *Proc. Inst. Mech. Eng. Part D J. Automob. Eng.* **1984**, *198*, 201–209. [CrossRef]
22. Grześ, P. Partition of heat in 2D finite element model of a disc brake. *Acta Mech. Autom.* **2011**, *5*, 35–41.
23. Yevtushenko, A.A.; Grześ, P. Finite element analysis of heat partition ratio in a pad/disc brake system. *Num. Heat Trans. Part A-Appl.* **2011**, *58*, 521–542. [CrossRef]
24. Qiu, L.Q.; Qi, H.S.; Wood, A. Two-dimensional finite element analysis of the heat partition ratio of a friction brake. *Proc. Inst. Mech. Eng. Part J J. Eng. Trib.* **2018**, *232*, 1489–1501. [CrossRef]
25. Afzal, A.; Abdul Mujeebu, M. Thermo-Mechanical and Structural Performances of Automobile Disc Brakes: A Review of Numerical and Experimental Studies. *Arch. Comput. Methods Eng.* **2019**, *26*, 1489–1513. [CrossRef]
26. Carslaw, H.C.; Jaeger, J.C. *Conduction of Heat in Solids*, 2nd ed.; Clarendon Press: Oxford, UK, 1959.
27. Hasselgruber, H. Der Schaltvorgang einer Trockenreibung Kupplung bei kleinster Erwärmung. *Konstruction* **1963**, *15*, 41–45.
28. Ginzburg, A.G. Coefficient of distribution of the heat flows during braking. In *Calculation and Testing of Friction Couples*; Mechanical Engineering: Moscow, Russia, 1974. (In Russian)
29. Yevtushenko, A.; Kuciej, M.; Topczewska, K. Effect of the Temporal Profile of the Friction Power on Temperature of a Pad-Disc Brake System. *J. Theoret. Appl. Mech.* **2019**, *57*, 461–473. [CrossRef]
30. Xiong, C.; Chen, M.; Yu, L. Analytical model and material equivalent methods for steady state heat partition coefficient between two contact discs in multi-disc clutch. *Proc. J. Mech. Eng. J. Automob. Eng.* **2019**, *234*, 857–871. [CrossRef]
31. Xia, Y.; Yano, A.; Hayashi, N.; Noraguchi, N.; Xie, G. Analysis of temperature and heat partitioning coefficient during friction between polymer and steel. *Tribol. Int.* **2022**, *171*, 107561. [CrossRef]
32. Yevtushenko, A.; Topczewska, K.; Zamojski, P. Influence of Thermal Sensitivity of Functionally Graded Materials on Temperature during Braking. *Materials* **2022**, *15*, 963. [CrossRef]
33. Yevtushenko, A.; Topczewska, K.; Kuciej, M. Analytical Determination of the Brake Temperature Mode during Repetitive Short-Term Braking. *Materials* **2021**, *14*, 1912. [CrossRef]
34. Sneddon, I.N. *The Use of Integral Transforms*; McGraw-Hill: New York, NY, USA, 1972.
35. Abramowitz, M.; Stegun, I. *Handbook of Mathematical Functions with Formulas, Graphs, and Mathematical Tables*; United States Department of Commerce, National Bureau of Standards (NBS): Washington, DC, USA, 1964.
36. Vashchenko-Zakharchenko, M.E. *Symbolic Calculus and Its Application to Integration of Linear Differential Equations*; University Press: Kiev, Ukraine, 1862.
37. Yevtushenko, A.; Topczewska, K.; Zamojski, P. The Effect of Functionally Graded Materials on Temperature during Frictional Heating: Under Uniform Sliding. *Materials* **2021**, *14*, 4285. [CrossRef]
38. Bateman, H.; Erdélyi, A. *Tables of Integral Transforms*; McGraw-Hill: New York, NY, USA, 1954; Volume 1.
39. Yevtushenko, A.; Topczewska, K.; Zamojski, P. The Effect of Functionally Graded Materials on Temperature during Frictional Heating at Single Braking. *Materials* **2021**, *14*, 6241. [CrossRef]
40. Özişik, N.M. *Heat Conduction*; John Wiley & Sons: New York, NY, USA, 1993.
41. Charron, F. *Partage de la Chaleur Entre Deux Corps Frottants*; Publications Scientifiques et Techniques; Blondel La Rougery: Paris, France, 1943; 182p.
42. Talati, F.; Jalalifar, S. Analysis of heat conduction in a disk brake system. *Heat Mass Transf.* **2009**, *45*, 1047–1059. [CrossRef]
43. Adamowicz, A.; Grześ, P. Analysis of disc brake temperature distribution during single braking under non-axisymmetric load. *Appl. Therm. Eng.* **2011**, *31*, 1003–1012. [CrossRef]
44. Loizou, A.; Qi, H.S.; Day, A.J. A fundamental study on the heat partition ratio of vehicle disk brakes. *J. Heat Transf.* **2013**, *135*, 121302. [CrossRef]
45. Yevtushenko, A.; Kuciej, M.; Topczewska, K. Analytical model for investigation of the effect of friction power on temperature in the disk brake. *Adv. Mech. Eng.* **2017**, *9*, 1–12. [CrossRef]
46. Topczewska, K. Influence of the Time of Increase in Contact Pressure in the Course of Braking on the Temperature of a Pad–Disc Tribosystem. *Mater. Sci.* **2018**, *54*, 250–259. [CrossRef]
47. Mao, J.J.; Ke, L.L.; Yang, J.; Kitipornchai, S.; Wang, Y.S. The coupled thermoelastic instability of FGM coatings with arbitrarily varying properties: In-plane sliding. *Acta Mech.* **2018**, *229*, 2979–2995. [CrossRef]

materials

MDPI

Article

Temperature during Repetitive Short-Term Operation of a Brake with Functionally Graded Friction Element

Aleksander Yevtushenko, Katarzyna Topczewska * and Przemysław Zamojski

Faculty of Mechanical Engineering, Bialystok University of Technology (BUT), 45C Wiejska Street, 15-351 Bialystok, Poland

* Correspondence: k.topczewska@pb.edu.pl

Abstract: The object of study is the temperature of a braking system, operating in repetitive short-term (RST) mode. One element of the considered friction pair is made of a functionally gradient material (FGM), and the other of a homogeneous material. To determine the temperature on the friction surfaces of both elements, the previously obtained, exact solution of the boundary value problem of heat conduction was adopted, with account of the heat generation due to the friction. A calculation scheme was proposed that takes into consideration thermal sensitivity of materials and variations of the friction coefficient under the influence of temperature. Calculations were performed for two-component FGM (ZrO_2–Ti-6Al-4V) in combination with gray cast iron (ChNMKh). It was found that for selected friction pair materials, consideration of their thermal sensitivity reduces the time of braking and the value of temperature achieved on the friction surfaces. At the same time, the whole process was characterized by a good stability of braking with a slight decrease in efficiency in each subsequent cycle.

Keywords: repetitive short-term braking operation; functionally graded materials; frictional heating; temperature

Citation: Yevtushenko, A.; Topczewska, K.; Zamojski, P. Temperature during Repetitive Short-Term Operation of a Brake with Functionally Graded Friction Element. *Materials* **2023**, *16*, 881. https://doi.org/10.3390/ma16020881

Academic Editor: Itzhak Green

Received: 22 November 2022
Revised: 30 December 2022
Accepted: 13 January 2023
Published: 16 January 2023

1. Introduction

The operating mode of vehicles may consist of successive cycles of braking and accelerating, which is typical for driving in mountainous or urban terrains. Braking systems of vehicles in such a driving mode also work periodically, so the friction pair elements are frictionally heated during braking applications and are convectively cooled during acceleration stages [1–3]. However, the unforced convective heat exchange with the environment is not enough to prevent the problem of overheated friction elements during braking actions, because it has insignificant influence on the temperature distribution [4,5]. Therefore, the characteristic feature of the repetitive short-term (RST) braking mode is that the temperature increases throughout the volumes of the friction couple elements with each subsequent braking cycle. Excessive temperature conditions occurring at repeated braking processes may lead to thermal instability of the friction couple and changes in material properties, and hence a significant reduction in braking effectiveness.

In most of the published studies concerning frictional heating processes in braking systems under the RST mode, numerical methods have been used to solve the thermal problems of friction [1–5]. Partly, this stems from the fact that they allow for direct application of the inhomogeneous temperature state of friction elements that has been found at the end of each cycle as the initial temperature distribution in the next cycle, which is impossible to perform by means of an analytical approach. The thermal behavior of the brake disc system during single as well as RST braking modes have been studied in [4,5], with consideration of the convective heat exchange on the free surfaces of a disc. The numerical solutions to the problem of heat conduction and the corresponding quasistatic thermoelasticity problem were obtained using the finite element method (FEM). The influence of the heat transfer coefficient on the temperature and thermal stress distributions in a brake disc during RST was investigated. It was concluded that free convection cooling of the disc has no pronounced effect on the temperature and thermal stresses during the single braking event, whereas by increasing the number of braking cycles, the heat exchange process has a higher effect on the thermal state of a brake system, particularly in the case of long-lasting cooling periods [5]. The temperature generated during repetitive braking slightly drops with the rise in the heat transfer coefficient, and this relationship has been found to be approximately linear in time. Another computational model for determining the transient temperature field in a brake disc during repeated braking is proposed in [1], with a special emphasis placed on the mutual dependence of velocity and maximum temperature. The calculations were carried out with consideration of the temperature-dependent coefficient of friction and thermal sensitivity of materials. The mean, flash, maximum, and volume temperature of the disc were determined based on the equations of heat dynamics of friction and wear. It was shown that the flash temperature had the highest values at the beginning of process, and gradually decreased with time in each successive braking [1]. This is consistent with the conclusion formulated in [4] that during repetitive braking, an increase in the number of brake cycles affects the local contact conditions, which leads to the growth of the real contact area between the friction elements, where the flash temperature appears. A similar coupled calculation scheme is proposed in [2], which allows one to take into consideration the interdependence of the friction coefficient and the maximum temperature achieved during each stage of the RST braking mode. Such a formulated nonlinear thermal problem of friction was solved using the finite difference method.

Besides numerical models, analytical methods are applied for simulating the frictional heating process during repetitive short-term braking [3,6,7]. Then, in order to establish the initial condition of the thermal problem of friction for the subsequent braking cycle, the volume-averaged temperature of friction components is involved. Comparative analysis of the temperature distribution in a ventilated disc brake system during repeated braking is carried out in [3] by means of the numerical (FEM), analytical, and experimental methods. The proposed numerical model simulated the mutual motion of the stationary pad and the rotating disc by applied moving heat source. The calculated temperature field was compared with the corresponding results obtained from analytical solutions to the problems of heat conduction, as well as with the experimental data achieved from the thermocouple's measurements. It was concluded that results determined from both theoretical methods are convergent with the experimental data [3]. Another analytical scheme to find the mean and volume temperature during repetitive short-term braking is proposed and successfully verified using experimental data in [6]. In [7], considering the thermal friction problem is considered during repetitive short-term braking mode by means of an analytical approach. A solution to the linear boundary value problem of heat conduction supported by empirical dependencies of material properties was used to calculate the mean temperature. Thermal sensitivity of the friction pair materials was partially taken into account by adjusting constant values of their properties and friction coefficient to the actual thermal state of elements in each braking application.

The above-mentioned studies concern the frictional heating of braking systems with friction elements made of homogeneous materials, or composites with spatially averaged properties. In this paper, an analytical calculation scheme is proposed to determine the temperature during repetitive short-term braking mode, in a tribosystem with a functionally graded friction element.

2. Scheme of Braking and Model Assumptions

Brake system operation during repetitive short-term (RST) mode is based on the successive performance of n cycles. Each of the full cycles $k = 1,2,\ldots,(n-1)$ consists of two stages—braking and accelerating—and the last, n-th interrupted cycle has only a braking period [7,8]. In the braking stages, with constant contact pressure p_0, the velocity of the system $V^{(k)}$ changes linearly from the initial value V_0 to the zero at the stop moment $t = t_s^{(k)}$ [7]:

$$V^{(k)}(t) = V_0\, V^{*(k)}(t),\ V^{*(k)}(t) = 1 - \frac{t}{t_s^{(k)}},\ 0 \le t \le t_s^{(k)}, \tag{1}$$

$$t_s^{(k)} = \frac{2W_0}{q_0^{(k)} A_a},\ k = 1, 2, \ldots, n, \tag{2}$$

where A_a—nominal area of contact between the friction elements, W_0—initial kinetic energy of the system, $q_0^{(k)}$—nominal value of the specific friction power. After a stop in each cycle, there is an acceleration stage, which consists of increasing the speed to the initial value, again V_0, in the time $t = t_c$, as follows [7]:

$$V^{(k)}(t) = V_0\, V^{*(k)}(t),\ t_s^{(k)} \le t \le t_k = t_s^{(k)} + t_c, \tag{3}$$

$$V^{*(k)}(t) = \frac{t - t_s^{(k)}}{t_c},\ k = 1, 2, \ldots, n-1. \tag{4}$$

The full duration of the RST brake mode is equal to:

$$t_b = t_s^{(1)} + t_s^{(2)} + \ldots + t_s^{(n)} + (n-1)t_c. \tag{5}$$

Braking stages are accompanied by intensive frictional heating of the friction elements. Before determining the resulting transient temperature field, the following simplifying assumptions were made:

1. Initial temperature of considered a friction pair at the beginning of the subsequent braking is equal to the volume-averaged temperature of the system;
2. As a result of the friction forces interaction, the heat is generated on the contact area of the elements and absorbed by them along the normal direction to the friction surface. The friction thermal contact of the elements during heating is perfect;
3. Unforced convection cooling of the system during braking stages is omitted.

According to the above assumptions, the friction pair can be formed by two semi-infinite bodies $z \ge 0$ and $z \le 0$, and the sought transient temperature field is one-dimensional, i.e., $T = T(z,t)$. The scheme of the considered tribosystem is illustrated below, in Figure 1.

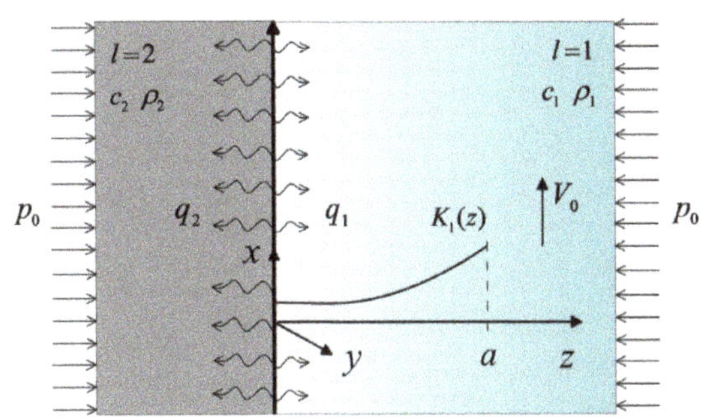

Figure 1. Scheme of the system.

Further, all variables and parameters related to the first semispace $z \ge 0$ are indicated by a subscript $l = 1$, and to the second semispace $z \le 0$ by a subscript $l = 2$. The first semispace $z \ge 0$ is made of the two-component (base and core) FGM. Suppose that $K_{1,m}$, $c_{1,m}$, $\rho_{1,m}$ are the thermal conductivity, specific heat, and density of the materials of the base ($m = 1$) and core ($m = 2$), respectively. The thermal conductivity of the FGM increases exponentially (with the gradient parameter $\gamma \ge 0$) in the direction normal to the working surface of the base. The homogeneous material of the second semispace $z \leq 0$ has the coefficient of thermal conductivity K_2, specific heat c_2, and density ρ_2.

Additionally, it was assumed that the materials of both elements and the friction coefficient f are thermally sensitive [2]:

$$K_{1,m}(T) = K_{1,m}^{(0)} K_{1,m}^{*}(T),\ c_{1,m}(T) = c_{1,m}^{(0)} c_{1,m}^{*}(T),\ \rho_{1,m}(T) = \rho_{1,m}^{(0)} \rho_{1,m}^{*}(T),\ m = 1,2, \tag{6}$$

$$K_{1,m}^{(0)} \equiv K_{1,m}(T_0),\ c_{1,m}^{(0)} \equiv c_{1,m}(T_0),\ \rho_{1,m}^{(0)} \equiv \rho_{1,m}(T_0),\ m = 1,2, \tag{7}$$

$$K_2(T) = K_2^{(0)} K_2^{*}(T),\ c_2(T) = c_2^{(0)} c_2^{*}(T),\ \rho_2(T) = \rho_2^{(0)} \rho_2^{*}(T), \tag{8}$$

$$K_2^{(0)} \equiv K_2(T_0),\ c_2^{(0)} \equiv c_2(T_0),\ \rho_2^{(0)} \equiv \rho_2(T_0), \tag{9}$$

$$f(T) = f_0 f^{*}(T),\ f_0 \equiv f(T_0), \tag{10}$$

where T_0—initial temperature of the system, and the corresponding dimensionless temperature functions are marked with the superscript '$*$'. Typically, dependencies of the type (6)–(10) are obtained as a result of processing appropriate experimental data [9].

3. Analytical Model

A key element of the proposed mathematical model is the solution to the appropriate thermal problem of friction for the considered system during execution of the individual braking processes. On the basis of the above qualitative assumptions of such a problem, a non-linear boundary value problem of heat conduction can be formulated with account of heat generation due to friction. The considered system consists of two semi-infinite bodies, one of which is made of a FGM and the other of a homogeneous material. Unfortunately, the significant nonlinearity of such a problem, caused by the thermal sensitivity of the friction coefficient as well as mechanical and thermal properties, means that the solutions can be obtained only by the numerical methods [1,2].

Another approach is also known, consisting of adapting an appropriate solution to the linear problem to determine the temperature of the thermally sensitive braking system [8]. It has been carried out for the case of a friction pair made of homogeneous materials [7]. This study presents an algorithm to establish the temperature of a thermally sensitive braking system with a friction couple consisting of a functionally graded element in combination with a homogeneous one. For such a friction pair, an exact solution of the linear, thermal problem of friction was obtained (with the coefficient of friction and thermophysical properties unchanged) during single braking with a constant deceleration [10]. Based on such a solution, the evolution of the temperature on the friction surfaces of the system operating in RST mode during the subsequent k–th braking can be calculated from the formulas [10]:

$$T^{(k)}(t) = \hat{T}^{(k)} + \Lambda^{(k)} T^{*(k)}(\tau),\ 0 \leq t \leq t_s^{(k)},\ k = 1,2,\ldots,n, \tag{11}$$

where

$$\Lambda^{(k)} = \frac{q_0^{(k)} a^{(k)}}{K_{1,1}^{(k)}}, \tag{12}$$

$$T^{*(k)}(\tau) = \frac{1}{\gamma^{*(k)}}\left[1 - \frac{4}{\pi}\int\limits_0^{\infty} G^{(k)}(x) P^{(k)}(\tau, x) dx\right], 0 \leq \tau \leq \tau_s^{(k)}, \tag{13}$$

$$G^{(k)}(x) = \frac{K_\varepsilon^{(k)} [J_1(x)]^2}{x^2 \left\{ [J_0(x)]^2 + [K_\varepsilon^{(k)} J_1(x)]^2 \right\}}, \tag{14}$$

$$P^{(k)}(\tau,x)=e^{-X_k\tau}-\frac{(1-e^{-X_k\tau})}{X_k\tau_s^{(k)}},\ X_k=\frac{(\gamma^{*(k)}x)^2}{4},\tag{15}$$

$$\tau=\frac{k_1^{(k)}t}{(a^{(k)})^2},\ \tau_s^{(k)}=\frac{k_1^{(k)}t_s^{(k)}}{(a^{(k)})^2},\ k_1^{(k)}=\frac{K_{1,1}^{(k)}}{\rho_1^{(k)}c_1^{(k)}},\ k_2^{(k)}=\frac{K_2^{(k)}}{\rho_2^{(k)}c_2^{(k)}},\tag{16}$$

$$c_1^{(k)}=\mathrm{v}c_{1,1}^{(k)}+(1-\mathrm{v})c_{1,2}^{(k)},\ \rho_1^{(k)}=\mathrm{v}\rho_{1,1}^{(k)}+(1-\mathrm{v})\rho_{1,2}^{(k)},\ 0\le\mathrm{v}\le1,\tag{17}$$

$$K_\varepsilon^{(k)}=\frac{K^{*(k)}}{\sqrt{k^{*(k)}}},\ K^{*(k)}=\frac{K_2^{(k)}}{K_{1,1}^{(k)}},\ k^{*(k)}=\frac{k_2^{(k)}}{k_{1,1}^{(k)}},\ \gamma^{*(k)}=\ln\left(\frac{K_{1,2}^{(k)}}{K_{1,1}^{(k)}}\right),\tag{18}$$

$$K_{1,m}^{(k)}\equiv K_{1,m}(\hat{T}^{(k)}),\ c_{1,m}^{(k)}\equiv c_{1,m}(\hat{T}^{(k)}),\ \rho_{1,m}^{(k)}\equiv\rho_{1,m}(\hat{T}^{(k)}),\ m=1,2,\tag{19}$$

$$K_2^{(k)}\equiv K_2(\hat{T}^{(k)}),\ c_2^{(k)}\equiv c_2(\hat{T}^{(k)}),\ \rho_2^{(k)}\equiv\rho_2(\hat{T}^{(k)}),\tag{20}$$

$$q_0^{(k)}=f^{(k)}p_0V_0,\ f^{(k)}\equiv f(\hat{T}^{(k)}),\tag{21}$$

$$a^{(k)}=\max(a_1^{(k)},a_2^{(k)}),\tag{22}$$

$$a_l^{(k)}=\begin{cases}d_l,\ a_{l,eff}^{(k)}\ge d_l\\ a_{l,eff}^{(k)},\ a_{l,eff}^{(k)}<d_l\end{cases},\ a_{l,eff}^{(k)}=\sqrt{3k_l^{(k)}t_s^{(k)}},\ l=1,2,\tag{23}$$

where $J_n(x)$, $n=0,1$—Bessel functions of the first kind; d_l, $l=0,1$—thickness of the friction elements (e.g., pad and disc).

The volume temperature $\hat{T}^{(k)}$ of the friction system before the start of the k–th braking was found as [7]:

$$\hat{T}^{(k)}=0.5(\hat{T}_0^{(k)}+\hat{T}_1^{(k)}),\ k=1,2,\ldots,n,\tag{24}$$

where

$$\hat{T}_i^{(k)}=T_0+\frac{\alpha_i^{(k)}W_0}{2G_{2,i}^{(k)}c_{2,i}^{(k)}}\left(\frac{e^{-\beta_i^{(k)}t_c}-e^{-k\beta_i^{(k)}t_c}}{1-e^{-\beta_i^{(k)}t_c}}\right),\ i=0,\ 1,\tag{25}$$

$$\beta_i^{(k)}=\frac{hA_{\text{vent}}}{2G_{2,i}^{(k)}c_{2,i}^{(k)}},\tag{26}$$

$$G_{2,i}^{(k)}=A_2d_2\rho_2^{(i,k)},\tag{27}$$

$$c_2^{(0,k)}=c_2^{(0)},\ c_2^{(1,k)}=c_2(\hat{T}_0^{(k)}),\ \rho_2^{(0,k)}=\rho_2^{(0)},\ \rho_2^{(1,k)}=\rho_2(\hat{T}_0^{(k)}),\tag{28}$$

where h—coefficient of the convective heat transfer from the surface of the disc with an area A_{vent} during the acceleration stages, $\alpha_i^{(k)}$—heat partition ratio (HPR). The methodology for determining HPR for the functionally graded friction couple is proposed in [11]. Based on this methodology, the heat transfer coefficient in formula (25) for the considered friction pair (FGM—homogeneous material) was found in the form [11]:

$$\alpha_i^{(k)}=\frac{\varepsilon_i^{(k)}}{1+\varepsilon_i^{(k)}},\ \varepsilon_i^{(k)}=0.625\frac{d^*K_i^{*(k)}}{\gamma_i^{*(k)}}\sqrt{\frac{\pi}{\tau_2^{(i,k)}}},\ k=1,2,\ldots,n,\ i=0,\ 1,\tag{29}$$

where

$$K_i^{*(k)}=\frac{K_2^{(i,k)}}{K_{1,1}^{(i,k)}},\ \gamma_i^{*(k)}=\ln\left(\frac{K_{1,2}^{(i,k)}}{K_{1,1}^{(i,k)}}\right),\ d^*=\frac{d_1}{d_2},\tag{30}$$

$$\tau_2^{(i,k)}=\frac{k_2^{(i,k)}t_s^{(k)}}{d_2^2},\ k_2^{(i,k)}=\frac{K_2^{(i,k)}}{\rho_2^{(i,k)}c_2^{(i,k)}},\tag{31}$$

$$K_{1,m}^{(0,k)} = K_{1,m}^{(0)},\ K_{1,m}^{(1,k)} = K_{1,m}(\hat{T}_0^{(k)}),\ m = 1,2,\ K_2^{(0,k)} = K_2^{(0)},\ K_2^{(1,k)} = K_2(\hat{T}_0^{(k)}). \quad (32)$$

It should be noted that from formulas (24) and (25), it follows that before the start of the first braking ($k = 1$), the volume temperature $\hat{T}^{(1)}$ is equal to the initial temperature of the system T_0. Before the start of subsequent braking, when determining the volume temperature $\hat{T}^{(k)}, k = 2, \ldots, n$ the first component $\hat{T}_0^{(k)}$ in formula (24) was also established using the properties of the materials at the initial temperature T_0, while the second component $\hat{T}_1^{(k)}$ was used to correct the result by taking into consideration the thermal sensitivity of the materials.

4. Numerical Analysis

The following scheme for determination of the temperature evolution on the working surfaces of selected friction pair elements was proposed:

1. Based on experimental data, finding the dependences of material properties and the friction coefficient on temperature in forms (6)–(10). Determining the value of material properties $K_{1,m}^{(0)}, c_{1,m}^{(0)}, \rho_{1,m}^{(0)}, m = 1,2$, (7), $K_2^{(0)}, c_2^{(0)}, \rho_2^{(0)}$ (9) and the coefficient of friction f_0 (10) at the initial temperature T_0;
2. Introduction of the input operational parameters: $p_0, V_0, T_0, W_0, n, A_a, A_{\text{vent}}, d_1, d_2, h, t_c$,v;
3. Start of the first braking: $k = 1$;
4. Determination of the volume temperature $\hat{T}^{(k)}$ (24)–(32);
5. Using the dependencies (6)–(10), establishment of the material properties values $K_{1,m}^{(k)}, c_{1,m}^{(k)}, \rho_{1,m}^{(k)}, m = 1,2$ (19), $K_2^{(k)}, c_2^{(k)}, \rho_2^{(k)}$ (20), the friction coefficient $f^{(k)}$, and specific friction power $q_0^{(k)}$ (21) at the volume temperature $\hat{T}^{(k)}$;
6. Determination of the stop time $t_s^{(k)}$ (2) and temporal profile of velocity $V^{(k)}(t)$, $0 \le t \le t_s^{(k)}$ (1);
7. Calculation of the temperature evolution $T^{(k)}(t), 0 \le t \le t_s^{(k)}$ (11)–(23);
8. Starting the next $k+1$ braking cycle and repeating starting from point (5) or ending the calculation process after reaching the equality $k = n$.

The above scheme was performed for a selected friction pair, which the first element is made of the two-component FGM: zirconium dioxide ZrO_2 (base, $m = 1$)—titanium alloy $Ti-6Al-4V$ (core, $m = 2$), and the second homogeneous element is made of the gray cast iron ChNMKh.

The properties (7) and (9) of these materials at the initial temperature $T_0 = 20°C$ are as follows [10,11]:

ZrO_2

$$K_{1,1}^{(0)} = 1.94\,\text{W}\,\text{m}^{-1}\,\text{K}^{-1},\ c_{1,1}^{(0)} = 452.83\,\text{J}\,\text{kg}^{-1}\,\text{K}^{-1},\ \rho_{1,1}^{(0)} = 6102.16\,\text{kg}\,\text{m}^{-3}, \quad (33)$$

$Ti-6Al-4V$

$$K_{1,2}^{(0)} = 6.87\,\text{W}\,\text{m}^{-1}\,\text{K}^{-1},\ c_{1,2}^{(0)} = 538.08\,\text{J}\,\text{kg}^{-1}\,\text{K}^{-1},\ \rho_{1,2}^{(0)} = 4431.79\,\text{kg}\,\text{m}^{-3}, \quad (34)$$

ChNMKh

$$K_2^{(0)} = 52.17\,\text{W}\,\text{m}^{-1}\,\text{K}^{-1},\ c_2^{(0)} = 444.6\,\text{J}\,\text{kg}^{-1}\,\text{K}^{-1},\ \rho_2^{(0)} = 7100\,\text{kg}\,\text{m}^{-3}. \quad (35)$$

Dependencies of material properties on the temperature have the forms:

ZrO_2 [12–15]

$$K_{1,1}(T) = 1.9365 + 0.7 \cdot 10^{-4} T + 0.5 \cdot 10^{-6}\, T^2 - 0.2 \cdot 10^{-9} T^3, \quad (36)$$

$$c_{1,1}(T) = 437.96 + 0.7767T - 0.17 \cdot 10^{-2} T^2, \quad (37)$$

$$\rho_{1,1}(T) = 6104.6 - 0.1212T - 0.4 \cdot 10^{-4}T^2 + 0.3 \cdot 10^{-7}T^3 - 0.1 \cdot 10^{-10}T^4, \quad (38)$$

Ti − 6Al − 4V [16,17]

$$K_{1,2}(T) = 6.6926 + 8.9177 \cdot 10^{-3}\,T + 6.8432 \cdot 10^{-6}T^2, \quad (39)$$

$$c_{1,2}(T) = 529.9316 + 0.4154T - 4.01646 \cdot 10^{-4}T^2 + 1.6364 \cdot 10^{-7}T^3, \quad (40)$$

$$\rho_{1,2}(T) = 4434 - 0.1088T - 0.8 \cdot 10^{-4}T^2 + 10^{-7}T^3 - 0.6 \cdot 10^{-10}T^4, \quad (41)$$

ChNMKh [18]

$$K_2(T) = 53.24 - 0.028T, \quad (42)$$

$$c_2(T) = 432.43 + 0.559T + 2.712 \cdot 10^{-4}T^2 + 1.657 \cdot 10^{-6}T^3 + 9.439 \cdot 10^{-10}T^4, \quad (43)$$

$$\rho_2(T) = \rho_2^{(0)}. \quad (44)$$

The dependence of the friction coefficient on temperature for the considered friction pair has the form (10), where [19]

$$f_0 = 0.27,\ f^*(T) = e^{-\kappa(T^*-1)},\ T^* = T/T_0,\ \kappa = 0.7 \cdot 10^{-4}. \quad (45)$$

Graphs of the corresponding dimensionless functions $K^*_{1,m}(T) = K_{1,m}(T)/K^{(0)}_{1,m}$, $c^*_{1,m}(T) = c_{1,m}(T)/c^{(0)}_{1,m}$, $\rho^*_{1,m}(T) = \rho_{1,m}(T)/\rho^{(0)}_{1,m}$, $m = 1,2$, $K^*_2(T) = K_2(T)/K^{(0)}_2$, $c^*_2(T) = c_2(T)/c^{(0)}_2$, $\rho^*_2(T) = \rho_2(T)/\rho^{(0)}_2$ and $f^*(T)$ are presented in Figure 2.

Calculations were made with the following operating parameters [7]: $p_0 = 1.47\,\text{MPa}$, $V_0 = 27.78\,\text{m}\,\text{s}^{-1}$, $W_0 = 392.1\,\text{kJ}$, $n = 5$, $A_a = 4.05 \cdot 10^{-2}\text{m}^2$, $A_{\text{vent}} = 4.44 \cdot 10^{-2}\text{m}^2$, $d_1 = 5.5\,\text{mm}$, $d_2 = 11\,\text{mm}$, $h = 100\,\text{W}\text{m}^{-1}\text{K}^{-1}$, $t_c = 5\,\text{s}$, $\text{v} = 0.5$.

Temporal profiles of the velocity $V^{(k)}(t)$ (1), (2) and specific friction power $q^{(k)}(t) = f^{(k)}p_0V^{(k)}(t)$, $0 \le t \le t_s^{(k)}$, $k = 1,2,\ldots,5$ are illustrated in Figure 3. A noticeable effect is the extension of the braking stage in each subsequent cycle of the process (Figure 3a). During each of the five braking applications, the intensity of the performed friction work (equal to the area under the graph) is the same (Figure 3b). This fact made it possible to compare the relevant temperature evolutions, demonstrated in Figure 4. This figure shows a comparison of friction surface temperature changes during braking $T^{(k)}(t)$ (11)–(24), found with (solid lines) and without (dotted lines) consideration of the dependencies (37)–(45) of material properties on the temperature. Results corresponding to the dotted curves were obtained for the properties of materials (34)–(36) at the initial temperature $T_0 = 20°\text{C}$. In both variants, temperature changes of the friction coefficient were taken into account in form (46).

With the exception of the first braking, consideration of the materials' thermal sensitivity resulted in a drop of the temperature on the friction surfaces. This effect is most noticeable in the last, fifth braking. Calculated values of the friction coefficient $f^{(k)}$ (21), time of braking $t_s^{(k)}$ (2), volume temperature $\hat{T}^{(k)}$ (24)–(32), and maximum temperature for each of the five braking actions, obtained with account of thermal sensitivity, are presented in Table 1. Corresponding data found for constant values of material properties are demonstrated in Table 2. Additionally, the data from Table 1 are presented in graphical form in Figure 5. With each successive braking, the coefficient of friction $f^{(k)}$ at the volume temperature $\hat{T}^{(k)}$ decreases, while the braking time $t_s^{(k)}$, volume temperature $\hat{T}^{(k)}$, and maximum temperature $T^{(k)}_{\max}$ increase. Consideration of materials' thermal sensitivity in the proposed analytical model results in greater stability of the friction coefficient value, shorter time of braking stages, lower values of volume, and maximum temperature, compared to the corresponding data found with unchanged material properties. The differences in the temperature values, obtained with and without account of the thermal sensitivity of the friction pair materials, increase with each successive braking.

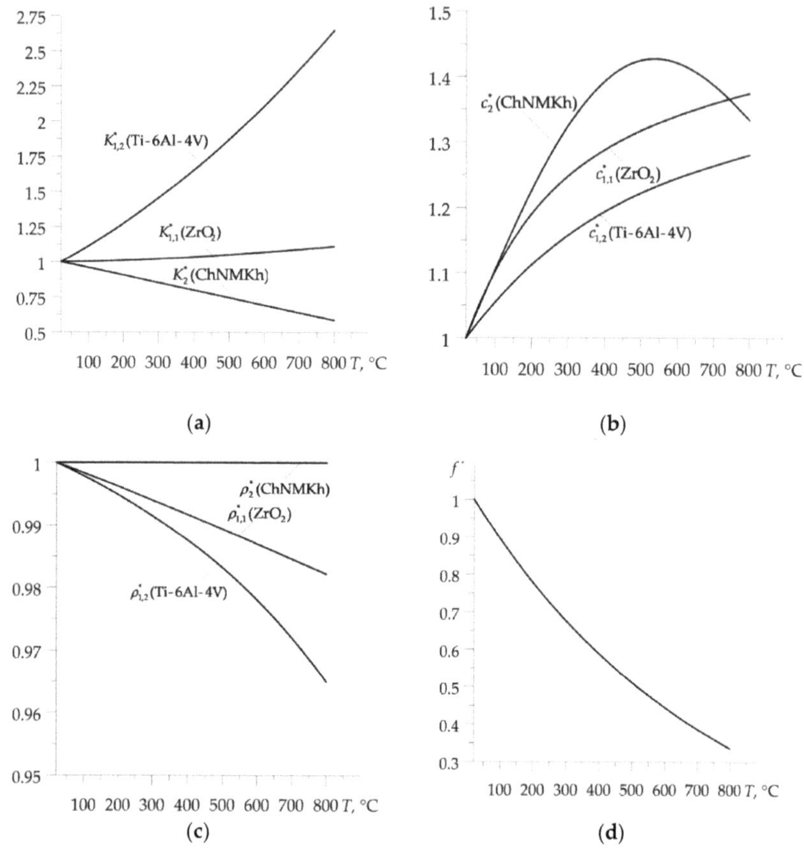

Figure 2. Temperature dependencies of the dimensionless material properties: (**a**) coefficient of thermal conductivity; (**b**) specific heat; (**c**) density; and (**d**) friction coefficient of the considered friction pair.

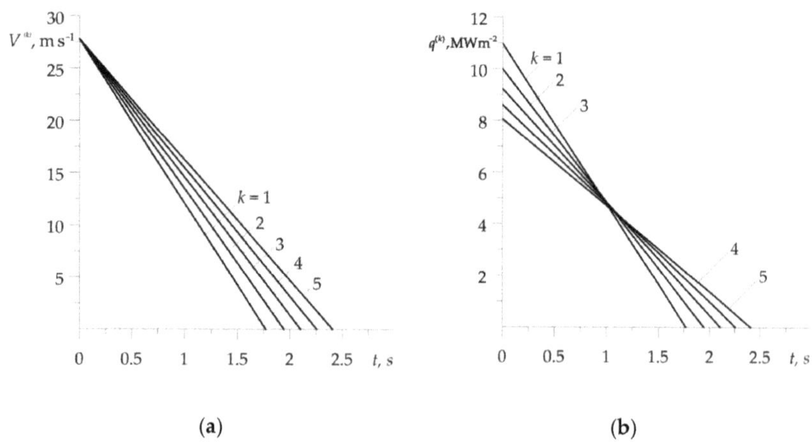

Figure 3. Evolutions of the: (**a**) velocity; (**b**) specific friction power, during each of the five braking applications.

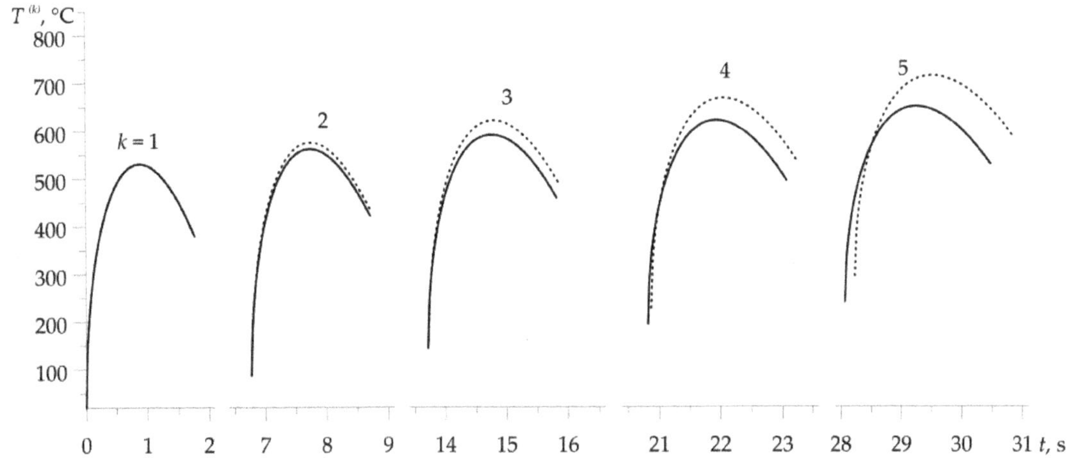

Figure 4. Evolutions of temperature on the friction surface during each of the five braking actions with (solid lines) and without (dotted lines) taking into account the thermal sensitivity of materials.

Table 1. Calculation results obtained with consideration of the materials' thermal sensitivity.

k	1	2	3	4	5
$f^{(k)}$	0.27	0.25	0.23	0.21	0.20
$t_s^{(k)}$, s	1.77	1.94	2.10	2.26	2.41
$\hat{T}^{(k)}$, °C	20	86.85	143.21	193.48	239.74
$T_{max}^{(k)}$, °C	530.27	561.29	591.46	621.11	650.49

Table 2. Calculation results obtained with constant properties of materials.

k	1	2	3	4	5
$f^{(k)}$	0.27	0.24	0.22	0.20	0.18
$t_s^{(k)}$, s	1.77	1.95	2.15	2.36	2.60
$\hat{T}^{(k)}$, °C	20	89.83	158.79	226.89	294.12
$T_{max}^{(k)}$, °C	530.27	575.76	621.75	668.19	715.04

Due to the fact that the curve of friction heat resistance (Figure 2d) is monotonically increasing the function of the temperature, the change in the friction coefficient during successive braking has the opposite form to the evolution of temperature (Figure 6). At the beginning of each braking cycle, the coefficient of friction is reduced until the maximum temperature $T_{max}^{(k)}$ is reached. In the subsequent period of the temperature drop, which lasts until the stop time $t_s^{(k)}$, the coefficient of friction slightly grows. With the increase in the number of braking applications, the minimum value of $f^{(k)}$ decreases.

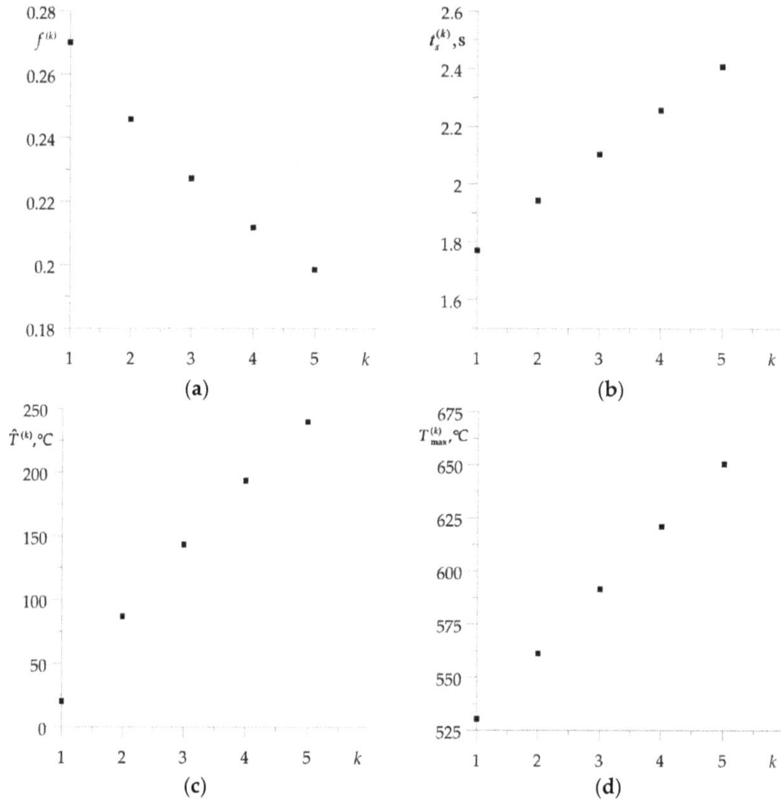

Figure 5. Dependencies on the number of braking applications k of: (**a**) friction coefficient $f^{(k)}$ (21); (**b**) braking time $t_s^{(k)}$ (2); (**c**) volume temperature $\hat{T}^{(k)}$ (24); (**d**) maximum temperature on the friction surface $T_{\max}^{(k)}$.

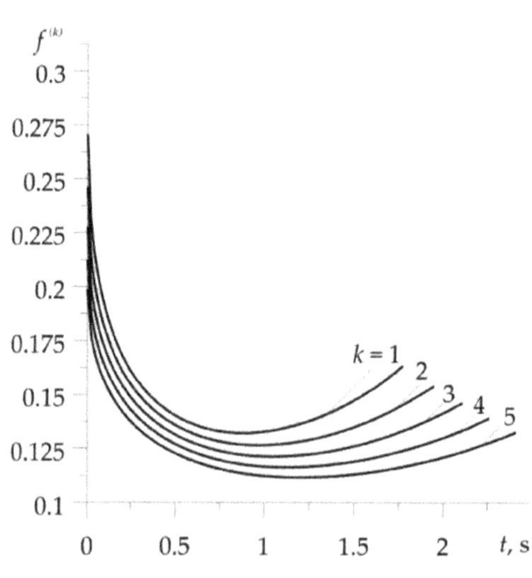

Figure 6. Evolution of friction coefficient $f^{(k)}$ (21) during five successive braking cycles.

On the basis of the results shown in Figure 6, the parameters characterizing the operation of the braking system during the subsequent cycle, such as the average value of the friction coefficient $f_m^{(k)}$, its stability $f_s^{(k)} = f_m^{(k)}/f_{\max}^{(k)}$, fluctuation $f_f^{(k)} = f_{\min}^{(k)}/f_{\max}^{(k)}$, and braking efficiency $f_{eff}^{(k)} = f_s^{(k)}/\left(t_s^{(k)}\right)^2$, were determined (Table 3). All braking cycles were characterized by good stability, and the first braking turned out to be the most effective for the selected friction pair. With each subsequent braking, the efficiency decreases.

Table 3. Parameters of the braking process evaluation.

k	1	2	3	4	5
$f_m^{(k)}$	0.149	0.142	0.135	0.129	0.123
$f_s^{(k)}$	0.553	0.576	0.593	0.608	0.620
$f_f^{(k)}$	0.489	0.515	0.534	0.550	0.563
$f_{eff}^{(k)}$, s^{-2}	0.177	0.153	0.134	0.119	0.107

5. Conclusions

An analytical scheme was proposed to determine the temperature during the repeated short-term (RST) operation mode of the braking system, in which one of the friction elements is made of a functionally gradient material (FGM). The proposed approach is a generalization of the authors' results concerning a single braking process [10–12]. Calculations were carried out for a friction pair made of a two-component FGM (base ZrO_2, core Ti–6Al–4V) and ChNMKh gray cast iron, for five braking actions. It was found that the braking time, the volume, and maximum temperature values increased almost linearly with the number of braking cycles. Involving the thermal sensitivity of materials into the calculation model causes a decrease in the maximum temperature value in relation to the results obtained for materials with invariant properties under temperature changes. This effect becomes more noticeable with each subsequent braking cycle. The coefficient of friction decreases rapidly at the beginning of each braking to a minimum value, then begins to increase slightly until standstill. The considered friction pair is characterized by good braking stability with sufficient efficiency, slightly decreasing with each successive braking.

It should be noted that the problem of determining the effect of FGM on temperature is currently intensively developed not only for bodies with unidirectional heat extension. An exact solution for transient heat conduction problem in an axisymmetric cylinder made of FGM whose thermal conductivity differs in two (radial and longitudinal) directions has been obtained [20]. Another analytical solution for steady-state heat transfer in a hollow sphere made of functionally graded material has been proposed [21].

Author Contributions: Conceptualization and methodology, A.Y.; software, P.Z.; validation, A.Y. and K.T.; formal analysis, P.Z.; investigation, A.Y., K.T., and P.Z.; writing—original draft preparation, A.Y. and K.T.; writing—review and editing, K.T.; visualization and figure preparation, P.Z.; supervision, A.Y.; project administration, K.T. All authors have read and agreed to the published version of the manuscript.

Funding: This investigation was performed within the framework of research project No. 2017/27/B/ST8/01249, funded by the National Science Centre, Poland, and project financing was through the program of the Minister of Education and Science of Poland named "Regional Initiative of Excellence" in 2019–2022, project No. 011/RID/2018/19; amount financed—PLN 12,000,000.

Institutional Review Board Statement: Not applicable.

Informed Consent Statement: Not applicable.

Data Availability Statement: No new data were created or analyzed in this study. Data sharing is not applicable to this article.

Conflicts of Interest: The authors declare no conflict of interest.

Nomenclature

a	Effective depth of heat penetration of the friction element (m)
A_a	Area of the nominal contact region (m^2)
A_{vent}	Area of the ventilated surface of the brake disc (m^2)
c	Specific heat (J kg^{-1}K^{-1})
d	Thickness of friction elements (m)
f	Coefficient of friction (dimensionless)
h	Coefficient of heat transfer (W m^{-2}K^{-1})
$J_k(\cdot)$	Bessel functions of the first kind of the kth order
k	Thermal diffusivity (m^{2}s^{-1})
K	Thermal conductivity (W m^{-1}K^{-1})
K_ε	Dimensionless coefficient of thermal activity of friction couple
n	Number of brakings in RST brake mode
p	Pressure on the contact surface (Pa)
p_0	Nominal value of the contact pressure (Pa)
q	Specific friction power (W m^{-2})
q_0	Nominal value of specific friction power (W m^{-2})
t	Time (s)
t_b	Time of performance of all RST mode of braking (s)
t_c	Cooling time during acceleration stage (s)
t_s	Stop time (s)
T	Temperature (°C)
T^*	Dimensionless temperature
$\hat{T}$	Volume temperature (°C)
T_0	Initial temperature (°C)
v	Volume fraction of the FGM components (dimensionless)
V	Velocity (m s^{-1})
V_0	Initial velocity (m s^{-1})
W_0	Initial kinetic energy (J)
z	Spatial coordinate in axial direction (m)
α	Heat partition ratio (dimensionless)
γ	Parameter of material gradient (m $^{-1}$)
Λ	Scaling factor of temperature (°C)
ρ	Density (kg m^{-3})
τ	Dimensionless time
τ_s	Dimensionless time of braking
superscript k	Number of a braking cycle
subscript l	Number of the friction pair element
subscript m	Number of the component material of functionally graded element

References

1. Grześ, P. Maximum temperature of the disc during repeated braking applications. *Adv. Mech. Eng.* **2019**, *11*, 1–13. [CrossRef]
2. Yevtushenko, A.; Kuciej, M. Calculation of friction characteristics of disc brakes used in repetitive short-term braking mode. *J. Frict. Wear* **2020**, *41*, 509–516. [CrossRef]
3. Yevtushenko, A.A.; Kuciej, M.; Grześ, P.; Wasilewski, P. Temperature in the railway disc brake at a repetitive short-term mode of braking. *Int. Commun. Heat Mass Transf.* **2017**, *84*, 102–109. [CrossRef]
4. Adamowicz, A.; Grześ, P. Analysis of disc brake temperature distribution during single braking under non-axisymmetric load. *Appl. Therm. Eng.* **2011**, *31*, 1003–1012. [CrossRef]
5. Adamowicz, A. Effect of convective cooling on temperature and thermal stresses in disk during repeated intermittent braking. *J. Frict. Wear* **2016**, *37*, 107–112. [CrossRef]
6. Dunaevsky, V.V. Prediction of railroad friction braking temperatures: Prediction of average bulk and average surface temperatures of railroad wheels and brake discs. *Tribol. Trans.* **1991**, *34*, 343–352. [CrossRef]
7. Yevtushenko, A.; Topczewska, K.; Kuciej, M. Analytical Determination of the Brake Temperature Mode during Repetitive Short-Term Braking. *Materials* **2021**, *14*, 1912. [CrossRef] [PubMed]
8. Chichinadze, A.V.; Braun, E.D.; Ginzburg, A.G.; Ignat'eva, Z.V. *Calculation, Testing and Selection of Friction Couples*; Nauka: Moscow, Russian, 1979. (In Russian)
9. Chichinadze, A.V. *Polymers in Friction Assembles of Machines and Devices: A Handbook*; Allerton Press Inc.: New York, NY, USA, 1984.

10. Yevtushenko, A.; Kuciej, M.; Topczewska, K.; Zamojski, P. Temperature in the Friction Couple Consisting of Functionally Graded and Homogeneous Materials. *Materials* **2022**, *15*, 3600. [CrossRef] [PubMed]
11. Yevtushenko, A.; Topczewska, K.; Zamojski, P. The Heat Partition Ratio during Braking in a Functionally Graded Friction Couple. *Materials* **2022**, *15*, 4623. [CrossRef] [PubMed]
12. Yevtushenko, A.; Topczewska, K.; Zamojski, P. Influence of Thermal Sensitivity of Functionally Graded Materials on Temperature during Braking. *Materials* **2022**, *15*, 963. [CrossRef] [PubMed]
13. Kingery, W.D.; Francl, J.; Coble, R.L.; Vasilos, T. Thermal Conductivity: X, Data for Several Pure Oxide Materials Corrected to Zero Porosity. *J. American Ceramic Soc.* **1954**, *37*, 107–110. [CrossRef]
14. Pankratz, L.B. *Thermodynamic Properties of Elements and Oxides-Bulletin 672*; U.S. Bureau of Mines: Washington, DC, USA, 1982.
15. Taylor, D. Thermal expansion data. II: Binary oxides with the fluorite and rutile structures, MO_2, and the antifluorite structure, M_2O. *Trans. J. British Ceramic Soc.* **1984**, *83*, 32–37.
16. Deem, H.W.; Wood, W.D.; Lucks, C.F. The relationship between Electrical and Thermal Conductivities of Titanium Trans. *Metall. Soc. AIME* **1958**, *212*, 520–523.
17. Cezairrliyan, A.; McClure, J.L.; Taylor, R.J. Thermophysical Measurements on 90Ti-6Al-4V Alloy Above 1450 K Using a Transient (Subsecond) Technique. *J. Res. Nat. Bur. Stand.–A Phys. Chem.* **1977**, *81*, 251–256. [CrossRef]
18. Chichinadze, A.V.; Matveevskii, R.M.; Braun, E.P. *Materials in Triboengineering of Unsteady Processes*; Nauka: Moscow, Russia, 1986.
19. Fu, P.; Zhao, J.; Zhang, X.; Kang, G.; Wang, P.; Kan, Q. Thermo-mechanically coupled sliding contact shakedown analysis of functionally graded coating-substrate structures. *Int. J. Mech. Sci.* **2022**, *222*, 107241. [CrossRef]
20. Delouei, A.A.; Emamian, A.; Karimnejad, S.; Sajjadi, H. A closed-form solution for axisymmetric conduction in a finite functionally graded cylinder. *Int. Commun. Heat Mass Transf.* **2019**, *108*, 104280. [CrossRef]
21. Delouei, A.A.; Emamian, A.; Karimnejad, S.; Sajjadi, H.; Jing, D. Two-dimensional analytical solution for temperature distribution in FG hollow spheres: General thermal boundary conditions. *Int. Commun. Heat Mass Transf.* **2020**, *113*, 104531. [CrossRef]

materials

MDPI

Article

Analytical Determination of the Brake Temperature Mode during Repetitive Short-Term Braking

Aleksander Yevtushenko, Katarzyna Topczewska and Michal Kuciej *

Faculty of Mechanical Engineering, Bialystok University of Technology (BUT), 45C Wiejska Street, 15-351 Białystok, Poland; a.yevtushenko@pb.edu.pl (A.Y.); k.topczewska@pb.edu.pl (K.T.)
* Correspondence: m.kuciej@pb.edu.pl

Abstract: An algorithm to determine the maximum temperature of brake systems during repetitive short-term (RST) braking mode has been proposed. For this purpose, the intermittent mode of braking was given in the form of a few cyclic stages consisting of subsequent braking and acceleration processes. Based on the Chichinadze's hypothesis of temperature summation, the evolutions of the maximum temperature during each cycle were calculated as the sum of the mean temperature on the nominal contact surface of the friction pair elements and temperature attained on the real contact areas (flash temperature). In order to find the first component, the analytical solution to the one-dimensional thermal problem of friction for two semi-spaces taking into account frictional heat generation was adapted. To find the flash temperature, the solution to the problem for the semi-infinite rod sliding with variable velocity against a smooth surface was used. In both solutions, the temperature-dependent coefficient of friction and thermal sensitivity of materials were taken into account. Numerical calculations were carried out for disc and drum brake systems. The obtained temporal variations of sliding velocity, friction power and temperature were investigated on each stage of braking. It was found that the obtained results agree well with the corresponding data established by finite element and finite-difference methods.

Keywords: repetitive short-term braking; frictional heating; temperature; thermal sensitivity of materials; friction coefficient

Citation: Yevtushenko, A.; Topczewska, K.; Kuciej, M. Analytical Determination of the Brake Temperature Mode during Repetitive Short-Term Braking. *Materials* **2021**, *14*, 1912. https://doi.org/10.3390/ma14081912

Academic Editor: Chuang Dong

Received: 15 February 2021
Accepted: 7 April 2021
Published: 11 April 2021

1. Introduction

Repetitive short-term mode of braking (RST) is a sequential performance of a certain number of cycles consisting of two stages: braking (heating) and acceleration (cooling) [1]. In analytical and numerical models, the main characteristics to describe the braking process (temporal profiles of velocity and friction power, time to standstill, maximum temperature, etc.) are found from solutions to the following problems [2,3]:

1. initial value problem for vehicle motion;
2. boundary-value problem of heat conduction, taking into account frictional heat generation (the so-called thermal problem of friction).

The schemes of the characteristics' determination, consisting of subsequent solutions to the above-mentioned problems, are the so-called uncoupled models. In this kind of scheme, to determine the sliding velocity profile, time to stop and resulting evolution of specific friction power, first, the initial value problem for the equation of motion of braking system has to be considered. After finding the specific friction power in this manner, its temporal profile is adapted to one of the boundary conditions during formulation of the boundary-value problem of heat conduction. The equality of the specific friction power and the sum of heat flux intensities, directed inside each element of the friction pair perpendicular to the contact surface, is required in this condition. The disadvantage of such an approach is that, when obtaining solutions, a constant, usually averaged with the braking time, the value of the friction coefficient is used. In consequence, the calculations do not take into account the mutual influence of sliding velocity and temperature.

Numerical uncoupled models of braking in the RST mode, based on the finite element method (FEM), were developed for automotive [4–7] and railway [8] brakes. The influence of convective cooling of free surfaces on temperature [4] and thermal stresses were studied in [5–7]. A comparative analysis of volumetric and mean temperature on the friction surface of the disc brake of the diesel multiple unit, obtained by means of: (a) numerical solution to the spatial thermal problem of friction for a disc-pad system, (b) analytical solution to the corresponding one-dimensional problem [9], and (c) experimental data using thermocouples [10] was carried out in [8]. To predict temperature, a computer model of an automotive disc brake operation during short-term multiple brakings was developed [11]. A theoretical and experimental methodology for determining the heat transfer coefficient during the RST braking mode has been proposed [12].

However, if for a selected friction pair the experimental data of frictional thermal stability are known, then the temperature calculations can be made according to the coupled models [13,14]. In this kind of model, the above-mentioned problems are related to the temperature-dependent coefficient of friction, and their solutions at each time step are obtained simultaneously. The implementation of the coupled calculation scheme allows to take into account the interdependence of sliding velocity and temperature during each stage of the RST braking mode. Since both of the above-mentioned problems in the coupled models are non-linear, their solutions were obtained numerically, using FEM [15] or the finite differences method (FDM) [16].

The use of numerical methods allows to take into account the actual dimensions of the braking system, including the contact area of the pad and disc, as well as the cooling of free surfaces, thermal sensitivity of materials, etc. On the other hand, the use of numerical methods requires researcher to have advanced skills when selecting the space–time grid, controlling the stability of calculations, performing verification of the obtained results with appropriate experimental data, etc. Therefore, analytical methods of solving thermal problems of friction during braking are still being investigated, simultaneously with numerical and experimental methods. These are mainly linear boundary value problems of heat conduction [17]. The exact obtained solutions to such problems during single braking cases are in the form of closed engineering formulas and allow for instant assessment of the temperature during braking under light conditions with sufficient accuracy; when the volumetric and the average temperatures of the friction surface do not exceed 100 °C and 200 °C, respectively [18–20]. Under heavy and medium braking conditions, when the volume temperature reaches values above 250 °C, the friction coefficient and properties of the friction pair materials change significantly due to the influence of temperature; therefore, numerical methods are used [21,22]. An attempt was made to use the exact solutions of linear thermal problems of friction to determine temperature during single braking of a system made of thermally sensitive friction materials [23–25].

It should be noted that, in the sense of methodology, the results contained in the monograph by A.V. Chichinadze et al. [9] and in [16] are closest to our study. The algorithm for determining the maximum temperature of a braking system in [9] is based on an analytical solution to the thermal problem of friction during single braking for two layers. This solution was obtained with two very important simplifying assumptions. The first one is the presumption that the temperature of any point on the axis, perpendicular to the friction surface, is directly proportional to the braking time. The second assumption is that the temperature increase, in this case, is equal to the increase in the mean volumetric temperature of the system. With such simplifications, this solution can be defined as approximate in an analytical form. A detailed comparative analysis of the temperature fields during single braking, obtained with the use of an approximate solution [9] and the exact solution, was carried out in [18]. The numerical FDM solution to the thermal problem of friction for the RST operating mode of a disc brake was obtained in [16]. This solution also belongs to the class of approximate solutions; whereas the main purpose of this study was to obtain exact engineering formulae to this problem and the development, on the basis of them, of a corresponding calculation scheme to find the maximum temperature in a braking system operating in RST mode.

2. Statement of the Problem

A braking system operating in a repetitive short-term (RST) mode is considered. This intermittent braking mode comprised the sequential performance of $n-1$ full cycles and the final (nth) discontinuous cycle (Figure 1). Each of the full cycles consisted of two stages. The first one was braking, where the linear velocity of sliding, $V^{(k)}$, decreased from the initial value, V_0, to zero at time moments $t = t_s^{(k)}$, $k = 1,2,\ldots, n-1$. Immediately after the standstill, the second phase of the cycle began, which was acceleration with the released brake. During this stage the velocity increased to initial value V_0 at time $t = t_c$.

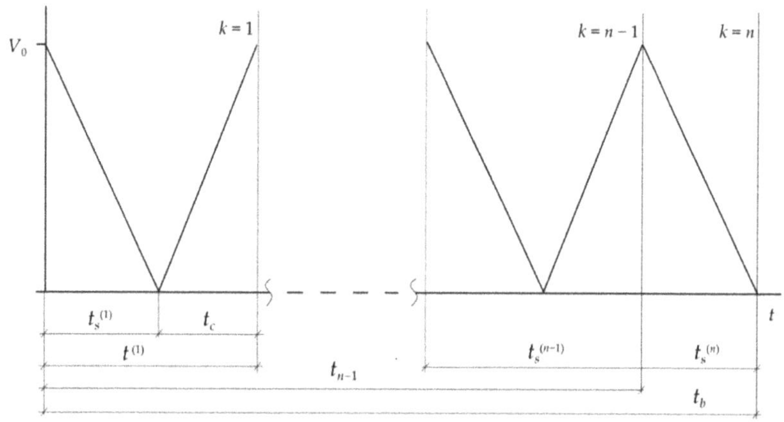

Figure 1. Scheme of sliding velocity variations during repetitive short-term (RST) braking mode.

Thus, the duration of one full cycle was $t_k = t_s^{(k)} + t_c$, $k = 1,2,\ldots, n-1$, and the overall time of all full braking cycles was $t_{n-1} = \sum\limits_{k=1}^{n-1} t_k$. The last interrupted cycle was performed with the applied brake, so speed $V^{(n)}$ dropped from initial value V_0 to zero at stop moment $t = t_s^{(n)}$. Thus, the total time of the RST operation mode of braking was $t_b = t_{n-1} + t_s^{(n)}$.

Braking phases are accompanied by intense frictional heating of the friction pair elements (disc-pad, drum-shoe, etc.). Because of high temperature gradients, the friction coefficient and thermo-mechanical properties of friction pair elements can differ significantly at the initial phase and after braking. Hence, the assumption is made that, during the kth braking, friction coefficient f, thermal conductivities K_l, specific heat capacities c_l, densities ρ_l and Brinell hardness HB_l of friction pair materials $l = 1, 2$ depend on temperature T in the form of, Equations (1) and (2):

$$f(T) = f_0 f^*(T), \tag{1}$$

$$K_l(T) = K_{l,0}K_l^*(T), c_l(T) = c_{l,0}c_l^*(T),\ \rho_l(T) = \rho_{l,0}\rho_l^*(T),\ HB_l(T) = HB_{l,0}B_l^*(T), \tag{2}$$

Where Equations (3)–(8),

$$f_0 = f(T_0),\ K_{l,0} = K_l(T_0),\ c_{l,0} = c_l(T_0),\ \rho_{l,0} = \rho_l(T_0),\ HB_{l,0} = HB_l(T_0), \tag{3}$$

$$f^*(T) = f_1 + \frac{f_2}{[f_3(T-f_4)]^2+1} + \frac{f_5}{[f_6(T-f_7)]^2+1}, \tag{4}$$

$$K_l^*(T) = K_{l,1} + \frac{K_{l,2}}{[K_{l,3}(T - K_{l,4})]^2 + 1} + \frac{K_{l,5}}{[K_{l,6}(T - K_{l,7})]^2 + 1}, \tag{5}$$

$$c_l^*(T) = c_{l,1} + \frac{c_{l,2}}{[c_{l,3}(T - c_{l,4})]^2 + 1} + \frac{c_{l,5}}{[c_{l,6}(T - c_{l,7})]^2 + 1}, \tag{6}$$

$$\rho_l^*(T) = \rho_{l,1} + \frac{\rho_{l,2}}{[\rho_{l,3}(T - \rho_{l,4})]^2 + 1} + \frac{\rho_{l,5}}{[\rho_{l,6}(T - \rho_{l,7})]^2 + 1}, \tag{7}$$

$$HB^*(T) = HB_1 + \frac{HB_2}{[HB_3(T - HB_4)]^2 + 1} + \frac{HB_5}{[HB_6(T - HB_7)]^2 + 1}, \tag{8}$$

where T_0—initial temperature, f_j, $K_{l,j}$, $c_{l,j}$, $HB_{l,j}$, $\rho_{l,j}$, $l = 1,2; j = 1,2,\ldots,7$ are parameters of the experimental data approximations [26]. Here and below, subscripts $l = 1$ and $l = 2$ refer to quantities associated with the primary element (disc, drum, etc.) and friction lining (pad, shoe, etc.), respectively.

The remaining assumptions of the calculation model are as follows:

1. At the initial moment of each braking phase, the friction element is pressed against the primary element contact surface with uniform pressure p, which exponentially increases with time t, from zero to nominal value p_0, Equation (9):

$$p(t) = p_0 p^*(t),\ p^*(t) = 1 - e^{-\frac{t}{t_i}},\ 0 \le t \le t_s^{(k)}, k = 1,2,\ldots,n, \tag{9}$$

   where t_i—time of pressure increase.
2. At the initial moment of the kth cycle of braking, the distribution of temperature in the tribosystem is homogeneous and equal to the averaged volumetric temperature of friction pair $T_0^{(k)}$;
3. As a result of the friction forces acting on the contact area of friction pair elements, heat is generated and absorbed by these elements in the normal directions of their friction surfaces;
4. The thermal contact of friction pair elements is perfect. In other words, the sum of heat flux intensities directed into friction elements, is equal to the specific friction power, and the temperatures of its contact areas are equal.
5. During the subsequent braking phases, the free surfaces of the brake system are adiabatic and during the acceleration stages, unforced convection cooling takes place.

The applied contact pressure (10) causes a reduction of sliding velocity $V^{(k)}$ during the kth braking, according to the following relation [25], Equation (10):

$$V^{(k)}(t) = V_0\, V^{*(k)}(t),\ V^{*(k)}(t) = 1 - \frac{t}{t_{s,0}^{(k)}} + p^*(t)\frac{t_i}{t_{s,0}^{(k)}},\ 0 \le t \le t_s^{(k)},\ k = 1,2,\ldots,n, \tag{10}$$

Where, Equation (11),

$$t_{s,0}^{(k)} = \frac{2W_0}{q_0^{(k)} A_a}, \tag{11}$$

W_0—initial kinetic energy of the system, which depends on the mass of the vehicle and its velocity at the moment just before the brake applications, $q_0^{(k)}$—nominal specific friction power, A_a—nominal contact surface area, usually determined by the dimensions of the contact surface of friction element. Braking time $t_s^{(k)}$ is obtained from the stop condition, Equation (12):

$$V^{*(k)}(t_s^{(k)}) = 0. \tag{12}$$

In case of the immediate ($t_i \to 0$) achievement of nominal pressure value p_0 in the Equation (9) from Equation (10) we obtain the linear drop in speed, Equation (13):

$$V^{*(k)}(t) = 1 - \frac{t}{t_{s,0}^{(k)}},\ 0 \le t \le t_{s,0}^{(k)}. \tag{13}$$

Thus, parameter $t_{s,0}^{(k)}$ (11) is the time of braking with constant deceleration.

In the phases of acceleration, the velocity increases linearly with time, Equation (14):

$$V^{*(k)}(t) = \frac{t - t_s^{(k)}}{t_c},\ t_s^{(k)} \le t \le t_k,\ k = 1,2,\ldots,n. \tag{14}$$

The evolution of maximum temperature $T_{\max}^{(k)}$ on the friction surface during kth braking was sought in the form of [9,15]:

$$T_{\max}^{(k)}(t) = T_m^{(k)}(t) + T_f^{(k)}(t),\ 0 \le t \le t_s^{(k)},\ k = 1,2,\ldots,n,. \tag{15}$$

where $T_m^{(k)}$—mean temperature of the nominal contact surface and $T_f^{(k)}$—average temperature of the real contact area (flash temperature).

3. Solution to the Problem

3.1. Heat Generation on the Nominal Contact Surface

Taking into account the above assumptions, to find component $T_m^{(k)}$ in the relation (16), we used the a calculation scheme of frictional contact of two different semi-infinite bodies, $0 \le z < \infty$ ($l = 1$) and $-\infty < z \le 0$ ($l = 2$), sliding against each other with velocity $V^{(k)}(t)$ (10)–(13). Initiated by frictional heating during kth braking, we found transient temperature field $T^{(k)}(z,t)$ in this system from the solution to the following boundary-value problem of heat conduction, Equations (16)–(20):

$$\frac{\partial^2 T^{(k)}(z,t)}{\partial z^2} = \frac{1}{k_{1,0}^{(k)}} \frac{\partial T^{(k)}(z,t)}{\partial t}, 0 < z < \infty, 0 < t \le t_s^{(k)}, \tag{16}$$

$$K_{2,0}^{(k)} \frac{\partial T^{(k)}(z,t)}{\partial z}\bigg|_{z=0^-} - K_{1,0}^{(k)} \frac{\partial T^{(k)}(z,t)}{\partial z}\bigg|_{z=0^+} = q^{(k)}(t), 0 < t \le t_s^{(k)}, \tag{17}$$

$$T^{(k)}(0^+,t) = T^{(k)}(0^-,t) \equiv T_m^{(k)}(t),\ 0 < t \le t_s^{(k)}, \tag{18}$$

$$T^{(k)}(z,t) \to T_0^{(k)},\ |z| \to \infty,\ 0 < t \le t_s^{(k)} \tag{19}$$

$$T^{(k)}(z,0) = T_0^{(k)},\quad |z| < \infty,\ k = 1,2,\ldots,n \tag{20}$$

where, Equations (21)–(23),

$$q^{(k)}(t) = q_0^{(k)} q^{*(k)}(t),\ q_0^{(k)} = f_0^{(k)} p_0 V_0,\ q^{*(k)}(t) = p^*(t) V^{*(k)}(t) \tag{21}$$

$$k_{l,0}^{(k)} = \frac{K_{l,0}^{(k)}}{\rho_{l,0}^{(k)} c_{l,0}^{(k)}},\ l = 1,2, \tag{22}$$

$$f_0^{(k)} = f(T_0^{(k)}),\ K_{l,0}^{(k)} = K_l(T_0^{(k)}),\ c_{l,0}^{(k)} = c_l(T_0^{(k)}),\ \rho_{l,0}^{(k)} = \rho_l(T_0^{(k)}). \tag{23}$$

Averaged volumetric temperature $T_0^{(k)}$ of a braking system before kth braking was calculated from the formulas, Equations (24)–(26) [9,16]:

$$T_0^{(k)} = \frac{T_{0,0}^{(k)} + T_{0,1}^{(k)}}{2},\ k = 1, 2, \ldots, n,. \tag{24}$$

$$T_{0,i}^{(k)} = T_0 + \frac{\gamma_i W_0}{2G\, c_{1,i}} \left(\frac{e^{-\alpha_i t_c} - e^{-k\alpha_i t_c}}{1 - e^{-\alpha_i t_c}} \right),\ i = 0,\ 1, \tag{25}$$

$$\gamma_i = \frac{\sqrt{K_{1,i}\rho_{1,i}c_{1,i}}}{\sqrt{K_{1,i}\rho_{1,i}c_{1,i}} + \sqrt{K_{2,i}\rho_{2,i}c_{2,i}}} \tag{26}$$

where, Equations (27)–(29),

$$\gamma_i = \frac{\sqrt{K_{1,i}\rho_{1,i}c_{1,i}}}{\sqrt{K_{1,i}\rho_{1,i}c_{1,i}} + \sqrt{K_{2,i}\rho_{2,i}c_{2,i}}} \tag{27}$$

$$\alpha_i = \frac{hA_{vent}}{G\, c_{1,i}} \tag{28}$$

$$K_{l,1} = K_l(T_{0,0}^{(k)}),\ c_{l,1} = c_l(T_{0,0}^{(k)}),\ \rho_{l,1} = \rho_l(T_{0,0}^{(k)}), \tag{29}$$

h—heat transfer coefficient, G, A_{vent}—weight and total surface area of the primary element of friction pair, respectively; values $K_{l,0}$, $c_{l,0}$, $\rho_{l,0}$, $l = 1,\ 2$ are defined by Equation (3). For the first braking cycle ($k = 1$)), from relations (25) and (26), it follows that $T_0^{(1)} = T_0$.

Exact solution to the boundary-value problem of heat conduction, (17)–(22), during single braking was obtained in [27]. Generalizing this solution to the considered case of repetitive braking, the sought temperature on the friction surface was written in the form, Equation (30):

$$T_m^{(k)}(t) = T_0^{(k)} + T_{m,0}^{(k)} T_m^{*(k)}(t)\ ,0 \le t \le t_s^{(k)}\ ,k = 1, 2, \ldots, n \tag{30}$$

where, Equations (31)–(37),

$$T_m^{*(k)}(t) = \gamma_0^{(k)} \sqrt{\frac{t}{t_{s,0}^{(k)}}} \left[\left(1 + \frac{t_i}{t_{s,0}^{(k)}} - \frac{2t}{3t_{s,0}^{(k)}} \right) \frac{2}{\sqrt{\pi}} - \left(1 + \frac{3t_i}{2t_{s,0}^{(k)}} - \frac{t}{t_{s,0}^{(k)}} \right) \mathrm{F}\left(\sqrt{\frac{t}{t_{s,0}^{(k)}}} \right) + \frac{t_i}{t_{s,0}^{(k)}} \mathrm{F}\left(\sqrt{\frac{2t}{t_{s,0}^{(k)}}} \right) \right] \tag{31}$$

$$\mathrm{F}(x) = \frac{2}{\sqrt{\pi}} \sum_{n=0}^{\infty} (-1)^n \frac{(2x^2)^n}{(2n+1)!!},\ 0 \le x \le 3,\ \mathrm{F}(x) = \frac{2}{\sqrt{\pi}} \sum_{n=0}^{\infty} \frac{(2n-1)!!}{(2x^2)^{n+1}},\ x > 3, \tag{32}$$

$$\gamma_0^{(k)} = \frac{\sqrt{K_{1,0}^{(k)}\rho_{1,0}^{(k)}c_{1,0}^{(k)}}}{\sqrt{K_{1,0}^{(k)}\rho_{1,0}^{(k)}c_{1,0}^{(k)}} + \sqrt{K_{2,0}^{(k)}\rho_{2,0}^{(k)}c_{2,0}^{(k)}}}, \tag{33}$$

$$T_{m,0}^{(k)} = \frac{q_0^{(k)} a^{(k)}}{K_{1,0}^{(k)}} \tag{34}$$

$$a^{(k)} = \max\left\{ a_l^{(k)} \right\} \tag{35}$$

$$a_l^{(k)} = \begin{cases} d_l,\ a_{l,eff}^{(k)} \ge d_l\ , \\ a_{l,eff}^{(k)},\ a_{l,eff}^{(k)} < d_l\ , \end{cases} \tag{36}$$

$$a_{l,eff}^{(k)} = \sqrt{3k_{l,0}^{(k)} t_{s,0}^{(k)}}, \quad (37)$$

$d_l, l = 1,\ 2$ thicknesses of the friction pair components.

The algorithm to find the mean temperature for the selected friction pair consisted in the subsequent performance of the following steps:

1. Based on the experimental data, by means of the approximation formulas (1)–(8), describe the thermal stability of friction $f(T)$ and temperature dependencies of thermal $K_l(T)$, $c_l(T)$ and mechanical $\rho_l(T)$, $HB_l(T)$ properties of friction materials $l = 1,2$;
2. Set the operation input parameters: p_0, V_0, T_0, W_0, n, A_a, h, G, t_i, t_c, d_l, $K_{l,0}$, $c_{l,0}$, $HB_{l,0}$, $\rho_{l,0}$, $l = 1,2$;
3. Begin the first ($k = 1$) braking cycle;
4. Establish the averaged volumetric temperature $T_0^{(k)}$ of the friction pair from Formulas (25)–(29);
5. Taking into account dependencies (1)–(8) calculate the values of friction coefficient $f_0^{(k)}$ and materials properties $K_{l,0}^{(k)}, \rho_{l,0}^{(k)}, l = 1,2$ (25) in temperature $T_0^{(k)}$;
6. Determine braking time $t_s^{(k)}$ and temporal profile of velocity $V^{(k)}(t), 0 \le t \le t_s^{(k)}$ from Equations (10)–(12) and (22);
7. Calculate the evolution of mean temperature on nominal contact surface $T_m^{*(k)}(t)$, $0 \le t \le t_s^{(k)}$ (33)–(38);
8. Start the subsequent $(k+1)$ cycle of braking and repeat the calculations, beginning from point 4). The calculation process ends when condition $k = n$ is met.

3.2. Temperature of the Real Contact Region

Real friction surfaces of braking system elements are not perfectly smooth, they are characterized by significant roughness and waviness. Therefore, these elements are not in contact on an entire nominal friction area, but only in some parts of one, and consists of roughness waves. These waves form the so-called contour surface of contact, the area of which ($A_c^{(k)}$) changes during kth braking, according to the law, Equations (38) and (39) [28]:

$$A_c^{(k)}(t) = A_a \left[\frac{p(t) b_0^{\nu-1}}{HB_{\min}^{(k)}(t)} \right]^{\frac{1}{\nu+1}}, \ 0 \le t \le t_s^{(k)}, \quad (38)$$

$$HB_{\min}^{(k)}(t) = \min\left\{ HB_1[T_m^{(k)}(t)], HB_2[T_m^{(k)}(t)] \right\} \quad (39)$$

where p, $T_m^{(k)}$—pressure (9) and mean temperature (30)–(37) on the nominal contact surface with area A_a, b_0, ν—parameters of the reference curve for the harder element of the friction pair materials. Usually this is the material of a primary element (disc, drum etc.). Pressure $p_c^{(k)}$, on contour area $A_c^{(k)}$ (38), (39) was found from the relation, Equation (40):

$$p_c^{(k)}(t) = p(t) \frac{A_a}{A_c^{(k)}(t)}, \ 0 \le t \le t_s^{(k)} \quad (40)$$

In a typical braking system, the friction linings are made of more deformable and less durable material than the primary element material. Thus, we assume that:

1. Asperities have the spherical shape and are located on the surface of the harder and stiffer primary element, while the friction lining surface is smooth.
2. Plastic roughness deformation mechanism takes place. This means that the contact of a single asperity with the friction lining surface lasts until its material becomes plastic due to a rapid increase in temperature and the appearance of significant thermal stresses.

3. Before coming into contact with the friction lining, the temperature of asperity does not change along its height and is equal to the mean temperature of nominal contact area $T_m^{(k)}$.

Contact of a single asperity with a smooth friction lining surface creates the real region of contact with a diameter as follows Equation (41) [29]:

$$d_r^{(k)}(t) = \left(\frac{8r_{av}h_{\max}}{\nu}\right)^{\frac{1}{2}}\left[\frac{p_c^{(k)}(t)}{HB_{\min}^{(k)}(t)b_0}\right]^{\frac{1}{2\nu}},\ 0 \le t \le t_s^{(k)}, \tag{41}$$

where r_{av}—averaged radius of the asperities rounding and $h_{\max}$—maximum height of the roughness. Total area $A_r^{(k)}$ of the real contact region is related to area A_a of the nominal contact surface according to the formula, Equation (42):

$$A_r^{(k)}(t) = A_a\frac{p(t)}{HB_{\min}^{(k)}(t)},\ 0 \le t \le t_s^{(k)}. \tag{42}$$

Evolution of temperature $T_f^{(k)}$ on the real contact region with area $A_r^{(k)}$ (42) (flash temperature) was determined from the following Equation (43) [26,30]:

$$T_f^{(k)}(t) = \left(1+\frac{1}{\sqrt{2}}\right)\frac{f_m^{(k)}(t)p(t)V^{(k)}(t)A_a d_r^{(k)}(t)}{A_r^{(k)}(t)[4K_{1,m}^{(k)}(t)+\sqrt{\pi V^{(k)}(t)\,d_r^{(k)}(t)\,K_{2,m}^{(k)}(t)\,c_{2,m}^{(k)}(t)\,\rho_{2,m}^{(k)}(t)}]},\ 0 \le t \le t_s^{(k)}, \tag{43}$$

Where Equation (44):

$$f_m^{(k)}(t) \equiv f\left[T_m^{(k)}(t)\right],\ K_{l,m}^{(k)}(t) \equiv K_l\left[T_m^{(k)}(t)\right],\ c_{l,m}^{(k)}(t) \equiv c_l\left[T_m^{(k)}(t)\right],\ \rho_{l,m}^{(k)}(t) \equiv \rho_l\left[T_m^{(k)}(t)\right],\ l = 1,2. \tag{44}$$

Summarizing, another scheme was proposed for determining the flash temperature evolution during *k*th braking:

1. On the basis of the friction surface profiles of the primary friction element, the average values of parameters b_0, ν, r_{av}, $h_{\max}$ characterizing the roughness shape and their distribution along the height were calculated in the longitudinal and transverse directions.
2. Knowing the temporal profile of mean temperature on nominal contact surface $T_m^{(k)}$ (30)–(37), by means of approximation functions (1)–(8), the evolutions of $HB_{l,m}^{(k)}$ (39) and $f_m^{(k)}$, $K_{l,m}^{(k)}$, $c_{l,m}^{(k)}$, $\rho_{l,m}^{(k)}$, $l = 1,2$ (44) were established.
3. Variations of contour contact area $A_c^{(k)}$ (38), (39) and contour pressure $p_c^{(k)}$ (40) during braking were determined, taking into account pressure profile p (9).
4. Changes of diameter $d_r^{(k)}$ (41) and total area of real contact $A_r^{(k)}$ (42) in time were established.
5. Evolution of flash temperature $T_f^{(k)}$ (43) was calculated, taking into account velocity temporal profile $V^{(k)}$ (10)–(12).

4. Numerical Analysis

Based on the analytical model proposed above, the temperature generated during RST braking mode was studied for two tribosystems: disc and drum brakes. The temperature field of the disc brake system was analyzed using the finite differences method in [16]. Additionally, the corresponding evolutions of temperature in the drum brake were presented in [9], where the calculations were performed on the basis of approximate solutions to the thermal problem of friction for two strips and experimental data. Results of calculations presented below were obtained for the same friction materials and operational parameters as the above-mentioned studies. In both systems, the RST braking mode was considered, which includes 4 cycles ($n = 4$): three full cycles (braking-acceleration) and the last interrupted cycle (braking). The brake disc is made of cast iron ChNMKh, and brake pads—were manufactured from cermet FMC-11. In the second system, the drum brake was made of 30KhHSA steel and the brake shoes were manufactured with retinax FC-16L. The sintered cermet friction material FMC-11 contains 64% Fe, 15% Cu, 3% SiO_2, 6% $BaSO_4$, 3% asbestos and 9% graphite. Retinax FC-16L is a composite based on phenol-formaldehyde resins and reinforced with brass shavings [21]. The thermophysical and mechanical properties of materials and friction coefficients of selected pairs at the initial temperature are included in Table 1. The methodology of calculations adopted in the present study is shown in Figure 2.

Table 1. Properties of friction pair materials at the initial temperature $T_0 = 20$ °C [26].

Material	f	K_l, Wm^{-1} K^{-1}	c_l, Jkg^{-1} K^{-1}	ρ_l, kg m^{-3}	HB_l, MPa
ChNMKh	0.45	52.17	444.6	7100	2100
FMC-11		35	479	4700	137
30KhHSA	0.39	38	490	7800	2050
FC-16L		0.79	961	2500	392

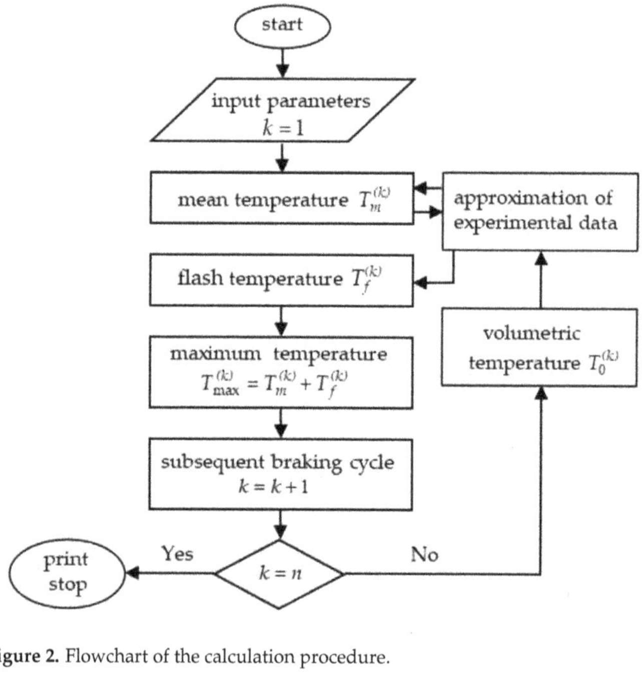

Figure 2. Flowchart of the calculation procedure.

Experimental data of frictional thermal stability and property variations under different temperature conditions of friction materials ChNMKh/FMC-11 and 30KhHSA/FC-16L are included in [9,26]. The values of coefficients in the functions (4)–(8), which approximate these data, are presented in Table 2. In this paper, it was assumed that density ρ_l of all selected materials changes slightly under temperature variations, so Equation (7) takes the form $\rho_l^*(T) = 1, l = 1, 2$. Graphs of functions $f(T)$ (4), $K_l(T)$ (5), $c_l(T)$ (6) and $HB_l(T)$, $l = 1, 2$ (8) are presented in Figure 3.

Table 2. Coefficients in approximation functions (4)–(6) and (8) for considered materials [31].

Coefficients	Material	$i = 1$	$i = 2$	$i = 3$, $°C^{-1}, \times 10^3$	$i = 4$, °C	$i = 5$	$i = 6$, $°C^{-1}, \times 10^3$	$i = 7$, °C
f_i	ChNMKh/FMC-11	0.01	1.07	1.5	–250	0	0	0
	30KhHSA/FC-16L	0	1.1	0.0014	300	0	0	0
$K_{l,i}$	ChNMKh	–2.37	4.22	0.196	–2543	0	0	0
	FMC-11	1.125	–0.64	2.3	900	0	0	0
	30KhHSA	2.455	–1.58	0.86	847	–1.05	6.3	–163
	FC-16L	1	0	0	0	0	0	0
$c_{l,i}$	ChNMKh	–0.85	6.6	0.57	4903	1.37	1.2	443
	FMC-11	0.78	0.74	3.5	1059	0.5	2.6	573
	30KhHSA	2.99	−1.4	$2 \cdot 10^{-6}$	859	–0.59	1.36	20
	FC-16L	1	0	0	0	0	0	0
$HB_{l,i}$	ChNMKh	–0.54	1	2	–50	1	1.7	500
	FMC-11	–0.93	0.83	2.34	546	2.02	2	–233
	30KhHSA	–0.55	1	3.3	0	1	2.5	400
	FC-16L	0.43	1.05	3.5	–250	0	0	0

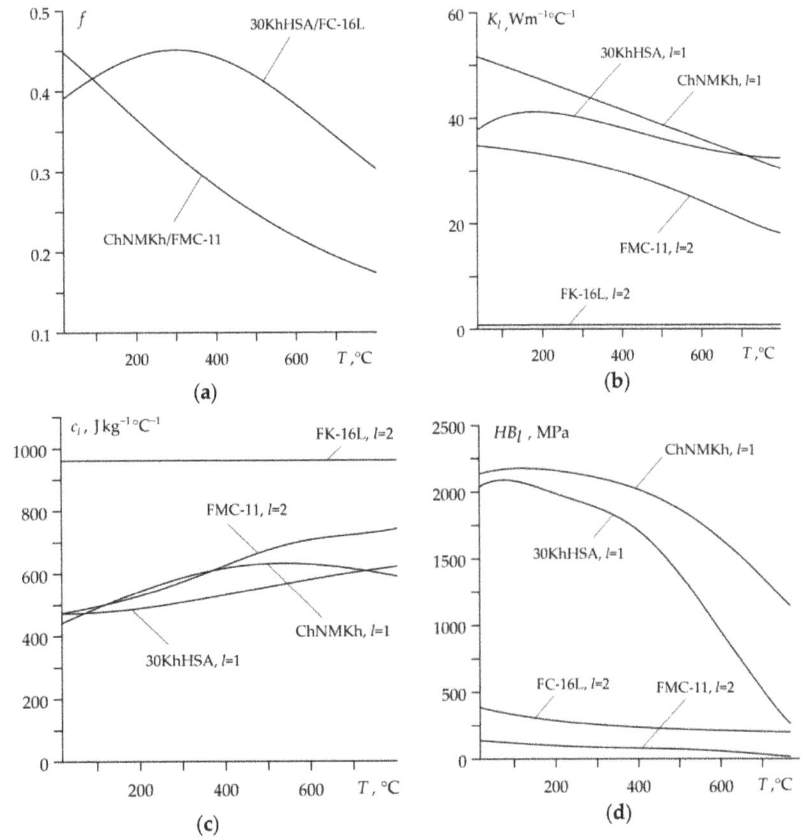

Figure 3. Graphs of functions approximating the experimental data of the dependence on temperature: (**a**) coefficient of friction f; (**b**) thermal conductivity K_l; (**c**) specific heat capacity c_l; (**d**) Brinell hardness HB_l, $l = 1, 2$.

4.1. Disc Brake System

The input values of operating parameters [15,16]: $p_0 = 1.47$ MPa, $V_0 = 27.78$ m s^{-1}, $W_0 = 392.1$ kJ, $A_a = 4.047 \cdot 10^{-2}$ m^2, $A_{vent} = 4.44 \cdot 10^{-2}$ m^2, $G = 1.58$ kg, $h = 100$ Wm^{-2}K^{-1}, $d_1 = 5.5$ mm, $d_2 = 10$ mm, $t_i = 0.5$ s, $t_c = 5$ s, $r_{av} = 450$ µm, $h_{\max} = 2.5$ µm, $b_0 = 1.0$, $\nu = 2.1$. Due to the geometric symmetry of the disc brake system with respect to the center plane of the disc, to determine the temperature field we consider the tribosystem, which consists of disc replacement and the pad with thicknesses of d_l, $l = 1,2$, respectively. The thickness of the disc replacement element d_1 is equal to the half of the real disc dimension.

The temporal profiles of velocity $V^{(k)}$ (10) and specific friction power $q^{(k)}$ (22) during four considered braking cycles are presented in Figure 4. Except for the short initial period of braking, when the pressure grows, the sliding speed linearly decreases with time and the stop time is longer with every subsequent cycle (Figure 4a). The specific friction power increases at the beginning of each braking cycle until reaching the maximum value, and then decreases to zero at the stop moment (Figure 4b), which is characteristic for the so-called rational braking mode [18,19]. It should be noted that the friction work performed at each braking cycle (the area under each of the four curves in Figure 4b) was the same and equal to the initial kinetic energy of the system $W_0 = 392.1$ kJ.

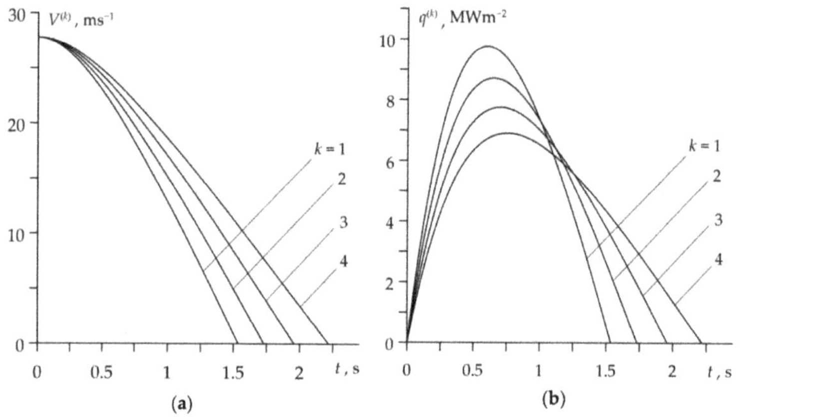

Figure 4. Changes in: (**a**) velocity $V^{(k)}$ and (**b**) specific friction power $q^{(k)}$ during $k = 1,2,3,4$ braking applications of a disc brake system.

The coefficient of friction $f_0^{(k)}$ (24) decreases with every subsequent braking from 0.45 during the first cycle to 0.28 during the fourth, last braking (Figure 5a). The drop of the friction coefficient value causes the elongation of braking stage $t_s^{(k)}$—its values are equal to 1.54 s, 1.73 s, 1.96 s and 2.22 s, respectively for $k = 1,2,3,4$ (Figure 5b).

Variations of mean temperature $T_m^{(k)}$ (30), flash temperature $T_f^{(k)}$ (43) and maximum temperature $T_{\max}^{(k)}$ (15) during each braking cycle of the process are shown in Figure 6.

Evolutions of mean temperature on the nominal contact area during every braking stage correspond to the temporal profiles of specific friction power, presented in Figure 4b. At the beginning of each braking cycle, the mean temperature $T_m^{(k)}$ growth takes place until maximum value is reached, and then the temperature drop to the standstill occurs. According to the adapted plastic roughness deformation mechanism, the flash temperature reaches significant values at the initial moments of the braking stages, when the contact area of the pad with the disc is relatively "cold". Subsequent heating of this area causes a rapid decrease in the flash temperature $T_f^{(k)}$. The mean temperature of the friction surface has a decisive influence on the time profile and the maximum temperature value. With each subsequent braking, the highest values of $T_m^{(k)}$ and $T_{\max}^{(k)}$ increase, while the $T_f^{(k)}$ decreases.

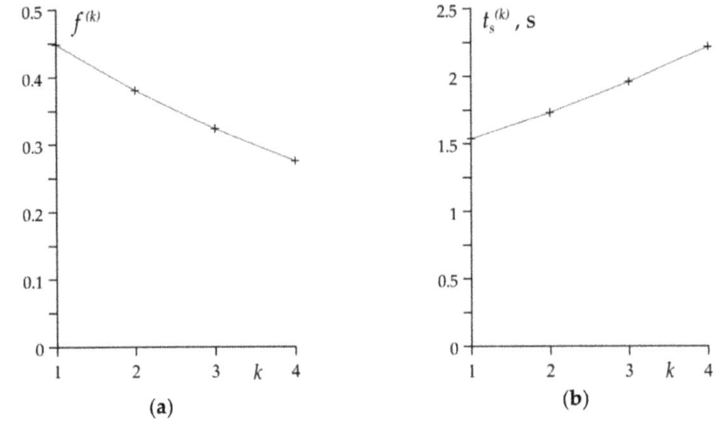

Figure 5. The values of: (**a**) friction coefficient $f_0^{(k)}$; (**b**) stop time $t_s^{(k)}$ for disc brake in each braking cycle $k = 1, 2, 3, 4$.

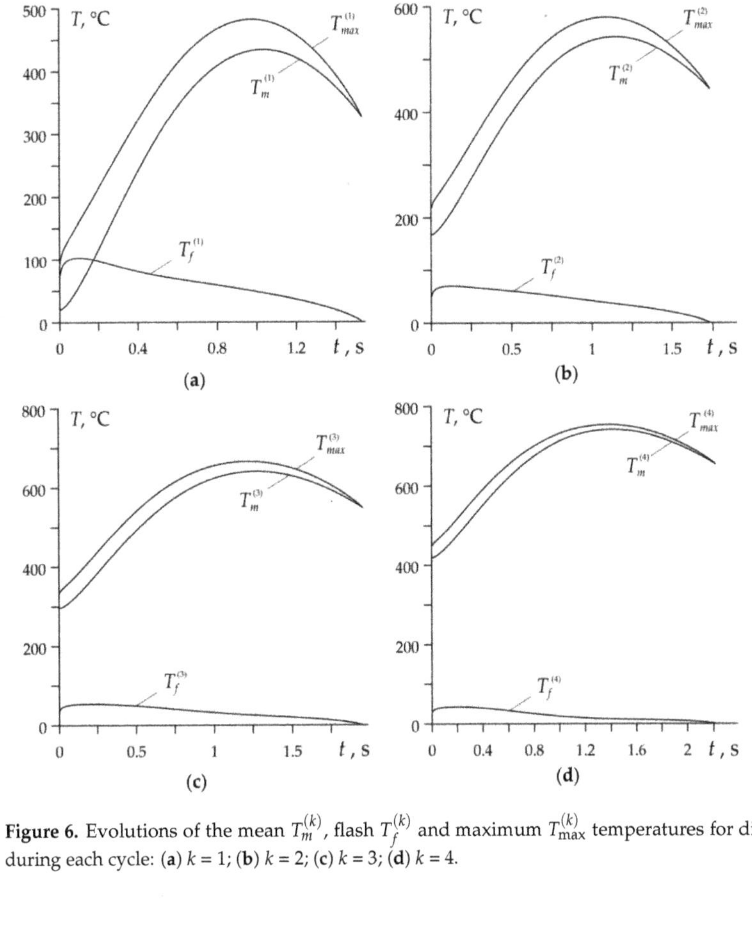

Figure 6. Evolutions of the mean $T_m^{(k)}$, flash $T_f^{(k)}$ and maximum $T_{\max}^{(k)}$ temperatures for disc brake during each cycle: (**a**) $k = 1$; (**b**) $k = 2$; (**c**) $k = 3$; (**d**) $k = 4$.

Temporal profiles of $T_m^{(k)}$, $T_f^{(k)}$ and $T_{\max}^{(k)}$ during the entire RST braking mode are presented in Figure 7. The combined time of all four braking cycles duration is 7.45 s and with account of the three stages of acceleration 15 s, gives the total time of the whole RST braking mode in the disc brake system $t_b = 22.45$ s.

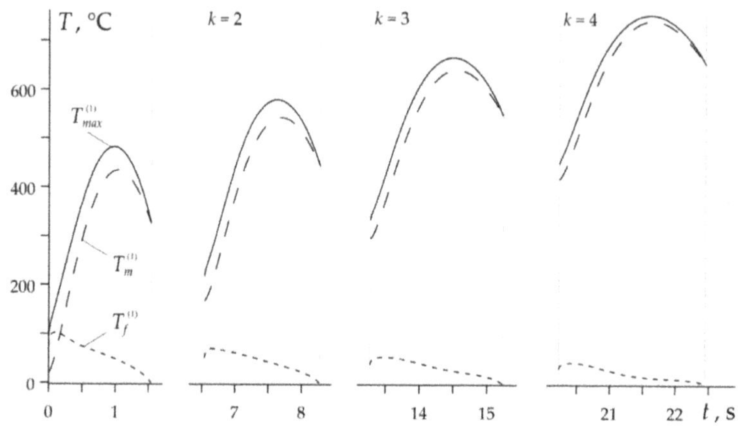

Figure 7. Variations of the maximum $T_{\max}^{(k)}$ (solid lines), the mean $T_m^{(k)}$ (dashed lines) and the flash $T_f^{(k)}$ (dotted lines) temperatures during disc brake RST mode.

4.2. Drum Brake System

The values of input parameters of braking in the RST mode by friction pair consisting of the drum ($l = 1$) and brake shoe ($l = 2$), were adapted [9]: $p_0 = 0.44$ MPa, $V_0 = 11.5$ m s^{-1}, $W_0 = 215.7$ kJ, $A_a = A_c = 3.85 \cdot 10^{-2}$ m^2, $A_{vent} = 15 \cdot 10^{-2}$ m^2, $G = 5.5$ kg, $h = 80$ Wm^{-2}K^{-1}, $d_1 = 10$ mm, $d_2 = 18$ mm, $t_i = 0.5$ s, $t_c = 25$ s, $r_{av} = 500$ μm, $h_{\max} = 4.5$ μm, $b_0 = 1.0$, $\nu = 2.2$.

The evolutions of sliding velocity and specific friction power for the drum brake are presented in Figure 8. There are some similarities with the corresponding profiles for a disc brake, which are shown in Figure 4. Sliding velocities also have short time of nonlinear drop (Figure 8a), and the profiles of specific friction power have a local maximum, shifted closer to the beginning of the braking period in this case (Figure 8b). Unlike the disc system, however, the duration of the braking stage in the drum brake is shortened and the maximum value of the specific friction power decreases with each subsequent braking.

In Figure 8a the shortening of the time of subsequent braking can be noticed, which is related to the corresponding changes in the friction coefficient: in the considered drum system its value during the first braking is 0.39 and increases linearly to 0.42 during the fourth braking (Figure 9a). The respective stop times are equal to 6.17 s and 5.77 s (Figure 9b).

Much smaller changes in the specific friction power time profiles during each braking (Figure 8b) than in the case of a disc brake result in the fact that the temperature evolution in the drum brake during each braking stage also differs much less (Figure 10). On the other hand, the contribution of the mean temperature $T_m^{(k)}$ and the flash temperature $T_f^{(k)}$ to the value of the maximum temperature $T_{\max}^{(k)}$ in the drum brake are completely different than in the disc brake. In the latter, as noted above, the time profile of the maximum temperature was primarily shaped by the mean temperature on the friction surface. In the drum brake, both components $T_m^{(k)}$ and $T_f^{(k)}$ show a significant influence on evolution and values of $T_{\max}^{(k)}$. The decisive factor in the initial braking phase is the flash temperature, and in the final stage—the mean temperature of the drum-brake shoe contact area.

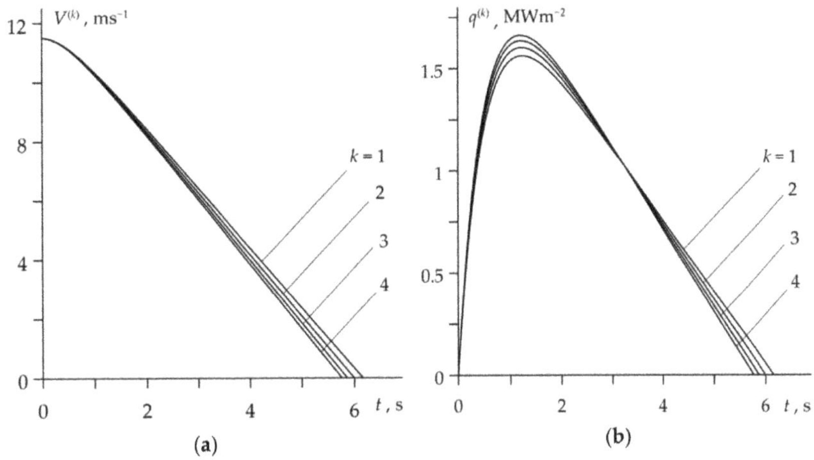

Figure 8. Changes in time during RST mode of the drum brake: (**a**) velocity $V^{(k)}$; (**b**) specific friction power $q^{(k)}$, $k = 1, 2, 3, 4$.

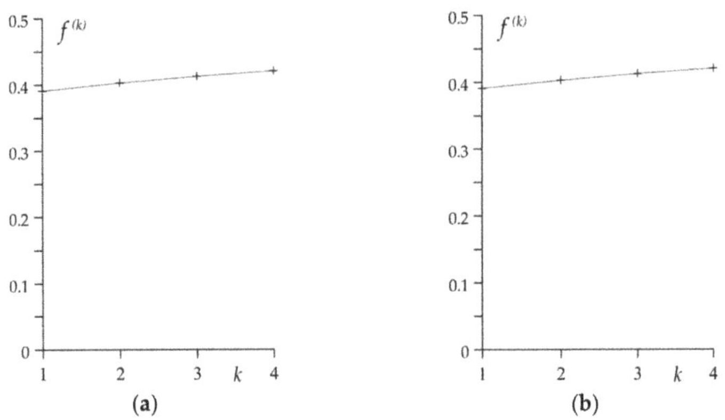

Figure 9. The values of: (**a**) coefficient of friction $f^{(k)}$; (**b**) stopping time $t_s^{(k)}$ for the drum brake in each braking cycle $k = 1, 2, 3, 4$.

The changes in the temperature with time during RST braking in the drum brake are presented in Figure 11. Duration of four braking cycles was equal to 23.81 s, three accelerations 75 s, so the total RST operation mode lasted 98.81 s. As in the case of the disc brake, with each subsequent braking, the highest values of the mean and the maximum temperatures increase, whereas the flash temperatures decrease. However, these relations are not as noticeable as in the disc brake. One of the reasons for such relatively small changes in temperature can result from much longer (five times) cooling phase during vehicle acceleration.

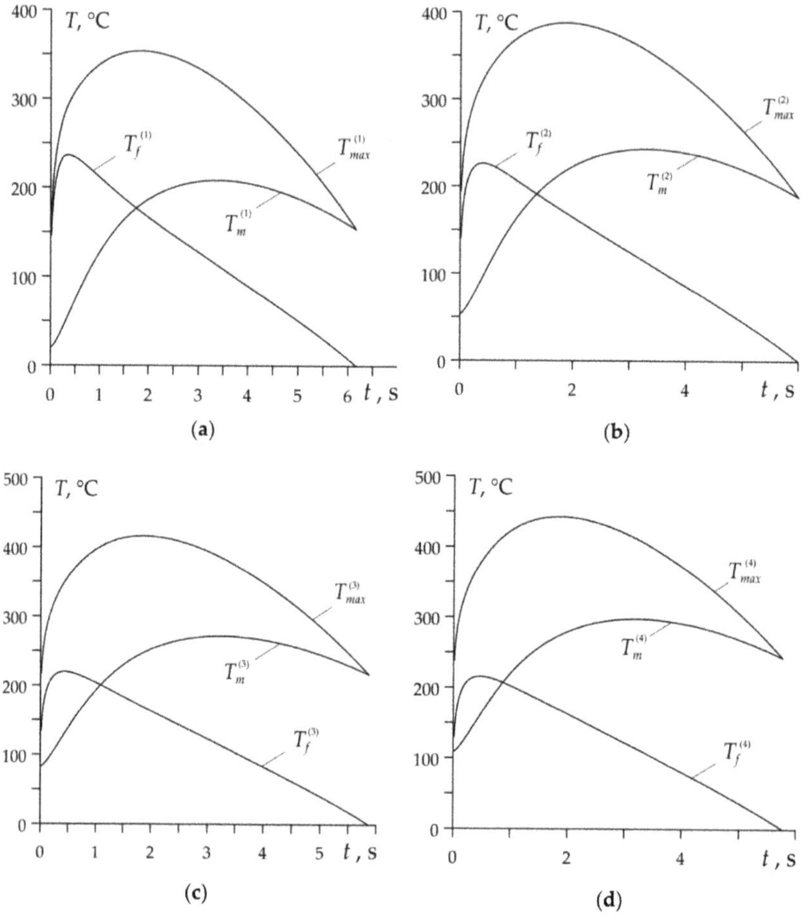

Figure 10. Evolutions of the mean $T_m^{(k)}$, flash $T_f^{(k)}$ and maximum $T_{\max}^{(k)}$ temperatures for drum brake during each cycle: (**a**) k = 1; (**b**) k = 2; (**c**) k = 3; (**d**) k = 4.

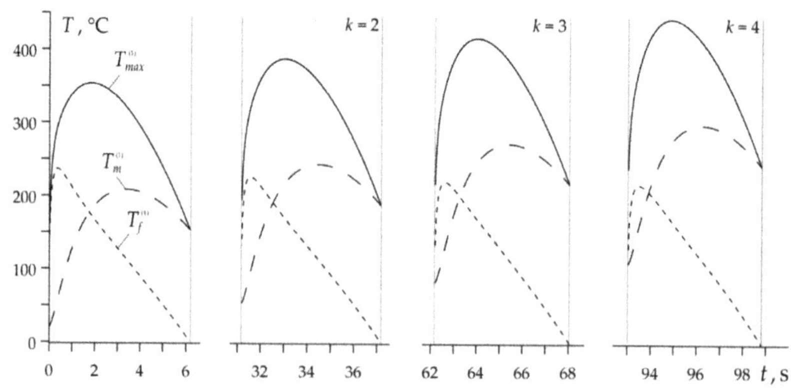

Figure 11. Evolutions of the maximum $T_{\max}^{(k)}$ (solid lines), mean $T_m^{(k)}$ (dashed lines) and flash $T_f^{(k)}$ (dotted lines) temperatures during drum brake RST mode.

Values of the friction coefficient $f^{(k)}$, braking time $t_s^{(k)}$, volumetric temperature $T_0^{(k)}$ and the highest values of the mean temperature $T_m^{(k)}$, flash temperature $T_f^{(k)}$ and maximum temperature $T_{\max}^{(k)}$ during performing each of the four cycles are included in Table 3.

Table 3. Calculated values of some characteristics at each braking cycle.

Characteristic	Brake System	$k = 1$	$k = 2$	$k = 3$	$k = 4$
$f^{(k)}$	Disc	0.45	0.38	0.32	0.28
	Drum	0.39	0.40	0.41	0.42
$t_s^{(k)}$, s	Disc	1.54	1.73	1.96	2.22
	Drum	6.17	6.00	5.87	5.77
$T_0^{(k)}$, °C	Disc	20	168	296	418
	Drum	20	53	83	109
$\max T_m^{(k)}$, °C	Disc	434	542	641	741
	Drum	208	243	272	298
$\max T_f^{(k)}$, °C	Disc	103	70	53	42
	Drum	237	226	220	216
$\max T_{\max}^{(k)}$, °C	Disc	482	560	666	753
	Drum	353	387	416	443

5. Summary of the Results and Discussion

The analytical model to determine the maximum temperature reached as a result of friction during the repetitive short-term operating mode of the braking systems, consisting of the n braking and acceleration cycles was proposed. The thermal sensitivity of the materials of the friction pairs and the temperature dependence of the friction coefficient were taken into account. According to the Chichinadze's summation hypothesis, the maximum temperature $T_{\max}^{(k)}$ during each braking ($k = 1, 2, \ldots, n$) was searched for as the sum of the mean temperature $T_m^{(k)}$ on the nominal contact area and the average temperature $T_f^{(k)}$ of the real contact area (flash temperature). The numerical analysis was carried out for two systems: disc brake (cast iron/cermet) and drum brake (steel/retinax) with $n = 4$:

- Dependence of the friction coefficient on temperature (thermal stability curve) shows a significant influence on the time profiles of the velocity, specific friction power and maximum temperature. The coefficient of friction, which decreases with increasing temperature in the disc brake system, results in elongation of each subsequent braking stage and growth of the maximum values of the specific friction power. The effect of the friction coefficient increase, under temperature increase to about 300 °C in the drum brake system, is the reduction of the braking time and the increase of the maximum values of the specific friction power.
- In the disc brake system operating in heavy mode, the evolution of temperature and its maximum values $T_{\max}^{(k)}$ are determined by the mean temperature $T_m^{(k)}$ on the nominal contact area. The contribution of the flash temperature $T_f^{(k)}$ to the maximum temperature is negligible.
- In the drum brake operating under light conditions, at the beginning of each braking stage, maximum temperature is determined mainly by the flash temperature, while at the end of braking it depends mostly from the mean temperature $T_m^{(k)}$.
- The results obtained by means of the proposed analytical model show satisfactory compliance with the relevant data obtained with the use of numerical methods, published in the scientific literature. In particular, the highest values of the maximum temperature $T_{\max}^{(k)}$ at the subsequent stages of braking $k = 1, 2, 3, 4$, found as a result of our calculations, are 482 °C, 560 °C, 666 °C and 753 °C (Table 3), and the corresponding data obtained in the article [16] are equal to 491 °C, 615 °C, 720 °C, 847

°C, respectively. The greatest relative percentage difference of the results occurred in the fourth stage and is equal to ≈ 11%. In the drum brake the maximum temperatures $T_{\max}^{(k)}$ determined by means of the proposed model, are equal to 353 °C, 387 °C, 416 °C and 443 °C (Table 3), and corresponding results presented in monograph [9] are 295 °C, 330 °C, 400 °C and 440 °C. The highest relative difference in outcomes occurs in the stage one and is equal to ≈ 21%. It should be noted that the mean temperature $T_m^{(k)}$ of the selected friction pair drum-brake shoe with similar input parameters during single braking ($k = 1$) was analyzed with the use of the finite difference method in the article [32]. The highest value of the mean temperature on the nominal contact area of the drum-brake shoe obtained was equal to 210 °C [32], which is in good agreement with the value 280 °C presented in Table 3.

6. Conclusions

The main advantage of the proposed approach is the demonstration of the possibility of adapting numerous exact solutions to the linear thermal problems of friction existing in the scientific literature, to determine the maximum temperature of the brake not only during a single, but also during a repetitive operation mode. This allows for express estimation with sufficient accuracy, not only of the maximum temperature, but also of important braking characteristics such as variations of the sliding velocity and specific friction power, stopping time and braking distance during each stage of RST braking mode. It should be emphasized that the proposed calculation model is coupled—it allows to take into account the mutual influence of all of the above-mentioned characteristics in the braking process, by the temperature-dependent coefficient of friction. From the point of view of application possibilities, it is important that this model can be used for calculations not only for materials with stable thermo-physical properties, but also for thermally sensitive materials.

Author Contributions: Conceptualization and methodology, A.Y and M.K.; software, K.T.; validation, A.Y.; formal analysis, K.T.; investigation, A.Y. and K.T.; writing—original draft preparation, A.Y. and K.T.; writing–review and editing, A.Y. and K.T.; visualization and figures preparation, K.T.; supervision, A.Y. and M.K.; project administration, M.K.; funding acquisition, M.K. All authors have read and agreed to the published version of the manuscript.

Funding: This study was performed within the framework of research project No. 2017/27/B/ST8/01249, funded by the National Science Centre, Poland and project financing through the program of the Minister of Education and Science of Poland named "Regional Initiative of Excellence" in 2019–2022, project No. 011/RID/2018/19, amount financing 12,000,000 PLN.

Institutional Review Board Statement: Not applicable.

Informed Consent Statement: Not applicable.

Data Availability Statement: No new data were created or analyzed in this study. Data sharing is not applicable to this article.

Conflicts of Interest: The authors declare no conflict of interest.

Nomenclature

A_a	area of the nominal contact surface (m^2)
A_c	area of the contour contact region (m^2)
A_r	area of the real contact region (m^2)
A_{vent}	area of the ventilated surface of the disc or drum (m^2)
b_0	parameter of the reference-surface curve (dimensionless)
c	specific heat (J $kg^{-1}K^{-1}$)
d	thickness (m)
d_r	diameter of an average spot of the real contact region (m)
f	coefficient of friction (dimensionless)
h	coefficient of heat transfer (W $m^{-2}K^{-1}$)
h_{max}	maximum roughness height on the friction surface of the disc or drum ((m))
HB	Brinell hardness (Pa)
K	thermal conductivity (W $m^{-1}K^{-1}$)
n	number of braking in RST brake mode
p	contact pressure (Pa)
p_0	nominal value of the contact pressure (Pa)
q	specific power of friction (W m^{-2})
q_0	nominal value of the specific power of friction (W m^{-2})
r_{av}	average rounding radius of roughness on the friction surface (r_{av})
t	time (s)
t_b	time of performance of all RST mode of braking (s)
t_c	cooling time at acceleration (s)
t_i	time of pressure increase (s)
t_s	time of braking (s)
T	temperature (°C)
T_0	initial (volumetric) temperature (°C)
T_f	flash temperature (°C)
T_m	mean temperature (°C)
T_{max}	maximum temperature (°C)
V	velocity (m s^{-1})
V_0	initial velocity (m s^{-1})
W_0	initial kinetic energy (J)
z	axial coordinate ((m))
Greek Symbols	
ν	parameter of the reference-surface curve (dimensionless)
ρ	density (kg m^{-3})
Index	
upper k	number of a stage of braking
lower l	number of the main ($l = 1$) and frictional ($l = 2$) elements of the friction couple

References

1. Chichinadze, A.V.; Eiss, N.S. *Polymers in Friction Assembles of Machines and Devices: A Handbook*; Allerton Press Inc.: New York, NY, USA, 1984.
2. Chichinadze, A.V. Processes in heat dynamics and modeling of friction and wear (dry and boundary friction). *Tribol. Int.* **1995**, *28*, 55–58. [CrossRef]
3. Day, A.J. *Braking of Road Vehicles*; Butterworth-Heinemann: Oxford, UK, 2014.
4. Adamowicz, A.; Grzes, P. Influence of convective cooling on a disc brake temperature distribution during repetitive braking. *Appl. Therm. Eng.* **2011**, *31*, 2177–2185. [CrossRef]
5. Adamowicz, A. Effect of convective cooling on temperature and thermal stresses in disk during repeated intermittent braking. *J. Frict. Wear* **2016**, *37*, 107–112. [CrossRef]
6. Adamowicz, A. Thermal stressed state of a disk in the process of multiple braking. *Mater. Sci.* **2016**, *51*, 814–820. [CrossRef]
7. Kang, S.S.; Cho, S.K. Thermal deformation and stress analysis of disk brakes by finite element method. *J. Mech. Sci. Technol.* **2012**, *26*, 2133–2137. [CrossRef]
8. Yevtushenko, A.; Kuciej, M.; Grzes, P.; Wasilewski, P. Temperature in the railway disc brake at a repetitive short-term mode of braking. *Int. Com. Heat Mass Trans.* **2017**, *84*, 102–109. [CrossRef]

9. Chichinadze, A.V.; Braun, E.D.; Ginzburg, A.G.; Ignat'eva, Z.V. *Calculation, Testing and Selection of Friction Couples*; Nauka: Moscow, Russia, 1979. (In Russian)
10. Ginsburg, A.G.; Romashko, A.M.; Titarenko, V.F. Calculation of temperature regime of a disc rail brake. In *Calculation and Modeling of Operation Mode of Breaking and Friction Devices*; Nauka: Moscow, Russia, 1974; pp. 21–25. (In Russian)
11. Lee, K. *Numerical Prediction of Brake Fluid Temperature Rise during Braking and Heat Soaking*; SAE Technical Paper 1999-01-0483; SAE International: Warrendale, PA, USA, 1999; pp. 1–9. [CrossRef]
12. Nosko, A.L.; Nosko, A.P. Cooling of braking devices of lifting-and-transport machines. *Her. Bauman Mosc. State Tech. Univ. Ser. Mech. Eng. Mach. Sci.* **2005**, *5*, 88–99. (In Russian)
13. Yevtushenko, A.A.; Grzes, P. The FEM-modeling of the frictional heating phenomenon in the pad/disc tribosystem (a review). *Num. Heat Transf. Part A* **2010**, *58*, 207–226. [CrossRef]
14. Wasilewski, P. Frictional heating in railway brakes: A review of numerical models. *Arch. Comput. Methods Eng.* **2020**, *27*, 45–58. [CrossRef]
15. Grzes, P. Maximum temperature of the disc during repeated braking applications. *Adv. Mech. Eng.* **2019**, *11*, 1–13. [CrossRef]
16. Yevtushenko, A.; Kuciej, M. Calculation of friction characteristics of disc brakes used in repetitive short-term braking mode. *J. Frict. Wear* **2020**, *41*, 687–697. [CrossRef]
17. Yevtushenko, A.A.; Kuciej, M. One-dimensional thermal problem of friction during braking: The history of development and actual state. *Int. J. Heat Mass Transf.* **2012**, *55*, 4148–4153. [CrossRef]
18. Yevtushenko, A.A.; Kuciej, M.; Topczewska, K. Analytical model for investigation of the effect of friction power on temperature in the disk brake. *Adv. Mech. Eng.* **2017**, *9*, 1–12. [CrossRef]
19. Yevtushenko, A.A.; Kuciej, M.; Topczewska, K. Effect of the temporal profile of the friction power on temperature of a pad-disc brake system. *J. Theoret. Appl. Mech.* **2019**, *57*, 461–473. [CrossRef]
20. Yevtushenko, A.A.; Kuciej, M.; Topczewska, K. Frictional heating during braking of the C/C composite disc. *Materials* **2020**, *13*, 2691. [CrossRef]
21. Yevtushenko, A.A.; Grzes, P. Initial selection of disk brake pads material based on the temperature mode. *Materials* **2020**, *13*, 822. [CrossRef]
22. Yevtushenko, A.A.; Grzes, P.; Adamowicz, A. The temperature mode of the carbon-carbon multi-disc brake in the view of the interrelations of its operating characteristics. *Materials* **2020**, *13*, 1878. [CrossRef] [PubMed]
23. Chichinadze, A.V.; Kozhemyakina, V.D.; Suvorov, A.V.; Strebezev, M.K.; Serik, A.B. Temperature field under model test of ring specimens at two side contact on new universal friction machine IM-58-T2. *Frict. Lubr. Mach. Mech.* **2007**, *7*, 25–33.
24. Chichinadze, A.V.; Kozhemyakina, V.D.; Suvorov, A.V. Method of temperature-field calculation in model ring specimens during bilateral friction in multidisc aircraft brakes with the IM-58-T2 new multipurpose friction machine. *J. Frict. Wear* **2010**, *31*, 23–32. [CrossRef]
25. Evtushenko, O.; Kuciej, M.; Topczewska, K. Determination of the maximal temperature of a pad–disc tribosystem during one-time braking. *Mater. Sci.* **2020**, *56*, 152–159. [CrossRef]
26. Chichinadze, A.V.; Matveevskii, R.M.; Braun, E.P. *Materials in Triboengineering of Unsteady Processes*; Nauka: Moscow, Russia, 1986. (In Russian)
27. Topczewska, K. Influence of the time of increase in contact pressure in the course of braking on the temperature of a pad-disc tribosystem. *Mater. Sci.* **2018**, *54*, 250–259. [CrossRef]
28. Demkin, N.B.; Izmailov, V.V.; Korotkov, M.A. Estimation of the deformation of rough spheres and cylinders in compression. *Wear* **1976**, *39*, 63–82. [CrossRef]
29. Demkin, N.B.; Ryzhov, E.V. *Surface Quality of Machine Contact Parts*; Mashinostroenie: Moscow, Russia, 1981. (in Russian)
30. Grzes, P. Finite element solution of the three-dimensional system of equations of heat dynamics of friction and wear during single braking. *Adv. Mech. Eng.* **2018**, *10*, 1–15. [CrossRef]
31. Grzes, P. Determination of the maximum temperature at single braking from the FE solution of heat dynamics of friction and wear system of equations. *Num. Heat Transf. Part A* **2017**, *71*, 737–753. [CrossRef]
32. Yevtushenko, A.; Kuciej, M.; Och, E.; Yevtushenko, O. Effect of the thermal sensitivity in modeling of the frictional heating during braking. *Adv. Mech. Eng.* **2016**, *8*, 1–10. [CrossRef]

materials

MDPI

Article

A Comparison of 3D and 2D FE Frictional Heating Models for Long and Variable Applications of Railway Tread Brake

Michal Kuciej [1,*], Piotr Grzes [1] and Piotr Wasilewski [2]

[1] Faculty of Mechanical Engineering, Bialystok University of Technology (BUT), 45C Wiejska Street, 15-351 Bialystok, Poland; p.grzes@pb.edu.pl

[2] Frimatrail Frenoplast S.A., 15 Watykańska Street, 05-200 Majdan, Poland; pwasilewski@frimatrail-frenoplast.pl

* Correspondence: m.kuciej@pb.edu.pl; Tel.: +48-85-746-92-00

Received: 25 September 2020; Accepted: 27 October 2020; Published: 29 October 2020

Abstract: The article proposes two 3D and 2D numerical FE models of frictional heating for the estimation of temperature distributions in railway tread brake in 1xBg configuration during repeated long-term braking. The results of computations were compared with the time courses of temperature measured using thermocouples throughout the duration of the tests on a full-scale dynamometer for two different brake shoe materials in combination with a steel wheel. The resulting temperature distributions calculated using the proposed models agreed well with the experimental measurements, and the maximum difference in temperature values does not exceed 20%. It has been proven that 2D FE model can be as efficient as 3D model to estimate the temperature distribution during long-term and variable braking in the considered friction node. The differences in the calculation of the temperature values using these models did not exceed 3%, and the calculation time for the 2D model, compared to the 3D model, was shorter approximately 85 times for the braking cycle lasting 5032 s, and approximately 45 times for the braking cycle lasting 3297 s.

Keywords: temperature; railway tread brake; 1xBg brake configuration; long-term braking; finite element method

1. Introduction

Despite the development of railway technology, tread brake, designed over 100 years ago, is still a widely used solution in railway vehicles, in particular in freight wagons and vehicles where available space is a significant limitation in the structure, e.g., in motor bogies of motive power units or in subway vehicles. One of the most important advantages of the tread brake is its simple design and high reliability. In a typical pneumatic railway brake, the force exerted on friction elements is generated by compressed air in the brake cylinder. In tread brake, the force from the brake cylinder is multiplied by brake rigging and presses the brake block (equipped with brake shoe) against the wheel running surface. The heat generated leads to rapid increase in temperature on the friction surfaces of brake components and other parts of the brake assembly.

The condition of tread-braked wheel running surface is significantly influenced by the material, from which the brake shoe is made. One of the important factors that led to the development of composite brake blocks in Europe was an attempt to reduce rolling noise of wheelsets, which is closely related to roughness of their running surface [1,2]. The use of composite brake shoes reduces the roughness of wheel running surface [2,3], and experimental studies have shown that in comparison to cast-iron, composite brake shoes also have less tendency to generate so-called hot spots [1].

Additionally, because of lower thermal conductivity, composite brake shoes increase the amount of heat absorbed by the wheel [4].

Manufacturers of railway braking systems verify friction pair selection on full-scale inertia dynamometer, i.e., such on which the friction pair is tested in its normal size. During the tests of this type, the actual operating conditions of the brake are simulated, including the routes of the proposed railway vehicles. The determined friction pair characteristics are evaluated, e.g., what is the relationship between the value of the coefficient of friction or wear and temperature, sliding velocity, pressure, or braked mass. The condition of the friction elements is also analyzed, i.e., whether signs of thermal degradation of the material or its damage are observed under given operating conditions. Dynamometer tests are also a stage of the process to receive approval granted by UIC (Union Internationale des Chemins de fer—International Union of Railways) to friction elements [5], and are also performed in the course of conformity assessment required by European law, among others, for brake shoes used in freight wagons [6,7].

Full-scale railway brake dynamometers are also used to determine temperature of friction elements in the course of brake application. In the article [8], Kim studied possible applications of infrared thermography in the characterization of railway components. It was successfully used to measure temperature of railway brake disc surface as well as to detect hot spots. It has been shown that the infrared technique allows for high-resolution measurement of the temperature field measured in real time and can complement the measurements with sliding thermocouples, the disadvantage of which is inaccurate measurement due to friction and the occurring heat generation. Paper by Hong et al. [9] is focused on investigating the mechanism of hot bands and hot spots formation in railway brake disc during high-energy brake applications. Experimental measurements, performed on a full-scale dynamometer, were employed to study so-called braking patterns which are observed throughout repetitive braking process. An attempt to simulate hot band formation in an FEM (finite element method) model has also been made by the authors and lead to the conclusion that taking into account one-sided wear of the pad would improve the agreement between the experiment and FE analysis.

Additionally, reduced-scale dynamometers are used to study tribological phenomena related to railway braking. In the article [10] Tang et al. modified contact interface of the friction block to analyze its influence on temperature fields formed during braking on the counter-face element as well as wear of the friction block and tendency to generate squeal. Rokhim et al. employed reduced-scale dynamometer to characterize frictional properties of composite railway brake shoe, namely dependence of the coefficient of friction on sliding velocity and contact pressure [11]. It has to be noted, however, that simulation of the operating conditions of the railway vehicle in reduced scale requires very careful design of the experiment, including geometry of the friction pair [12].

Due to the cost of conducting tests on dynamometer stands, and bearing in mind how important it is to determine the operating temperature and thermal loads of the friction pair for given parameters of railway vehicle operation, the object of interest of both scientific research and industry is modeling of the frictional heating process. This is evidenced by the multitude of papers published so far, the subject of which are experimental studies of temperature fields as well as analytical and numerical modeling of the frictional heating process.

A two-dimensional axisymmetric FE model to determine the mean temperature on the surfaces of the brake shoe and wheel was proposed in the papers [4,13]. The partition of heat flux between friction pair elements depended on the thermal resistance. The major advantage of the model is that it accounts for wheel-to-rail heat flow.

The model presented by Vernersson in the article [13] was calibrated by Teimourimanesh et al. using the results of temperature measurement obtained during the dynamometer test and passage of a metro vehicle [14]. Teimourimanesh used it afterwards to perform calculations of the friction pair temperature evolution for a metro vehicle [15]. The material of the wheel (ER7 steel) was thermosensitive, i.e., thermophysical properties depended on temperature. A constant and equal deceleration value was established for all braking cases, and—as a consequence—a constant value of friction generated at the wheel-brake shoe interface. The results of calculations of temperature changes were the basis for calculations of thermomechanical loads on the wheel. The analysis of the wheel mechanical loads was used to estimate the wheel designed service life (predicted mileage requiring replacement).

Petersson developed a two-dimensional model of the railway tread brake with a single brake shoe using FEM [16]. The model was calibrated on the basis of temperature measurements made during the test on a full-scale dynamometer. Using the model, the author examined the effect of heat flux distribution between the wheel and the brake shoe, pressure distribution and heat capacity of the brake shoe on the calculated temperature fields. Increasing the value of brake shoe heat capacity set in the model did not significantly affect the calculated value of the wheel temperature, but had a significant impact on the value of the brake shoe temperature

In the paper [17] Milošević et al. used FEM to develop a three-dimensional model of a frictionally heated wheel with a tread brake in 2xBgu configuration (bilateral configuration, two brake shoes per side of the wheel). The temperature field changes were calculated numerically for six variants of continuous brake applications with different values of maintained velocity and pressure exerted on the brake shoes. It was assumed that the coefficient of friction of the wheel-brake shoe system is constant during all brake applications and independent of the velocity and pressure. In the developed brake numerical model, perfect thermal contact between the wheel and the brake shoes was established (equal temperature on the contact surfaces).

In the article [18] Suchánek et al. employed FEM to calculate temperature of railway wheel in the course of two consecutive applications of a tread brake. The calculated temperature values were subsequently used to determine equivalent stress in the wheel. One of the major simplifications of the FE model presented in this study was that rotational motion of the wheel was not accounted for.

Handa et al. combined full-scale dynamometer test and FE model to investigate hollow wear of wheel running surface characteristic of tread-braked railway wheels [19]. It was concluded that the main factor contributing to this specific wear pattern is plastic deformation caused by rolling contact on the part of the wheel profile which is heated during brake application.

Additionally, disc brakes are the subject of scientific research concerning mathematical modeling of frictional heating. A few papers are shortly summarized below.

Sayeed Ahmed and Algarni [20] examined the relationship between the temperature change of the disc and the number of modifications in the form of grooves and holes. The heat flux entering the disc did not depend on the temperature in the lining area, which was possible due to the introduction of heat partition coefficient which was constant throughout the braking process. Its value depended on thermal effusivity of the disc and the brake pad material as well as the area of the entire rubbing path on the disc-pad friction surface. Hence, by changing the number of grooves (change in contact surface area), the value of the friction power directed to the disc also changed. The temperature results obtained on the basis of numerical calculations (ANSYS) were experimentally verified using the infrared thermometer.

The analytical model of a multi-disc braking system with the discs made of carbon friction composite material Termar-ADF was proposed in the article by Yevtushenko et al. [21]. The authors calculated temperature of a single brake disc with different length of the fiber bundles of the composite taking into account convection heat transfer with the surrounding air. Based on the developed model with micro, meso, and macro levels, the changes in temperature distributions and their correlations with coefficient of volumetric concentration of bundles as well as the length of the bundles were shown and discussed.

The methodology for selection of materials for disc brake pads was studied using 3D numerical model of the disc brake [22]. Six pads materials were tested in combination with cast iron disc. A number of parameters characterizing the frictional heating process during braking (maximum temperature, braking time, temperature-dependent coefficients of friction and wear rate, braking efficiency, etc.)

were accounted for allowing the selection of the optimum combination of materials in the light brake operation mode.

Tribological characteristics (dependence of the coefficient of friction and wear rate on the sliding distance, velocity, and contact pressure) of copper-graphite composites reinforced with Cu-coated or uncoated SiO_2 were studied in the article [23]. The measurements were carried out on a pin-on-ring tribometer.

A broader review regarding experimental research as well as analytical and numerical modeling of frictional heating in braking systems is presented in the articles [24–26].

Comprehensive analysis of braking systems which find application in railway vehicles was presented by Günay et al. [27]. The authors described in detail the method of assessing braking performance of railway vehicles. It is worth noting that an empirical formula, where coefficient of friction is dependent simultaneously on sliding velocity and contact pressure, was given for vehicles equipped with cast iron brake shoes. On the other hand, it is often assumed in numerical braking simulations that the friction coefficient remains constant during braking, regardless of the operating conditions [15–17,28]. One of the assumptions made in the present study is that taking into account frictional characteristics dependent on operating conditions may be vital to reliably model frictional heating during braking of railway vehicles equipped with composite brake shoes.

Apart from introduction, the paper is divided into five sections. First, the experimental part of the study is discussed, including description of the full-scale dynamometer, materials used in the research, test program, and results of braking torque measurement. Then, the statement of the boundary value heat conduction problem is given and finite element models of the wheel-brake shoe system are presented. Subsequent section contains results of numerical analysis, namely calculated temperature fields, their discussion and comparison with the experimental measurements. Finally, the study is summarized, and conclusions are drawn.

Experimental tests on inertia test benches (including measurements of temperature, wear, coefficient of friction, etc.), for braking systems of rail vehicles, are usually complicated, labor-intensive, and expensive. Numerical modeling has the potential to be a good alternative for dynamometer tests at the initial stage of the railway braking system design process. Therefore, the main purpose of this article is to evaluate accuracy of temperature fields estimated by numerical calculations concerning long-lasting (order of 103 s) braking process, performed for different friction materials, using 2D and 3D FE models, where variable operating conditions and frictional characteristics are accounted for.

2. Experimental Research on a Dynamometer Test Stand for Railway Brakes

The experimental part of this work was performed on a full-scale inertia dynamometer for testing of brake friction pairs at the Railway Research Institute in Warsaw (Figure 1a–c). The test bench was described extensively in the article by Konowrocki et al. [29].

The simplified schematic drawing of the dynamometer is presented in Figure 1e. The drive shafts (2) are coupled to the electric motor (1), flywheels (3), and an axis equipped with railway wheel (4). Flywheels are used to represent the mass of a railway vehicle, the braking of which is simulated during the test (the kinetic energy of the motor-shafts-flywheels-railway wheel system rotational motion corresponds to the kinetic energy of the translational motion of the railway vehicle). If solely combinations of flywheels were used, it would be possible to reach the moment of inertia corresponding only to a few specific railway vehicle masses. To obtain moment of inertia intermediate to the basic values, it is necessary to control the electric motor so as to reduce or supplement the moment of inertia generated by rotating flywheels [30]. Angular velocity of the system is controlled by the rotary encoder (Heidenhain ROD 436, Traunreut, Germany).

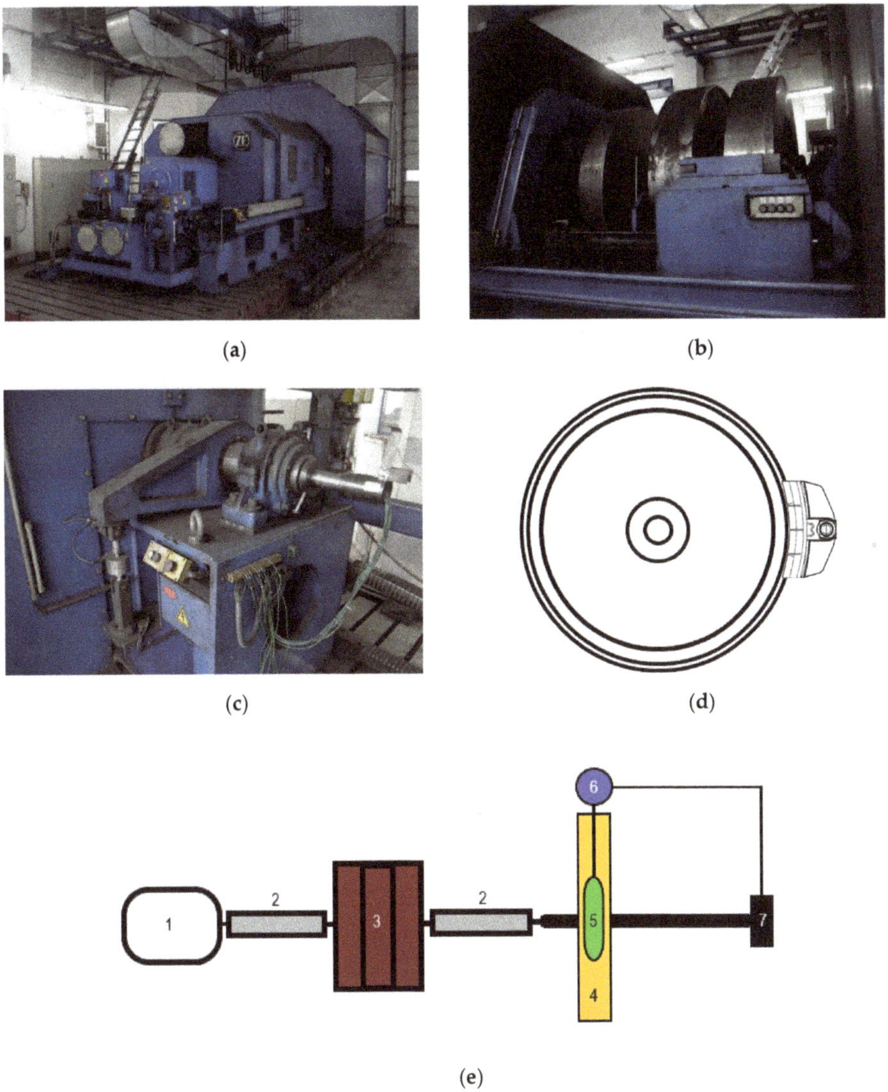

Figure 1. A full-scale inertia dynamometer for testing of brake friction pairs at the Railway Research Institute in Warsaw (**a**–**c**) (pictures courtesy of the Railway Research Institute in Warsaw) and (**d**) 1xBg railway tread brake configuration, (**e**) scheme of the test rig with main components: 1—electric motor; 2—drive shaft; 3—flywheels; 4—railway wheel; 5—brake block; 6—brake cylinder; 7—load cell.

The tested friction pair comprised railway wheel (4) and composite brake shoe installed in the holder, part of the brake block (5). Brake block is fixed to the frame which has axis of rotation independent from the motor-shafts-flywheels-railway wheel system but coaxial to it. This way, it is possible to measure friction generated at the wheel-brake shoe interface using load cell (7) (HBM-U2A). The principle of braking torque measurement is presented in Appendix B to UIC Code 548 [30]. Normal force exerted on the brake block comes from compressed-air brake cylinder (6) and was determined using load cell HBM-U2B.

Three K-type thermocouples (NiCr-NiAl, TTE426-K-2 × 0.22-850-1-QQP, Termoaparatura Wrocław, Wrocław, Poland) were installed below wheel running surface to measure wheel temperature in the course of the test. Thermocouples were located on average 5.8 mm below the working surface every 120° along the circumference and at the distance of 110 mm, 85 mm, and 55 mm from the edge of the wheel in the *z*-axis direction, which gives coordinates in the cylindrical system $T_{1E} = (r = 430.62, \theta = 210°, z = 110)$, $T_{2E} = (r = 430.24, \theta = 330°, z = 85)$, $T_{3E} = (r = 430.48, \theta = 90°, z = 55)$.

Two friction pairs were tested in 1xBg configuration [5,13], i.e., unilateral configuration with a single brake shoe (Figure 1d). ER7 steel wheel and two composite materials were subject of the experimental study, denoted Material A and Material B, respectively. The basic dimensions of the 1xBg brake shoe-wheel system are given in Table 1. The composition of the organic composites under study is given in Table 2, and an exemplary view of the brake shoes after the tests is shown in Figure 2. The materials chosen for the study differ in reinforcing fiber used in the formulation, which has significant impact on their mechanical, thermophysical, and tribological properties. Considerable amount of data acquired in the experimental part of this work allowed verification of the numerical models described in the paper. The results of comprehensive tribological tests of the abovementioned friction pairs, on reduced- and full-scale inertia dynamometer, are presented in the articles [31,32]. Braking time, wheel angular velocity, braking distance, brake cylinder pressure, braking torque, temperature under the wheel running surface, and cooling air flow velocity were measured throughout the test.

Table 1. Dimensions of the braking system.

Parameter	Value
nominal diameter of the wheel d_{eq}, m	0.87
nominal surface area A_a, mm^2	25,337.91
thickness of the brake shoe, m	0.08
width of the brake shoe, m	0.32

Table 2. Formulation of the organic composites, wt% [31,32].

	Base Formulation	Glass Fibre	Steel Fibre
Material A	65–75%	25–35%	0%
Material B		0%	25–35%

(**a**) (**b**)

Figure 2. View of the brake shoes made of: (**a**) Material A; (**b**) Material B after tests on a full-scale dynamometer.

The brake shoes were bedded-in before the execution of the main test program by braking at 120 km/h (contact force 30 kN, mass per wheel 7.5 t). The bedding-in procedure was repeated until reaching a minimum of 85% contact area between the brake shoe and the wheel. The main test on a full-scale dynamometer consisted of accelerating the wheel to angular velocity corresponding with the translational velocity of a rail vehicle in real conditions, i.e., equal to 160 km/h, 120 km/h and 80 km/h (for Material A—Figure 2a) and 80 km/h, 120 km/h, and 160 km/h (for Material B—Figure 2b), then braking to a stop. Cooling of the friction pair was performed by rotating the wheel at the angular velocity simulating the translational velocity of 100 km/h. Subsequent braking process commenced when the average temperature measured by three thermocouples below the surface dropped to less than 60 °C. The complete test program is presented in Table 3.

Table 3. Full-scale dynamometer test program for 1xBg configuration.

Number of Braking	Initial Velocityof the Vehicle V_0, km/h	Contact Force F_c, kN	Initial Temperature T_a, °C	Mass per Wheel m, t	Comments
1	160	10	ambient	2.5	Material A/ER7
2	120		50–60		
3	80				
4	160	30			
5	120				
6	80				
7	160	10		10	
8	120				
9	80				
1	80	10	ambient	2.5	Material B/ER7
2	120		50–60		
3	160				
4	80	30			
5	120				
6	160				
7	80	10		10	
8	120				
9	160				

Contact force F_c time courses and braking torque M measured during tests on a full-scale dynamometer for the considered friction systems are presented in Figure 3. The changes in vehicle velocity during braking are shown in Figure 4.

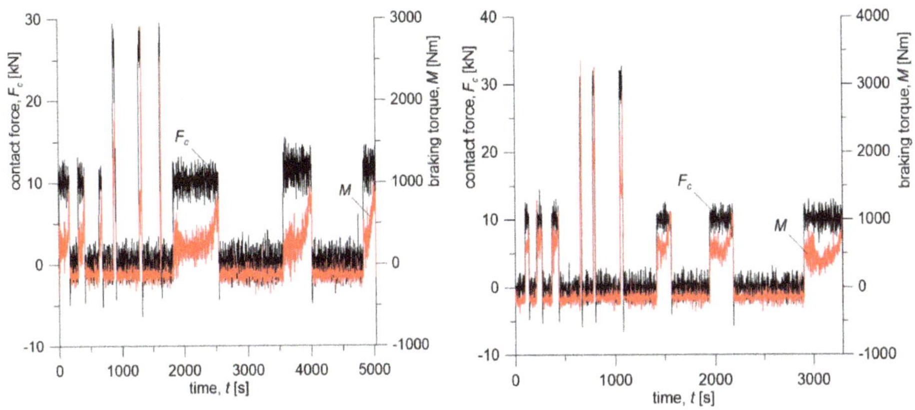

(a) (b)

Figure 3. Changes in contact force F_c and braking torque M during braking for: (**a**) Material A; (**b**) Material B.

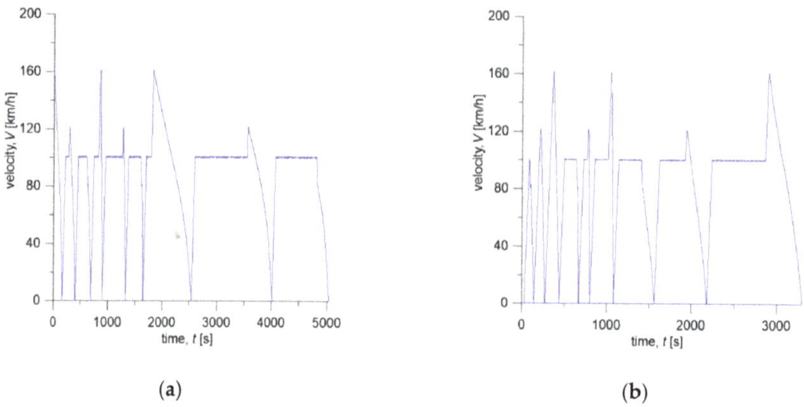

Figure 4. Changes in simulated velocity V of the vehicle during braking for: (**a**) Material A; (**b**) Material B.

3. Statement of the Problem

At the initial moment of time $t = 0$, wheel of the considered braking system rotates at constant angular velocity $\omega_0 = V_0 R_w^{-1}$, where V_0 is the linear velocity on the wheel rolling radius R_w, equal to the simulated vehicle velocity. Pressed by contact force F_c, the brake shoe comes into frictional sliding contact with the wheel, which leads to reduction of angular velocity of the wheel until it is stopped. After stopping, the brake is released, and the wheel is accelerated again. The above sequence was repeated in accordance with the test program carried out on the dynamometer (Table 3). The single braking time varied depending on the input parameters (initial velocity, contact force) and material (coefficient of friction, material properties) changing over time.

Assumptions made in the calculations:

1. kinetic energy of rotating masses is entirely converted into heat;
2. deformation of the system components due to mechanical forces and temperature is neglected—only thermal problem is considered;
3. materials of the friction elements are isotropic and their properties depend on temperature;
4. geometrical model of the wheel does not account for holes drilled for thermocouples due to very small diameter of the tip of the thermocouple; such dimension would imply very fine distribution of the finite element mesh in the vicinity of the holes;
5. the outer regions of the brake shoes were not modelled since the temperature in that location does not change during the entire analyzed process;
6. the coefficient of friction changes during braking, and these changes are known *a priori* on the basis of measurements on a full-scale dynamometer;
7. on the free surfaces, convective cooling takes place with the constant heat transfer coefficient, defined separately for specific areas on the surface of the wheel and brake shoes;
8. thermal radiation was not accounted for.

Two computational models of frictional heating for 1xBg wheel-brake shoe system were developed and tested: spatial (denoted 3D) and two-dimensional, axisymmetric (denoted 2D)—Figure 5. In the case of the spatial model, changes in the transient temperature field $T(r,\theta,z,t)$ of the wheel and brake shoes were determined from the solution of the following nonlinear boundary value heat conduction problem:

$$\frac{1}{r}\frac{\partial}{\partial r}\left(K_s(T)r\frac{\partial T}{\partial r}\right)+\frac{1}{r^2}\frac{\partial}{\partial \theta}\left(K_s(T)\frac{\partial T}{\partial \theta}\right)+\frac{\partial}{\partial z}\left(K_s(T)\frac{\partial T}{\partial z}\right)=\rho_s c_s(T)\frac{\partial T}{\partial t},\quad 0<t\le t_s,\quad (r,\theta,z)\in\Omega_s, \qquad (1)$$

$$\frac{1}{r}\frac{\partial}{\partial r}\left(K_w(T)r\frac{\partial T}{\partial r}\right)+\frac{1}{r^2}\frac{\partial}{\partial \theta}\left(K_w(T)\frac{\partial T}{\partial \theta}\right)+\frac{\partial}{\partial z}\left(K_w(T)\frac{\partial T}{\partial z}\right)=\rho_w c_w(T)\left[\frac{\partial T}{\partial t}+\omega(t)\frac{\partial T}{\partial \theta}\right], \quad 0<t\le t_s,\ (r,\theta,z)\in\Omega_w, \tag{2}$$

where $K_{s,w}$—thermal conductivity of materials, t—time, t_s—total braking time, T—temperature, ρ—mass density of materials, c—specific heat capacity of materials.

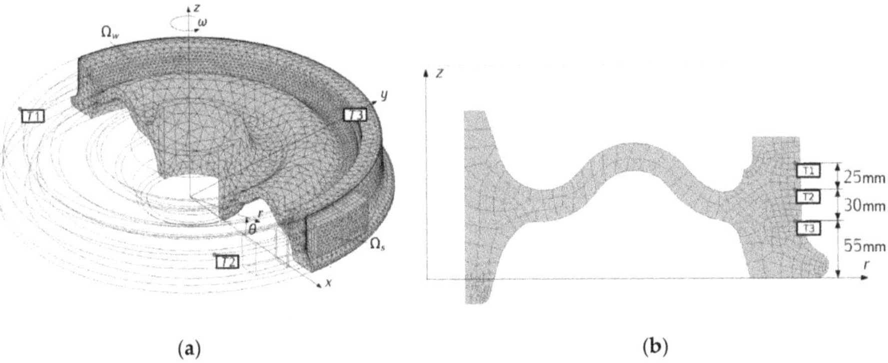

Figure 5. Finite element meshes of the tread brake used in braking simulation: (**a**) spatial; (**b**) two-dimensional (axisymmetric) with measurement points; r—radial coordinate, T_1, T_2, T_3 —temperature at specific location inside the wheel, θ—circumferential coordinate, Ω_s —region within the volume of the brake shoe, Ω_w —region within the volume of the wheel, ω —angular velocity of the wheel.

In the brake shoe-wheel contact area, denoted Γ, the conditions of perfect thermal contact were assumed:

$$K_w(T)\left.\frac{\partial T}{\partial r}\right|_{r=r_{eq}}-K_s(T)\left.\frac{\partial T}{\partial r}\right|_{r=r_{eq}}=q(r,\theta,z,t),\ (r,\theta,z)\in\Gamma,\ 0<t\le t_s, \tag{3}$$

$$T_w(r,\theta,z,t)=T_s(r,\theta,z,t),\ (r,\theta,z)\in\Gamma,\ 0<t\le t_s, \tag{4}$$

where:

$$q(r,\theta,z,t)=f(t)p(t)V(t),\ (r,\theta,z)\in\Gamma,\ 0<t\le t_s, \tag{5}$$

$q(r,\theta,z,t)$ is change in time of the total specific power of friction, $p(t)=F_c(t)/A_a$ is change in time of the contact pressure distributed uniformly on the friction surface A_a, $f(t)$—is change in time of friction coefficient, $V(t)$—is change in time of velocity on the equivalent radius of the wheel.

In the case of a two-dimensional model, friction power generated in the brake shoe-wheel contact area should be averaged using coverage factor $\eta=\theta_0/2\pi$. In consequence, calculations for the wheel and brake shoe had to be carried out separately. In addition, taking into account that 3D spatial simulations were performed for both friction elements, only wheel was analyzed in the 2D model. The axisymmetric temperature field $T(r,z,t)$ of the wheel was determined from the following boundary value heat conduction problem:

$$\frac{1}{r}\frac{\partial}{\partial r}\left(K_w(T)r\frac{\partial T}{\partial r}\right)+\frac{\partial}{\partial z}\left(K_w(T)\frac{\partial T}{\partial z}\right)=\rho_w c_w(T)\frac{\partial T}{\partial t},\ 0<t\le t_s,\ (r,z)\in\Omega_w. \tag{6}$$

Part of the friction power density directed towards the wheel was equal to:

$$q_w(r,z,t)=\gamma(t)\eta f(t)p(t)\,V(t),\ (r,z)\in\Gamma,\ 0<t\le t_s, \tag{7}$$

where the heat partition coefficient was calculated using Charron's formula [33]:

$$\gamma = \frac{\sqrt{K_w \rho_w c_w}}{\sqrt{K_s \rho_s c_s} + \sqrt{K_w \rho_w c_w}}. \tag{8}$$

Initially the wheel and the brake shoes were at ambient temperature $T_0 = T_a$.

Time profile of the friction power density $q(t)$ was calculated from braking torque and translational velocity changes determined on the dynamometer during braking. The product of the coefficient of friction and the contact pressure, appearing in Equation (5), was obtained from the measured braking torque $M(t) = f(t)p(t)A_a r_{eq}$. Knowing the velocity change (Figure 4), the friction power density is $q(t) = M(t)V(t)A_a^{-1}$ (Figure 6).

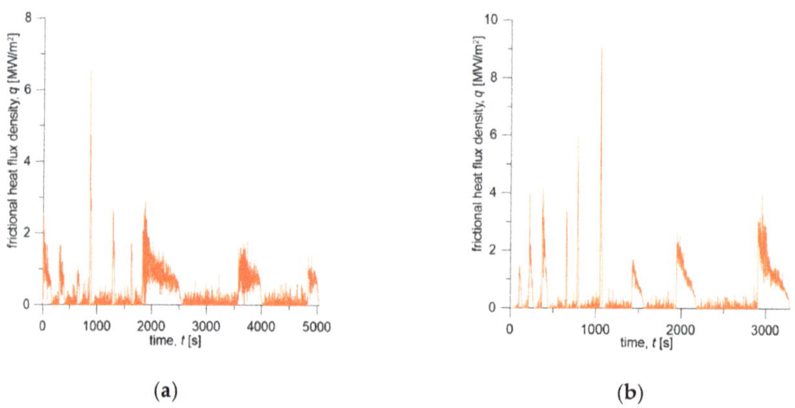

Figure 6. Changes in frictional heat flux density generated in the wheel-brake shoe contact area during braking: (**a**) Material A—ER7; (**b**) Material B—ER7.

4. Description of the 3D and 2D Finite Element Models of the Wheel-Brake Shoe System

The purpose of computer simulations of frictional heating was to develop computational models that allow determination of temperature field changes in the volume of the wheel and the brake shoe. Particularly important was the area near the contact surfaces of both objects, including the location of measuring points from the dynamometer tests. Due to the nature of frictional heating, the highest temperature gradients occur in the normal direction from the working surfaces of the contact elements to the outer cylindrical surfaces. In the case of the brake shoe (only 3D model), linear distribution of 8 finite elements in the 1/100 ratio was set in this direction. For the wheel, number of finite elements on the entire circumference of the working surface was increased (relative to the remaining area), with a view to minimizing the total number of finite elements (Figure 5a). Additionally, a thin (of the order of millimeters) surface layer was created.

To ensure accuracy of the results, several finite element meshes were built, starting from coarse grid and increasing the total number of model elements, including elements of the wheel's surface layer. Due to the complex shape of the wheel and partial contact on the friction surface with the brake shoe, a mesh consisting of tetrahedral, prismatic, and pyramidal elements was used.

When creating the mesh, it was necessary to merge opposite nodes of the brake shoe and wheel surfaces, hence the shape of the elements from the outside of both parts is consistent. These criteria forced a specific approach to dividing the area into spatial elements. The best solution—accuracy-wise—was to use hexagonal elements characterized by regular shapes—the edges are parallel to the direction of the largest temperature gradients. In the analyzed case it was difficult due to the fact that the surface of the brake shoe contacting the wheel is separated by slots. Therefore, the assumed contact area was smaller than determined from nominal width and length of the brake shoe. The first stage in the construction of the mesh was creating of triangular surface elements, then their extrusion from the contact surface to the outer surface of the brake shoe. In this way, prismatic elements were created. In the case of the wheel, a mesh of tetrahedral elements was generated in the entire volume. In the last stage, a layer replacing tetrahedral elements with prismatic ones was created at the working surface. Due to the largest temperature changes in the contact area, variable thickness distribution of the elements in both objects was introduced. The number of boundary layers was equal to 4, the boundary layer stretching factor 1.2, the thickness adjustment factor 1.2, thickness of first layer—automatic.

Preliminary calculations were carried out to verify the approach described above. It was found that further increasing the number of elements and changing their order to a higher one did not affect the temperature distribution. Finally, Lagrange second order finite elements were used to create the model's mesh (in the case of linear elements temperatures were significantly underestimated). The model comprised 91,794 tetrahedral, 796 pyramid, 19,202 prism, 24,872 triangular, and 1526 quadrilateral elements. The total number of degrees of freedom (DOFs) was 228,138 (plus 67,290 internal DOFs).

By contrast, the two-dimensional model (Figure 5b) was developed using axisymmetric (2D) quadrilateral elements (340 elements). The number of degrees of freedom was 1503.

5. Numerical Analysis

Numerical calculations of the space-time temperature distribution in the friction elements of railway tread brake in 1xBg configuration were carried out, using 3D and 2D models described in Section 4, for two combinations of brake shoe-wheel friction pairs (organic composite Material A/ER7 steel and organic composite Material B/ER7 steel).

Initial conditions and operating parameters for the numerical simulations were adopted from the test program carried out on a full-scale dynamometer (Table 3) and are consistent with changes in contact force F_c (Figure 3) and braking torque M during repeated braking (Figure 4). Knowing the time courses of F_c and M for both friction pairs, changes in friction heat flux densities q (Figure 6) generated on the brake shoe-wheel contact surface (Materials A or B and ER7 steel) were determined.

The thermophysical properties of the considered friction pair materials, which were used for calculations using 2D and 3D numerical models, are shown in Tables 4 and 5. Heat transfer coefficient with the environment h assumes different values in specific areas of the wheel (Figure 7) according to the data listed in Table 6.

Table 4. Properties of the brake shoe materials [31].

		Specific Heat Capacity c_s, J/(kg K)		Thermal Conductivity K_s, W/(m K)		Mass Density ρ_s, kg/ m^3	Thermal Diffusivity k_s, m^2/s
at temperature		30 °C	100 °C	30 °C	100 °C	20 °C	
brake shoe	material A	870	1040	1.18	1.47	1930	$7.013 \cdot 10^{-7}$
	material B	730	860	1.41	1.74	2350	$8.594 \cdot 10^{-7}$

Table 5. Thermal properties of the wheel material (ER7 [1] steel) [34].

Temperature, °C	Thermal Conductivity K_w, W/(m K)	Specific Heat Capacity c_w, J/(kg K)
0	47.3	440
20	44.1	510
400	39.3	570

[1] Density of the wheel $\rho_w = 7850\,\text{kg m}^{-3}$.

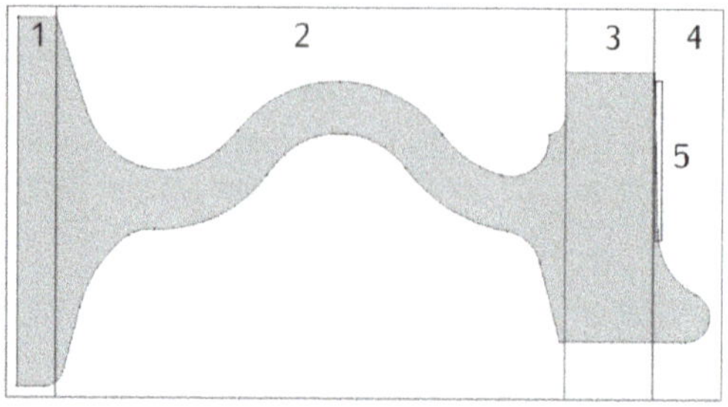

Figure 7. Areas of different convection coefficient h applied in braking simulations [34].

Table 6. Convection coefficient values in the various wheel surface zones [34].

Zone	Heat Transfer Coefficient h, W/(m^2 K)
1	32.6
2	55.9
3	55.9
4, 5	65.3

As established in Section 2, the temperature measurement inside the steel wheel was made using three K type thermocouples (NiCr-NiAl) located 5.8 mm below the working surface displaced every 120°. The same arrangement of measuring points was reproduced in the 3D spatial model (Figure 5a) and in the simplified 2D axisymmetric model (Figure 5b).

The curves of experimentally measured and calculated time courses of temperature based on the proposed numerical models are shown in Figures 8 and 9. The temperature values are average values measured by thermocouples (T_{1-3E}) and calculated at the corresponding points with the computational models (T_{1-3FEM}) at every time step.

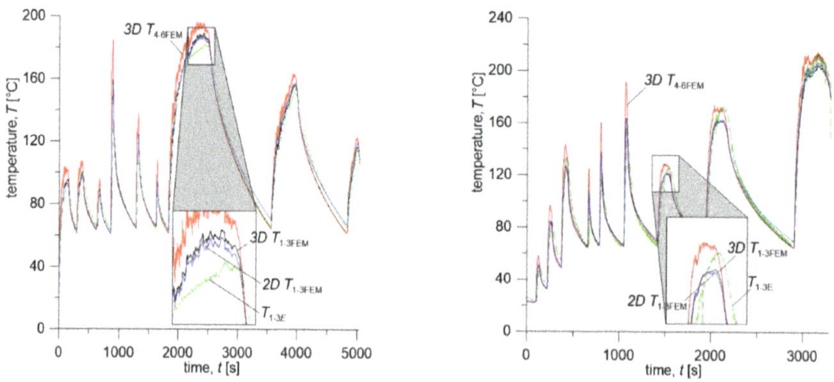

(**a**) (**b**)

Figure 8. Mean temperature changes: (**a**) Material A; (**b**) Material B; average values from three thermocouples (dashed lines) T_{1-3E}, three points from 3D and 2D numerical calculations under the surface of the wheel T_{1-3FEM} and three points on the surface of the wheel 3D T_{4-6FEM}.

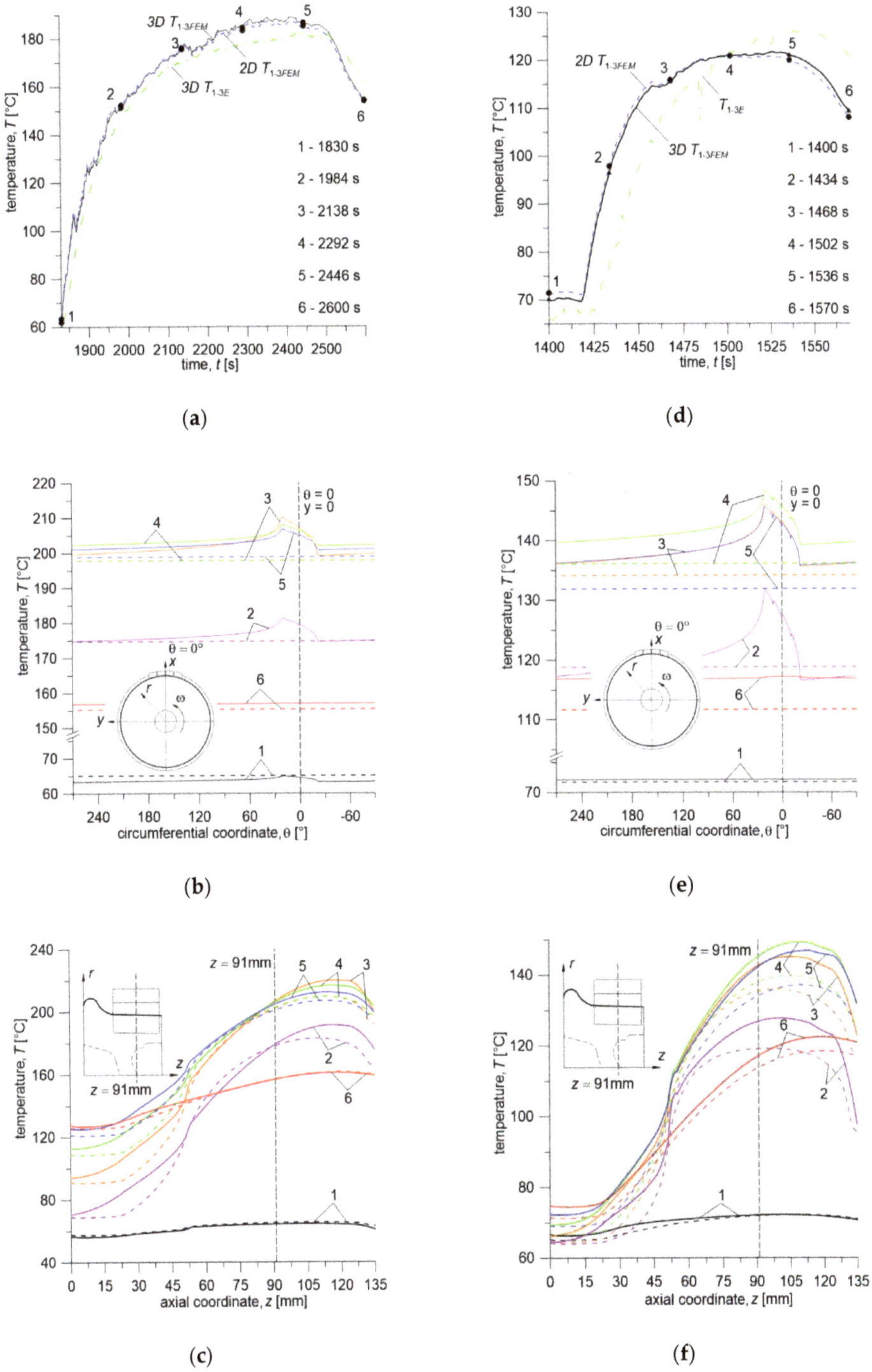

Figure 9. "7-th braking": (**a–c**) Material A/ER7; (**d–f**) Material B/ER7.

Temperature evolutions measured 5.8 mm ($T_1 = 5$ mm, $T_2 = 6$ mm, $T_3 = 6.5$ mm) below the wheel running surface (T_{1-3E}) and calculated at respective points with 3D and 2D models (T_{1-3FEM}) are presented in Figure 8. Additionally, average temperature on the wheel running surface T_{4-6FEM} was calculated with the 3D model keeping identical distance of the data collection points from the outer face of the wheel rim. The arrangement of points nos. 4–6 located on the wheel running surface is shown on the diagram in Figure 5a.

According to the test program from Table 3, for the friction pair comprising Material A (brake shoe) and ER7 steel (wheel), the first three brake applications (no. 1–3) were carried out successively from initial velocities of 160 km/h, 120 km/h, and 80 km/h, at a contact force of 10 kN and mass per wheel of 2.5 t. During braking no. 1 (160 km/h, 10 kN, 2.5 t), the average temperature measured 5.8 mm below the wheel surface reached maximum value of $T_{1-3E} = 92.3$ °C, for braking no. 2 (120 km/h, 10 kN, 2.5 t)—$T_{1-3E} = 99.2$ °C, and for brake application no. 3 (80 km/h, 10 kN, 2.5 t)—$T_{1-3E} = 83.2$ °C (Figure 8a). The total time of three braking cycles was $t_{1-3} = 866$ s. It should be noted that braking no. 1 began when the friction elements initial temperature was $T_0 = 23$ °C (cooling air (ambient) temperature was 10 °C), while subsequent brakings until the end of the tests began when the average temperature, measured by three thermocouples 5.8 mm below the wheel running surface, dropped to the value in the range of $T_{1-3E} = 62.7 \div 64.9$ °C.

The next stage of the study was the sequence of three brake applications (no. 4–6) initiated, as previously, at velocities of 160 km/h, 120 km/h, and 80 km/h and mass per wheel of 2.5 t, while the contact force was increased to 30 kN (Table 3). Due to the increase in the contact force, the time of each single braking (t_4, t_5, t_6) shortened, while the total time t_{4-6} including cooling time extended to 966 s. However, due to the more rapid braking process, the maximum temperature values are much higher, especially for brakings no. 4 and 5 at velocities of 160 km/h and 120 km/h—they are equal to $T_{1-3E} = 132.2$ °C, and 108.7 °C, respectively (Figure 8a).

In the final stage of the test (brakings no. 7–9) mass per wheel was increased to 10 t, and contact force was reduced to 10 kN, while the order of brake applications remained unchanged. Due to the fourfold increase in mass per wheel, the time of the entire sequence significantly lengthened, as compared to the previous two braking cycles 1–3 and 4–6 and is equal to $T_{1-3E} = 3686$ s. Additionally, the maximum values of the measured temperature are much higher and reach values of $T_{1-3E} = 182.5$ °C, 154.5 °C for individual brake applications (Figure 8a).

The first three braking sequences according to Figure 8a during the time from $t = 0$ to $t = 3297$s were carried out at the contact force $F_c = 10$ kN, which at the nominal area $A_a = 25250$ mm^2, gave the nominal contact pressure value of 0.396 MPa. The subsequent three brake applications were performed at $F_c = 30$ kN (pressure of about 1.188 MPa) and the last three again at 10 kN.

The test program presented in this article for the second friction pair, consisting of the brake shoe made of Material B and ER7 steel wheel, differs from the previous one (i.e., Material A/ER7 system) in the order of initial braking velocities, i.e., 80 km/h, 120 km/h, and 160 km/h (Table 3). Considering the thermophysical properties of Material B and Material A (Table 4) and the change in the performed braking sequence, time course of temperature for this friction pair is significantly different (Figure 8b). The difference is pronounced especially in the total braking cycle time, which is equal to $t = 5032$ s for Material A/ER7 pair and t = 3297 s for Material B/ER7. This is mainly due to a completely different gradual heating of the wheel, and thus with smaller values of the maximum temperature in the wheel in the initial stage of testing and its faster decrease after subsequent braking to the programmed initial temperature. That difference can also be observed by comparing the time courses of friction power density corresponding to each of the friction pairs, shown in Figure 6.

In the first part of the program consisting of three brake applications no. 1–3, from 80 km/h, 120 km/h, and 160 km/h, with contact force of 10 kN and mass per wheel of 2.5 t, the maximum average temperature measured by three thermocouples located 5.8 mm below the wheel running surface, reached values T_{1-3E} = 49 °C, 83.2 °C, 132.7°, respectively, where its duration was t_{1-3} = 652 s. During the subsequent braking sequence (brakings no. 4–6 from 80 km/h, 120 km/h and 160 km/h, at F_c = 30 kN and m = 2.5 t) average temperature reached maximum values of T_{1-3E} = 89.5 °C, 120 °C, 158.4 °C, respectively, where t_{4-6} = 766 s. Throughout the final part of the program (brake applications no. 7–9 from 80 km/h, 120 km/h, and 160 km/h, at F_c = 10 kN and m = 10 t), maximum values of the average temperature are T_{1-3E} = 126 °C, 173 °C, 213.8 °C, respectively, for t_{7-9} = 1879 s.

The time courses of temperature calculated with the 2D and 3D FE models show very good agreement with measured values (Figure 8a,b). It can be seen that simplifications assumed in the proposed numerical models, i.e., the condition of equality of temperature on the brake shoe-wheel contact surface, introduction of the heat partition coefficient in 2D system, as well as convective heat transfer coefficient which is constant in time, but different in particular zones of the wheel surface, (see Figure 7), are acceptable and do not significantly affect accuracy of temperature estimation, even for such long and variable brake applications. Both the temperature values calculated from the spatial (3D T_{1-3FEM}) and two-dimensional (2D T_{1-3FEM}) models slightly increase the temperature compared to the experimental measurement results. The difference between these values is not more than 20% (braking no. 4, Material A and B). Comparing the 3D T_{1-3FEM} and 2D T_{1-3FEM} temperature values calculated at each time step, it can be stated that their difference is negligible (max less than 3%, however, it should be added that for both materials it is higher during cooling and is equal to 7÷9% after braking no. 8). It can be important when choosing the calculation model of frictional heating, especially in terms of the required computing power and "time consumption" concerning model preparation and calculation itself.

In addition, time courses of the average 3D T_{4-6FEM} temperature on the brake shoe-wheel contact surface calculated using the three-dimensional FE model at the condition of temperature equality on that surface were shown in Figure 8a,b.

In the case of friction pairs of the same type as presented 1xBg brake configuration, it must be stated that the initial estimation of the time course of temperature can be carried out using a 2D model, achieving a satisfactory approximation. The temperature calculations were performed on a workstation with following parameters: CPU Intel® Xeon® E5-2698 v4 @ 2.20GHz; RAM 64 GB (DDR4). The calculation times for the Material A/ER7 pair performed were equal to 1h 35 min for the 2D model and 114 h 50 min for the 3D model. However, in the case of the Material B/ER7 friction pair, calculation times were 1 h 13 min and 51 h 32 min, respectively.

In order to perform a detailed analysis of the time courses of temperature (measured—dash dotted lines, and calculated by 2D model—dashed lines and 3D model—solid lines) in the braking system discussed in this work, braking no. 7 for both friction pairs was considered (Figure 9). For Material A/ER7 friction pair, it is braking from 160 km/h at mass per wheel of 10 t and contact force of 10 kN, while for Material B/ER7 pair—80 km/h at mass per wheel of 10 t and contact force of 10 kN. The analysis of single brakings for both cases (plots in Figure 9a,d) confirms the earlier observation that the calculations made using both numerical models approximate the actual time course of temperature very well. Small percentage differences (max 3.37% for Mat. A and 3.3% for Mat. B) between the measured and calculated temperature values may stem both from simplifications imposed on numerical models and measurement errors resulting, among others, from the delay in response of the thermocouples used in experimental part of this study.

The analyzed multiple braking processes are relatively long—they differ significantly from the typical cases of single braking (order of 10^{-1} s) occurring e.g., in motor vehicles. The long duration of the processes, in addition to the large size of the objects such as the wheel of a rail vehicle, with an outer diameter of 870 mm (assumed rolling radius was 437 mm), require a significant computing power and lead to many longer calculations. Therefore, the results obtained in this study, apart from their correctness verified by experimental research, are also intended to justify the previously formulated statement that the 2D axisymmetric model is sufficient to provide results at a similar level of accuracy. In order to perform such analysis, temperature points 1–6 are marked on the time courses of temperature obtained from calculations made with the 2D and 3D models presented in Figure 9a and d. For the abovementioned time points the temperature distributions are shown in Figure 9b and c (Material A/ER7 pair) and Figure 9e and f (Material B/ER7 pair)—for the curve lying on the circumference of the circle in the middle of the brake shoe thickness (Figure 9b,e), and in the axial direction (Figure 9c,f)—for a profile marked in the schematic drawing with a thick line. As can be seen from the time courses of temperature in Figure 9b,e, during braking no. 7 for both considered friction pairs at time points no. 2–5, when the temperature rises and reaches maximum values, this distribution is non-uniform. The maximum temperature fluctuations around the circumference at these points are at a level of 5.5% (Material A) and 13% (Material B), which confirms the assumption that the axisymmetric model will be sufficient to obtain similar results. By contrast, the temperature distribution in the axial direction shown in Figure 9c,f, for both friction pairs, differ by a maximum of 171% for Material A and 115% for Material B. The temperature along the analyzed curve for two friction pairs is higher on the right-hand side of the line $z = 91$ mm due to the lower width of the area absorbing heat through conduction.

Temperature distributions inside the brake shoe and wheel during braking no. 7, corresponding to the time points indicated in Figure 9a,d, are presented for the considered friction pairs in Figure 10. In addition to presenting the subsequent stages of temperature change in the wheel and its gradual heating with the braking time, it is also worth noting that the temperature distribution in the wheel at the analyzed time points in both friction pairs is of an axisymmetric nature. The above observation also confirms the possibility of using the 2D FE model to analyze time courses of temperature in 1xBg braking systems during repeated, variable, and prolonged braking.

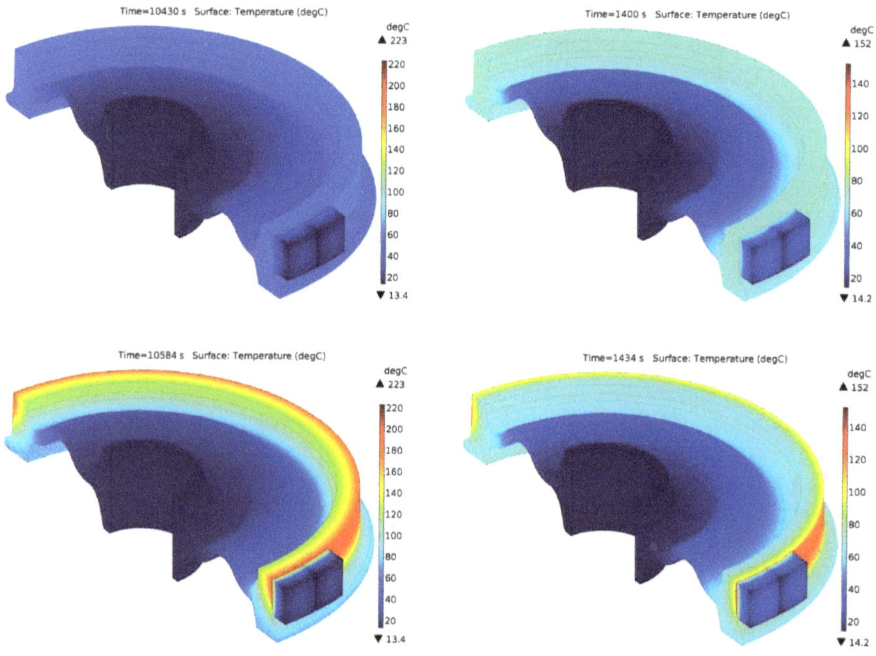

Figure 10. *Cont.*

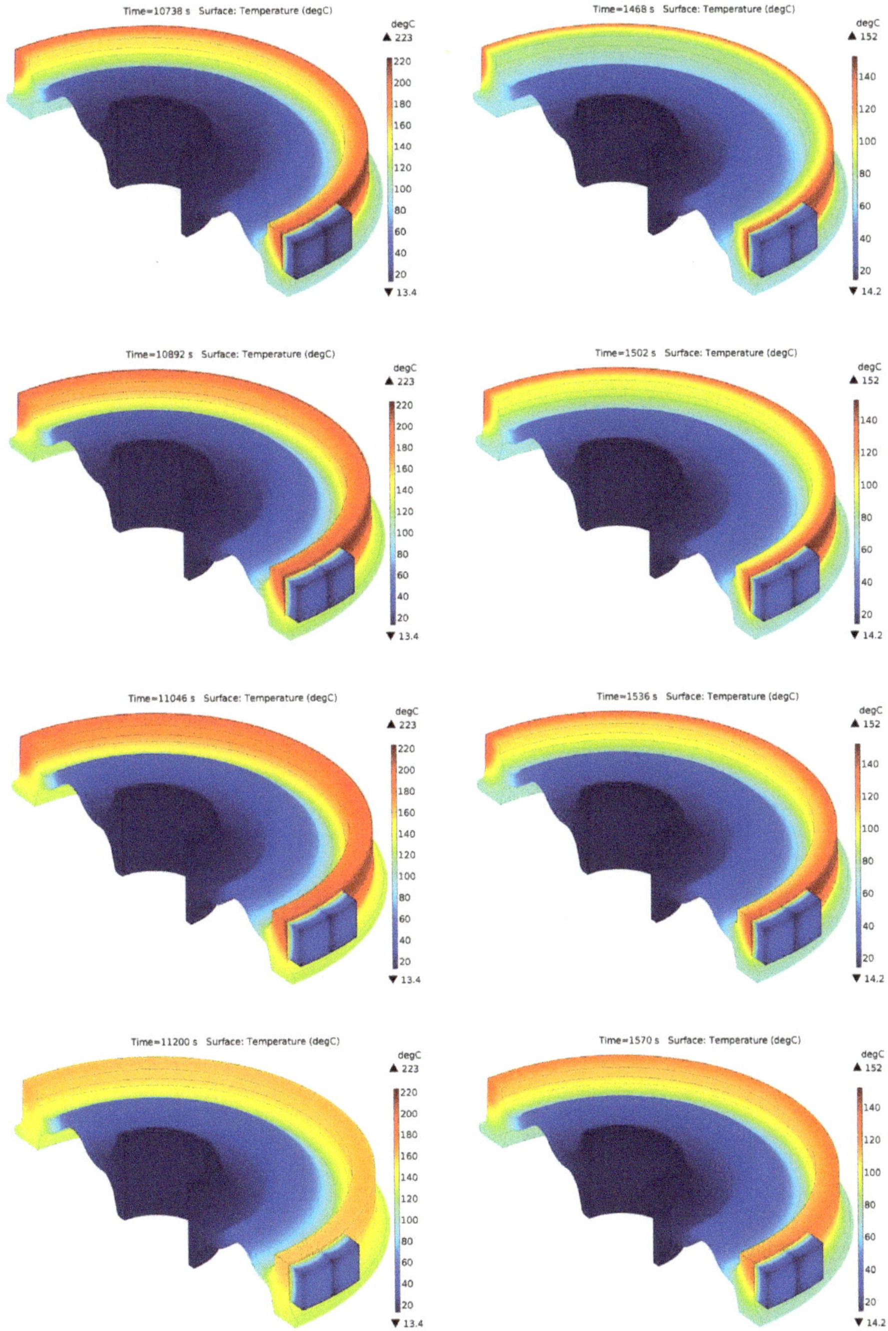

(**a**) (**b**)

Figure 10. Temperature distribution in the wheel and brake shoe for selected time points marked in Figure 9a,d, for braking no. 7: (**a**) Material A/ER7; (**b**) Material B/ER7.

6. Summary and Conclusions

Two 3D and 2D FE numerical models of frictional heating were proposed to calculate temperature fields in the railway tread brake in 1xBg configuration. The consistency of the results obtained for two different friction pairs composed of the brake shoes made of organic composite materials (denoted Materials A and B) and the steel wheel (ER7) demonstrates the usefulness of the developed computational models.

Temperature values calculated from 3D and 2D models are slightly overestimated as compared to the results of experimental measurements and the difference between these values does not exceed 20%. For this reason, it can be stated that both the 3D and 2D model may be used to successfully estimate the temperature distribution in the friction elements of 1xBg braking system, which is beneficial in the design of new friction pairs in terms of their thermophysical and tribological properties.

Comparison of the temperature calculated using 3D and 2D models at each time step shows that the difference is not greater than 3% for the frictional heating processes and not more than 9% during the cooling steps between successive brake applications. The comparison of models presented in the article proved that 2D frictional heating model can be successfully used to estimate the temperature distributions in the discussed friction node. The computation time (workstation CPU Intel® Xeon® E5-2698 v4 @ 2.20GHz; RAM 64 GB (DDR4)) performed with the 2D model compared to the more general 3D model is reduced by approximately 85 times for the braking cycle lasting 5032 s, and approximately 45 times for the braking cycle lasting 3297 s.

Both proposed numerical models of frictional heating have certain limitations of applicability. In this article speed and heat flux density determined from experimentally measured braking torque were used as input values. The next stage in the future development of the presented models will be, among others, the mutual coupling of the speed and the friction coefficient through the average temperature on the contact surface of the friction node.

Despite some limitations of the proposed models, they can be used not only to estimate the temperature level during repeated and long-term braking, but also, for example, to study the effect of convective heat exchange on the average wheel temperature or detection of potential wheel overheating related to extreme operating conditions.

The above numerical modeling approach can also be successfully applied to a different type of railway tread brake, e.g., 2xBgu.

The authors of this work did not find any studies where 3D and 2D models of railway tread brake were compared and verified with experimental data, which is also a major asset of this work.

Author Contributions: Conceptualization, M.K.; methodology, M.K.; software, P.G.; validation, M.K.; formal analysis, M.K. and P.W.; investigation, P.G.; resources, P.W.; data curation, P.G. and P.W.; writing—original draft preparation, M.K., P.G. and P.W.; writing—review and editing, M.K., P.G. and P.W.; visualization, P.G.; supervision, M.K.; project administration, M.K.; funding acquisition, M.K. and P.G. All authors have read and agreed to the published version of the manuscript.

Funding: This study was performed within the framework of research project No. 2017/27/B/ST8/01249, funded by the National Science Centre, Poland and project financing through the program of the Minister of Science and Higher Education of Poland named "Regional Initiative of Excellence" in 2019–2022, project number 011/RID/2018/19, amount of financing 12,000,000 PLN.

Acknowledgments: The authors would also like to gratefully acknowledge Jacek Kukulski at the Railway Research Institute in Warsaw for his valuable technical support.

Conflicts of Interest: The authors declare no conflict of interest.

Nomenclature

A_a	nominal area of the contact region between brake shoe and wheel (m^2)
c	specific heat capacity (J/(kg K))
d_{eq}	nominal wheel outer diameter $d_{eq} = 2r_{eq}$ (m)
f	coefficient of friction (dimensionless)
F_c	contact force (N)
h	heat transfer coefficient (W/(m^2 K))
k	thermal diffusivity (m^2/s)
K	thermal conductivity (W/(m K))
m	braking mass per one wheel (kg)
M	braking torque (N m)
p	contact pressure (MPa)
q	specific friction power (W/m^2)
r	radial coordinate (m)
r_{eq}	equivalent radius of the contact region (m)
R_w	equivalent radius of the wheel $R_w = r_{eq}$ (m)
t	time (s)
t_s	total braking time (s)
T	temperature (°C)
T_a	ambient temperature (°C)
T_0	initial temperature (°C)
T_{1-3}	average temperature from T_1, T_2 and T_3 (°C)
T_{4-6}	average temperature from T_4, T_5 and T_6 (°C)
T_1, T_2, T_3	temperature at specific location inside the wheel (°C)
T_4, T_5, T_6	temperature at specific location on the contact surface of the wheel (°C)
V	velocity on the equivalent radius of the wheel (vehicle velocity) (m/s)
V_0	initial vehicle velocity (m/s)
z	axial coordinate (m)
Greek symbols	
γ	heat partition coefficient (dimensionless)
Γ	contact region of the brake shoe and wheel (m^2)
η	coverage factor (dimensionless)
θ	circumferential coordinate (rad)
ρ	mass density (kg/m^3)
ω	angular velocity of the wheel (rad/s)
ω_0	initial angular velocity of the wheel (rad/s)
Ω_s	region within the volume of the brake shoe (m^3)
Ω_w	region within the volume of the wheel (m^3)
Subscripts	
s, w	brake shoe, wheel
E	experiment

References

1. Petersson, M.; Vernersson, T. Noise-related roughness on tread braked railway wheels-experimental measurements and numerical simulations. *Wear* **2002**, *253*, 301–307. [CrossRef]
2. de Vos, P. *Railway Noise in Europe: State of the Art Report*; Union Internationale des Chemins de fer: Paris, France, 2016.
3. Heutschi, K.; Bühlmann, E.; Oertli, J. Options for reducing noise from roads and railway lines. *Transp. Res. Part A Policy Pract.* **2016**, *94*, 308–322. [CrossRef]
4. Vernersson, T. Temperatures at railway tread braking. Part 2: Calibration and numerical examples. *Proc. Inst. Mech. Eng. Part F J. Rail Rapid Transit* **2007**, *221*, 429–441. [CrossRef]
5. *UIC Code 541-4 Brakes—Brakes with Composite Brake Blocks—General Conditions for the Certification of Composite Brake Blocks*, 5th ed.; Union Internationale des Chemins de fer: Paris, France, 2018.

6. European Commission, Commission Regulation (EU) No 321/2013 concerning the technical specification for interoperability relating to the subsystem 'rolling stock—Freight wagons' of the rail system in the European Union and repealing Decision 2006/861/EC. *Off. J. Eur.* **2013**, *OJ L 104*, 1–56.
7. Friction Elements for Wheel Tread Brakes for Freight Wagons (ERA/TD/2013- 02/INT v 3.0). Available online: https://www.era.europa.eu/sites/default/files/activities/docs/era_td_2013_02_int_en.pdf (accessed on 15 December 2014).
8. Kim, J. Non-destructive characterization of railway materials and components with infrared thermography technique. *Materials* **2019**, *12*, 4077. [CrossRef] [PubMed]
9. Hong, H.; Kim, M.; Lee, H.; Jeong, N.; Moon, H.; Lee, E.; Kim, H.; Suh, M.; Chung, J.; Lee, J. The thermo-mechanical behavior of brake discs for high-speed railway vehicles. *J. Mech. Sci. Technol.* **2019**, *33*, 1711–1721. [CrossRef]
10. Tang, B.; Mo, J.L.; Xu, J.W.; Wu, Y.K.; Zhu, M.H.; Zhou, Z.R. Effect of perforated structure of friction block on the wear, thermal distribution and noise characteristics of railway brake systems. *Wear* **2019**, *426*, 1176–1186. [CrossRef]
11. Rokhim, Y.A.; Surojo, E.; Muhayat, N.; Raharjo, W.W. Frictional Characteristic Evaluation of Composite Brake Block Using a Reduced-Scale Brake Dynamometer. In Proceedings of the 6th International Conference and Exhibition on Sustainable Energy and Advanced Materials, Surakarta, Indonesia, 16–17 October 2019; Sabino, U., Imaduddin, F., Prabowo, A., Eds.; Springer: Singapore, 2020; pp. 61–67. [CrossRef]
12. Desplanques, Y.; Roussette, O.; Degallaix, G.; Copin, R.; Berthier, Y. Analysis of tribological behaviour of pad–disc contact in railway braking: Part 1. Laboratory test development, compromises between actual and simulated tribological triplets. *Wear* **2007**, *262*, 582–591. [CrossRef]
13. Vernersson, T. Temperatures at railway tread braking. Part 1: Modeling. *Proc. Inst. Mech. Eng. Part F J. Rail Rapid Transit* **2007**, *221*, 167–182. [CrossRef]
14. Teimourimanesh, S.; Vernersson, T.; Lundén, R.; Blennow, F.; Meinel, M. Tread braking of railway wheels —Temperatures generated by a metro train. *Proc. Inst. Mech. Eng. Part F J. Rail Rapid Transit* **2014**, *228*, 210–221. [CrossRef]
15. Teimourimanesh, S.; Vernersson, T.; Lundén, R. Thermal capacity of tread-braked railway wheels. Part 1: Modelling. *Proc. Inst. Mech. Eng. Part F J. Rail Rapid Transit* **2016**, *230*, 784–797. [CrossRef]
16. Petersson, M. Two-dimensional finite element simulation of the thermal problem at railway block braking. *Proc. Inst. Mech. Eng. Part C J. Mech. Eng. Sci.* **2002**, *216*, 259–273. [CrossRef]
17. Milošević, M.S.; Stamenković, D.S.; Milojević, A.P.; Tomić, M.M. Modeling thermal effects in braking systems of railway vehicles. *Therm. Sci.* **2012**, *16*, 515–526. [CrossRef]
18. Suchánek, A.; Harušinec, J.; Loulová, M.; Strážovec, P. Analysis of the distribution of temperature fields in the braked railway wheel. In Proceedings of the MATEC Web of conferences, Machine Modelling and Simulations, Sklené Teplice, Slovak Republic, 5–8 September 2017; Vasko, M., Handrik, M., Jakubovičová, L., Kopas, P., Blatnická, M., Baniari, V., Štalmach, O., Sapietová, A., Sága, M., Eds.; EDP Sciences: Paris, France, 2018; Volume 157. [CrossRef]
19. Handa, K.; Ikeuchi, K.; Morimoto, F. Temperature-dependent wear of tread-braked railway wheels. *Wear* **2020**, *452–453*, 1–9. [CrossRef]
20. Sayeed Ahmed, G.M.; Algarni, S. Design, Development and FE thermal analysis of a radially grooved brake disc developed through direct metal laser sintering. *Materials* **2018**, *11*, 1211. [CrossRef]
21. Yevtushenko, A.; Kuciej, M.; Topczewska, K. Frictional heating during braking of the C/C composite disc. *Materials* **2020**, *13*, 2691. [CrossRef]
22. Yevtushenko, A.A.; Grzes, P. Initial selection of disc brake pads material based on the temperature mode. *Materials* **2020**, *13*, 822. [CrossRef]
23. Zou, H.; Ran, X.; Zhu, W.; Wang, Y.; Zhan, S.; Hao, Z. Tribological Behavior of Copper–Graphite Composites Reinforced with Cu-Coated or Uncoated SiO_2 Particles. *Materials* **2018**, *11*, 2414. [CrossRef]
24. Wasilewski, P. Frictional heating in railway brakes: A review of numerical models. *Arch. Comput. Method E.* **2020**, *27*, 45–58. [CrossRef]
25. Yevtushenko, A.A.; Kuciej, M. One-dimensional thermal problem of friction during braking: The history of development and actual state. *Int. J. Heat Mass. Tran.* **2012**, *55*, 4148–4153. [CrossRef]
26. Yevtushenko, A.A.; Grzes, P. The FEM-modeling of the frictional heating phenomenon in the pad/disc tribosystem (a review). *Numer. Heat Transf. Part A Appl.* **2010**, *58*, 207–226. [CrossRef]

27. Günay, M.; Korkmaz, M.E.; Özmen, R. An investigation on braking systems used in railway vehicles. *Eng. Sci. Technol. Int. J.* **2020**, *23*, 421–431. [CrossRef]
28. Suresh Babu, A.; Siva Prasad, N. Coupled field finite element analysis of railway block brakes. *Proc. Inst. Mech. Eng. Part F J. Rail Rapid Transit* **2009**, *223*, 345–352. [CrossRef]
29. Konowrocki, R.; Kukulski, J.; Walczak, S.; Groll, W. Dynamic interaction of cleansing brake insert for high speed train—Experimental investigation. *WUT J. Transp. Eng.* **2013**, *98*, 279–289. (In Polish)
30. *UIC Code 548 Brakes—Requirements of Friction Test Benches for the International Certification of Brake Pads and Brake Blocks*, 2nd ed.; Union Internationale des Chemins de fer: Paris, France, 2016.
31. Wasilewski, P.; Kuciej, M. Comparative study on the effect of fibre substitution on the properties of composite railway brake shoe. In Proceedings of the 9th International Scientific Conference "BALTTRIB 2017", Kaunas, Lithuania, 16–17 November 2017.
32. Wasilewski, P. Full-scale dynamometer test of composite railway brake shoes—Study on the effect of the reinforcing fibre type. *Acta Mech. Autom.* **2018**, *12*, 204–208. [CrossRef]
33. Charron, F. *Partage de la Chaleur Entre Deux Corps Frottants*; Publications Scientifiques et Techniques du Ministere de L'air: Paris, France, 1943.
34. Faccoli, M.; Ghidini, A.; Mazzù, A. Experimental and numerical investigation of the thermal effects on railway wheels for shoe-braked high-speed train applications. *Metall. Mater. Trans. A* **2018**, *49*, 4544–4554. [CrossRef]

materials

MDPI

Article

Comparative Analysis of Temperature Fields in Railway Solid and Ventilated Brake Discs

Aleksander Yevtushenko [1], Michal Kuciej [1,*], Piotr Grzes [1] and Piotr Wasilewski [2]

[1] Department of Mechanics and Applied Computer Science, Faculty of Mechanical Engineering, Bialystok University of Technology (BUT), 45C Wiejska Street, 15-351 Bialystok, Poland; a.yevtushenko@pb.edu.pl (A.Y.); p.grzes@pb.edu.pl (P.G.)

[2] Frimatrail Frenoplast S.A., 15 Watykańska Street, 05-200 Majdan, Poland; pwasilewski@frimatrail-frenoplast.pl

* Correspondence: m.kuciej@pb.edu.pl; Tel.: +48-85-746-9200

Abstract: A new approach to numerical simulation using the finite element method (FEM) for the rotational motion of discs for railway vehicle disc brake systems was proposed. For this purpose, spatial models of transient heating due to the friction of such systems with solid and ventilated discs were developed. The performed calculations and the results obtained allowed justification of the possibility of simplifying the shape of the ventilated brake disc through elimination of ventilation channels. This contributes to a significant reduction in computational time, without compromising the accuracy of the results. The spatial and temporal temperature distributions in the ventilated and the solid disc of the same mass were analyzed. The share of energy dissipated due to convection and thermal radiation to the environment in relation to the total work done during a single braking was investigated. The maximum temperature values found as a result of computer simulations were consistent with the corresponding experimental results.

Keywords: railway disc brake; frictional heating; moving heat source; temperature; finite element method

Citation: Yevtushenko, A.; Kuciej, M.; Grzes, P.; Wasilewski, P. Comparative Analysis of Temperature Fields in Railway Solid and Ventilated Brake Discs. *Materials* **2021**, *14*, 7804. https://doi.org/10.3390/ma14247804

Academic Editor: Juraj Gerlici

Received: 2 November 2021
Accepted: 14 December 2021
Published: 16 December 2021

1. Introduction

Carrying out a computer simulation with the use of FEA of the transient temperature field of the brake consists of several steps. In addition to the precise determination of the operating parameters as well as the shape and dimensions of the friction elements, it requires simplifying assumptions while developing the calculation models. Modern computational units make it possible to include a number of interdependent quantities (e.g., coefficient of friction, thermophysical properties, heat transfer coefficient, hardness, average temperature of the nominal contact area, flash temperature, maximum temperature, etc.) in numerical calculations of the thermal problems of friction. These values correspond with the complex shapes of the brake pads and discs, and even entire assemblies of parts in the vicinity of the brake [1–3]. Such an approach, aimed at a comprehensive assessment of the temperature state, is based on a system of equations of thermal dynamics of friction and wear [4]. The basic problems of this system concern (1) the determination of the time profiles of pressure, velocity, and friction power density; (2) obtaining experimental and then analytical relations of the coefficients of friction and wear rate on temperature; (3) the experimental dependency of the temperature-dependent material properties; (4) the determination of the average temperature of the nominal contact area and the temperature of the real contact area (flash temperature) and eventually their sum—the maximum temperature [4].

Along with the increase in the complexity of the computational model of the process by generalizing the basic equations of the HDFW (heat dynamics of friction and wear) system to different braking modes and brake systems, the computation time may be significantly extended. In order to shorten that time, analytical or numerical models that do not take into account temperature changes in the circumferential direction (axisymmetric or 2D models) are used [5]. Omission of the circumferential coordinates allow us to obtain reliable results with a solid disc, associated with the pads in the shape of a ring segment. Then, by introducing the heat partition coefficient, the temperature is set separately for the disc (2D model) and the pads (3D model). In the case of a ventilated disc with an irregular shape of the nominal contact area (due to external contours and internal cuts), only 3D calculation models should be used to determine the temperature mode of the brake.

Changes in transient temperature fields of a passenger vehicle brake disc with exponentially increasing pressure and non-linearly decreasing disc angular velocity were analyzed in the article [6]. The cyclic heating of the surface of the friction ring (rubbing path) of the disc was modeled by means of a moving surface heat source with alternately changing areas of heating and convection cooling. The changes in the temperature, braking time, and heat flux densities on the working surface of the disc were presented. A similar approach was proposed in [5]. During the numerical simulation of a moving heat source, a program code developed by the authors was proposed, generating boundary conditions on the working surface of the disc in the MES MSC.Patran/MSC.Nastran environment. For this purpose, functions describing the change of the heat flux density heating the disc were developed separately for each of the finite elements uniformly distributed in the circumferential and radial direction. This approach provides an accurate representation of the frictional power density portion in the area of displacement of the pad relative to the stationary disc. A significant advantage of the proposed method is the possibility of adapting this technique to a ventilated disc, provided that its outer surface is consistent and homogenous in the area of heating. The disadvantage, however, is that it is not possible to perform calculations using the contact model of heating the brake, taking into account the simultaneous mutual influence of the pads and the disc on the temperature.

The approach proposed in [6] was adapted to the thermal-structural coupling model in order to determine not only the temperature field, but also to carry out transient stress analysis and to describe the phenomenon of hot spot formation [7]. The main purpose of the work was to develop a methodology for identifying thermal fatigue cracks. The calculations were carried out for the regular shapes of the solid disc and pads. It was shown that at a fixed point of the disc, the temperature evolution takes the form of cyclically repeating stages of rising and falling with each revolution of the wheel.

The temperature fields of the ventilated disc brake elements during a single braking were determined numerically in the software environment based on the ANSYS Workbench platform [8]. In the developed contact thermo-structural coupled computational model, the stopping time as well as the exponential increase in contact pressure and the nonlinear velocity change were known a priori before the analysis. This also includes the variable heat transfer coefficient from the surfaces of the pads and the disc. The simulations were carried out using the direct coupling method, which is an iterative calculation of temperature and stress fields. The temperature changes over time at selected points on the disc obtained by the calculations were consistent with the corresponding results obtained using thermocouples embedded in these points. The temperature time profiles on the working surface of the disc revealed distinct and regular oscillations, declining with the distance from the friction surface.

Transient temperature fields of the pad and ventilated disc were found using the finite element method, adapted in COMSOL Multiphysics® [9]. Using the heat partition coefficient, simulations of frictional heating were carried out separately for the disc and the pads. Although the calculations of the temperature of the pads did not require interference with the standard tools available in the commercial FEM software environment, special modules (Mathematics and Deformed Geometry) were used to determine the temperature of the disc, to represent the movement of the heating area of a complex shape on the friction surface of the stationary disc. In fact, the change in the rotational velocity of the ventilated disc and the deformation of the surface layer of its geometric model were related, allowing for an accurate representation of the heat flux density proportional to the power of friction forces during braking.

The temperature fields of the disc and pads during a single braking of a rail vehicle were analyzed in the article [10]. The process of frictional heating during braking was simulated on a dynamometer test bench. When developing the calculation model, the ventilated disc was replaced with a solid disc with the same outer dimensions and mass. The same mass was maintained by increasing the thickness of the heated layer parallel to the friction surface as well as by simplifying the area of the disc-hub interface. In the contact area of the pad with the disc, perfect thermal friction conditions were assumed, i.e., the temperature of the opposite surfaces at each point was equal, and the sum of the heat flux densities directed from the friction surface to the inside of the disc and the pads was equal to the friction power density. In each of the ten tested variants of the single braking, a high consistency of the maximum disc temperature value, calculated by means of FEM and measured using thermocouples, was achieved. It was found that replacing a ventilated disc with a solid disc of the same mass during a single braking provides a sufficient level of accuracy in finding the temperature.

Currently, two types of brake discs are used in railway vehicles: solid and ventilated (in radial or tangential direction). The discs are manufactured as monoblocks or as individual friction rings. According to [11], taking into account the dimensions of the axle-mounted railway discs (typically 640 mm in diameter) and thicknesses from 80 (steel) to 110 mm (cast iron), high air pumping leads to aerodynamic loses and energy consumption. As a result, operating costs and the negative impact on the environment increase. In the above-mentioned work, the disc was developed with both radial vanes and circumferential pillars, the so-called radial vanes/pillared design. As was shown, such a structure of the disc does not differ from the traditional one in terms of the amount of energy dissipated by convection, and at the same time it shows much better (about 50%) equivalent aerodynamic (air pumping) losses. It was shown that a parameter that must be taken into account when designing discs is the ratio of convective power dissipation to aerodynamic power losses. Using the computational fluid dynamics (CFD) method, the changes during braking of the parameters, such as aerodynamic (pumping) power loss, convective-heat dissipation (power loss), and the aerodynamic efficiency ratio dependent on the angular velocity of the disc, were analyzed. It was emphasized that the commonly used discs for rail vehicles were produced for years without significant changes, before the introduction of the advanced CFD methods.

At a given time, there are many different approaches, methods, and simplifications in the numerical modeling of temperature fields of a rotating disc and stationary brake pads. Their analyses are provided in review articles [12,13]. Undoubtedly, one of the most important requirements when developing such models, apart from the accurate determination of the maximum temperature and flash temperature occurring in the real contact areas of the pads with the disc, the variability of friction coefficients and wear intensity, etc., was to shorten the computational time. Recent achievements in the modeling of disc temperature fields, the main goal of which was to develop a model that is universal and does not require significant time to obtain results, are presented in [14]. The authors proposed a calculation scheme that they described as a uniformly distributed heat source method, abbreviated as the UDHS (uniformly distributed heat source) method [14]. The method, unlike those in the literature, aimed at finding a way to obtain accurate results in a short time. The basic simplification assumption of the UDHS method consists of adopting a homogeneous distribution of heat flux intensity on the friction surface and approximating the mutual cyclic motion of the disc and the pad with the cosine function. The obtained results were verified on the basis of temperature measurements using thermocouples on a full-scale test bench.

A similar but more advanced approach, based on the Gaussian mixture function (GMHS—Gaussian mixture heat source Method), was proposed in [15]. The method, as in [14], is based on the Gaussian function approximation of the heat flux intensity distribution at a specific point on the working surface of the disc. At a constant angular velocity, these cycles were evenly distributed in time, and at a linearly decreasing velocity, the maximum value of the flux decreased with each subsequent cycle. It should be emphasized that this method has a significant simplification. At a given moment, the entire friction surface of the disc (in the circumferential and radial direction) was simultaneously heated with a heat flux of one and the same value.

A similar approach, albeit for the 2D axisymmetric model, simplifying the 3D ventilated disc model, was proposed in [16]. Calculations were performed for the braking parameters, disc dimensions, and material properties, as adapted from [17]. The function approximating the successive passes of the pad relative to the disc was developed so as to accurately map the temporal profile of the heat flux density (rectangular waveform), determined on the basis of the exponentially increasing pressure and the nonlinearly decreasing angular velocity of the disc. The calculated temperature values were compared with the corresponding measurement data from [17], and a very good agreement was obtained.

Initiated by an axisymmetric heat load, the spatial temperature field of the ventilated disc was determined numerically by means of FEM and CFD in [18]. The wheel-mounted brake disc R920K for the ER24PC locomotive was considered. First, distributions of the heat transfer coefficient were found in the FLUENT environment, and then the obtained results were used as boundary conditions for thermal analysis with the heat partition coefficient. It was found that the ventilation channels are heated after a certain time (a few seconds) from the beginning of the braking process. This delay in a temperature rise resulted in a small effect of convection cooling. Parallel laboratory tests were used to verify the calculations. The determined temperature evolution during braking agreed well with the data from the experimental studies.

The developed numerical model of the disc brake with axisymmetric heating of the ventilated disc [18] was then used to determine the thermal stress fields formed in the disc during emergency braking and during braking when riding downhill with a constant velocity [19].

In this paper, a comparative analysis of the temperature mode of the ventilated and solid discs of a rail vehicle was carried out, considering a single brake application. For this purpose, two techniques for modeling the rotational motion of the disc or the displacement of the pad relative to a fixed disc are proposed. The problems related to the optimization of the selection of a temporal step of such a size that would allow for an accurate mapping of the oscillating temperature changes at a specific point of the working surface of the disc are discussed in detail. During the calculations, the condition of equality of the total friction work found numerically and the initial kinetic braking energy obtained during experimental tests on a full-scale dynamometer test bench are verified. The calculated temperature values are shown to be consistent with the corresponding thermocouple measurement data.

2. Statement of the Problem

The subject of the study is the temperature field generated due to friction in a disc brake of a railway vehicle (Figure 1). The friction pair consists of the tangential vane type brake disc (640 mm × 110 mm), with the reduced ventilation and a set of organic composite brake pads typically used in passenger coaches.

During a single braking, the pressure p was the same at each point of the contact area, increasing linearly in time t from zero at the initial time $t = 0$ to the value p_1 at $t = t_1$, then increasing stepwise to the nominal value p_s and remaining at this value until the end of the process $t = t_s$:

$$p(t) = \begin{cases} p_1 t/t_1, \, 0 \le t \le t_1, \\ p_s, \, t_1 \le t \le t_s, \end{cases} \quad (1)$$

where $p_1 = 0.5F_1/A_a$, $p_s = 0.5F_s/A_a$, F_1—the value of the clamping force $t = t_1$, $F_s = 1.05F_1$, and A_a—surface area of the nominal contact of the pad with the disc. The change of pressure in time (Equation (1)) corresponds to the following time profile of the angular velocity of the disc:

$$\omega(t) = \begin{cases} \omega_0 + (\omega_1 - \omega_0)t/t_1, & 0 \le t \le t_1, \\ \omega_1 + (\omega_s - \omega_1)(t - t_1)/(t_s - t_1), & t_1 < t < t_s. \end{cases} \tag{2}$$

where $\omega_0 \equiv \omega(0)$, $\omega_1 \equiv \omega(t_1)$, and $\omega_s \equiv \omega(t_s)$.

Figure 1. An example of a ventilated disc brake system of a railway vehicle mounted on the inertial test stand in the Railway Research Institute in Warsaw, Poland (picture courtesy of the Railway Research Institute in Warsaw).

As a result of friction, heat is generated in the contact area between the pad and the disc, causing them to heat up. When developing the calculation model, the following assumptions were made to determine the brake temperature field:

1. The pads are positioned symmetrically on both sides of the disc. Therefore, due to the existing load and geometric symmetry, only the half-thickness disc in combination with one pad is analyzed;
2. The thermal contact of friction between the brake pads and the disc is perfect, i.e., the temperature in the contact area is the same, and the sum of the heat flux densities directed normally from the friction surface to the inside of each element is equal to the friction power density;
3. The cooling of the free surfaces of the pads and the disc proceeds due to convection and thermal radiation to the surrounding air;
4. Thermophysical properties of materials as well as friction and heat transfer coefficients do not change under the influence of temperature.

The parameters and quantities relating to the pad and the disc are denoted by the subscripts "p" and "d", respectively, and the pad-disc friction pair is related to the cylindrical coordinate system of spatial variables (r, θ, z). On the basis of assumptions 1–3, the transient temperature field $T(r, \theta, z, t)$ of the brake was found from the solution of the following heat conduction equations of the parabolic type:

$$\frac{\partial^2 T}{\partial r^2} + \frac{1}{r}\frac{\partial T}{\partial r} + \frac{1}{r^2}\frac{\partial^2 T}{\partial \theta^2} + \frac{\partial^2 T}{\partial z^2} = \frac{1}{k_p}\frac{\partial T}{\partial t}, \ (r, \theta, z) \in \Omega_p, \ 0 < t \le t_s, \tag{3}$$

$$\frac{\partial^2 T}{\partial r^2} + \frac{1}{r}\frac{\partial T}{\partial r} + \frac{1}{r^2}\frac{\partial^2 T}{\partial \theta^2} + \frac{\partial^2 T}{\partial z^2} = \frac{1}{k_d}\left[\frac{\partial T}{\partial t} + \omega(t)\frac{\partial T}{\partial \theta}\right], \ (r, \theta, z) \in \Omega_d, \ 0 < t \le t_s, \tag{4}$$

where

$$\Omega_p = \left\{(r,\theta,z) \in R^3 : r_p \le r \le R_p,\ -\theta_0 \le \theta \le \theta_0,\ 0 \le z \le \delta_p\right\}, \quad (5)$$

$$\Omega_d = \left\{(r,\theta,z) \in R^3 : r_d \le r \le R_d,\ -\pi \le \theta \le \pi,\ 0 \le z \le \delta_d\right\}, \quad (6)$$

$k_{p,d}, r_{p,d}$ and $R_{p,d}$ are the coefficients of thermal diffusivity and the inner and outer radii of the elements, respectively, and $2\theta_0$—cover angle of the pad.

Taking into account assumption 4, in the area of contact of the pad with the disc,

$$\Gamma = \left\{(r,\theta) \in R^2 : r_p \le r \le R_p,\ -\theta_0 \le \theta \le \theta_0,\ z = 0\right\}, \quad (7)$$

The following conditions should be met [20]:

$$T(r,\theta,0^+,t) = T(r,\theta,0^-,t),\ (r,\theta) \in \Gamma,\ 0 < t \le t_s, \quad (8)$$

$$K_d \frac{\partial T}{\partial z}\bigg|_{z=0^-} - K_p \frac{\partial T}{\partial z}\bigg|_{z=0^+} = q(r,\theta,t),\ (r,\theta) \in \Gamma,\ 0 < t \le t_s, \quad (9)$$

where $K_{p,d}$—thermal conductivity of materials, and the friction power density was determined from the equation:

$$q(r,\theta,t) = fp(t)r\omega(t),\ (r,\theta) \in \Gamma,\ \ 0 < t \le t_s, \quad (10)$$

in which f—coefficient of friction, and $p(t)$ and $\omega(t)$—pressure (Equation (1)) and angular velocity (Equation (2)) time profiles, respectively.

Based on Assumption 2, the time profiles of the heat flux densities directed normally from the free surface $\Gamma_{p,d}$ of the pad and the disc to the environment were written as

$$q_{p,d}^{diss}(t) = q_{p,d}^{conv}(t) + q_{p,d}^{rad}(t),\ \ 0 < t \le t_s, \quad (11)$$

where

$$q_{p,d}^{conv}(t) = h(T_a - T),\ q_{p,d}^{rad}(t) = \varepsilon_{p,d}\sigma(T_a^4 - T^4), \quad (12)$$

h—heat transfer coefficient, $\varepsilon_{p,d}$—emissivity of materials, $\sigma = 5.67 \cdot 10^{-8}\ \mathrm{W\,m^{-2}K^{-4}}$—Stefan-Boltzmann constant, and T_a—ambient temperature.

Initially, the pad and the disc were at temperature T_0.

Taking into account the form of the friction power density q (Equation (10)), the time course of the friction work during braking W was determined from the following equation:

$$W(t) = 2A_a \int_0^t q(r_{eq},0,\tau)d\tau,\ \ 0 < t \le t_s, \quad (13)$$

where

$$r_{eq} = \frac{2(R_p^3 - r_p^3)}{3(R_p^2 - r_p^2)}, \quad (14)$$

A_a—surface area of contact of the pad on one side of the disc.

On the other hand, taking into account Equations (11) and (12), the changes in the braking time of the amount of heat $W_{p,d}^{diss}$, dissipated from the free surfaces to the environment, were determined from the equations:

$$W_{p,d}^{diss}(t) = 2[W_{p,d}^{conv}(t) + W_{p,d}^{rad}(t)],\quad 0 < t \le t_s, \quad (15)$$

where

$$W_{p,d}^{conv}(t) = A_{p,d} \int_0^t q_{p,d}^{conv}(\tau)d\tau,\ \ W_{p,d}^{rad}(t) = A_{p,d} \int_0^t q_{p,d}^{rad}(\tau)d\tau, \quad (16)$$

$A_{p,d}$—half of the free surface areas $\Gamma_{p,d}$ of two pads and one disc.

3. Development of the 3D CAD Geometric Model and Generation of the Finite Element Mesh

The solution of the boundary value problem of heat conduction (Equations (3)–(10)) was obtained numerically using FEM, adapted in the COMSOL Multiphysics® software package [21]. The calculations were carried out on the basis of the three models, with a different degree of simplification of the shape of the disc, consisting mainly of taking into account or not taking into account the ventilation channels of the disc. These are models where:

I. The ribs forming the channels were replaced with a solid area of reduced thickness so as to keep the mass of the disc unchanged. In this model, the relative rotational movement of the disc and the pad takes place by assigning a variable velocity field at each point of the disc area at a fixed pad [10];
II. The actual shape of the disc with ventilation channels was taken into account. The rotational movement of the disc was replaced by the movement of the pad relative to the stationary disc. This is done by changing the outer contours of the disc in the vicinity of the contact surface according to the simulated displacement of the pad;
III. The solid disc from model I was considered, but with the relative displacement of the components according to model II.

One of the goals of this work was to carry out a comparative analysis of the brake temperature fields with a ventilated disc and a solid disc, representing a ventilated disc in a simplified form. A high convergence (relative maximum temperature difference did not exceeded 0.01%) of the results for the solid disc, obtained on the basis of the well-known and well-approved model I, was established with the corresponding temperature values obtained with the use of the new model III. This allowed us to be sure of the correct choice of the proposed modelling approach and the COMSOL Multiphysics® tools applied in model III, and then II. Therefore, only the results obtained from model III (solid disc) and model II (ventilated disc) will be presented and discussed (Figure 2).

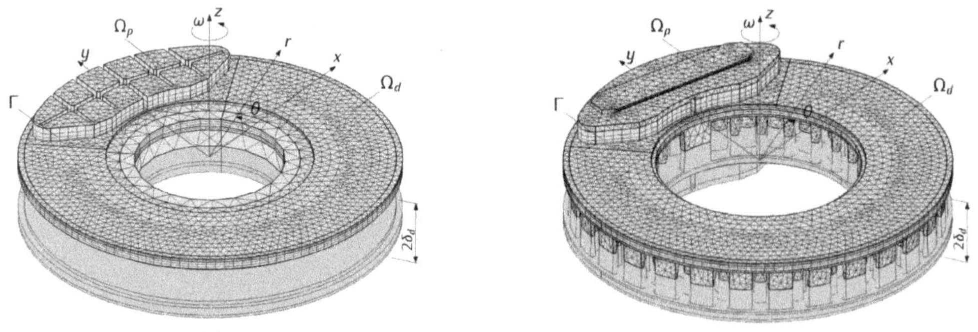

Figure 2. FE meshes and geometric 3D CAD models of the solid (model III) (**a**) and ventilated (model II) (**b**) types of disc brakes.

The 3D CAD geometric models of the tested braking system used to generate the finite element grids were created using the SOLIDWORKS® software (Version 2020). The construction of the CAD geometric model of the disc required taking into account the specificity of assigning properties and parameters of the Deformed Geometry tools of the COMSOL Multiphysics® program. In the geometric model of the pad, the wall inclinations were ignored due to the considerable distance from the friction surface (more than 10 mm). The backing plate, which strengthens the lining and allows it to be properly mounted in the holder, was replaced with the pad material. Holes were also omitted to improve the stability of the connection between the sheet and the pad. The brake pad holder was also not taken into account. Due to the considered short-term braking mode, all these simplifications did not affect the obtained results, but allowed us to reduce the total number of finite elements and thus significantly (several times) shortened the computational time. Considering the possible practical application of the proposed numerical FE models, this is of key importance.

The geometrical models of the ventilated and solid discs were transferred from SOLIDWORKS® to COMSOL Multiphysics® using and the Live Link™ tool (Version 5.3). At the same time, all modifications made in SOLIDWORKS® could be constantly updated in COMSOL Multiphysics®. Then, a second-order finite element mesh (quadratic Lagrange) was generated. The overall final FE mesh was created in a few steps. An important criterion for changing the size of the element of the parts of the assembly was the expected high temperature gradients resulting from the frictional heating process corresponding with the properties of materials. It was necessary to include several times lower thermal conductivity of the pad compared to the steel disc. Therefore, after creating the 2D triangle elements on the plane friction surface (15 mm was the maximum element size), the sweep feature with a predefined distribution type (arithmetic sequence) was created. In model III, the number of the finite elements within the area of the solid disc was equal to 179 tetrahedral and 24,036 prism elements, and the pad was divided into 2048 prism elements. Whereas in model II, the disc consisted of 33,518 tetrahedral, 50 pyramid, 14,595 prism and 140 hexagonal elements. In this model, the pad consisted of 4669 tetrahedral and 2048 prism elements.

4. Modeling the Rotational Motion of the Disc

In the case of a solid disc with no shape changes in the circumferential direction, the numerical modeling of temperature fields in FEM-based programs usually requires the following:

1. Separating the pad friction surface and thus defining the nominal contact area (heating area);
2. Formulating the boundary conditions on each surface (heat flux density, convection cooling, thermal insulation, etc.);
3. Defining a mathematical relationship for the velocity of each point of the disc during braking in order to obtain the effect of its rotation with respect to the stationary heating area.

The difficulty of describing the pad-disc frictional heating process arises when the disc is ventilated. In this case, its shape changes in the circumferential direction, caused by the free spaces alternating with elements constituting the ribs. This requires writing in-house program code to generate the function of changing the heat flux density separately for each element in the area of the surface of the rubbing path of the disc [5,22].

This article proposes the adaptation of special tools, available in the COMSOL Multiphysics® software, to deform the geometry in such a way that it corresponds to the rotational motion of the ventilated disc in relation to the stationary pad. This task was carried out for the disc, specially divided into objects for this purpose. The contours of the isolated heating area of the disc with a complex shape, described by the boundaries of the pad's friction surface, were deformed during braking. In spite of the rotation of the disc, the conditions of perfect thermal friction contact, with the fixed pad remaining in the initial position, were maintained. The described deformation occurred only for selected elements, constituting a solid layer of a certain thickness (the condition of the possibility of deformation is the continuity of the material). Due to the presence of ribs, the adjacent layer towards the inside of the disc, parallel to the contact surface, remained stationary (like the pad); see Figure 3a,b.

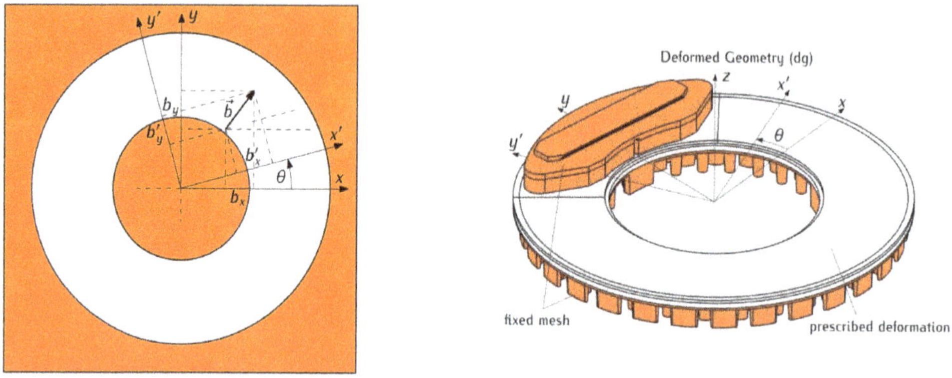

(**a**) (**b**)

Figure 3. Motion scheme for the pad-disc system adapted in COMSOL Multiphysics® software: (**a**) deformation of the layer of the disc in plane; (**b**) selection of the 3D objects of the braking system in Deformed Geometry (dg).

The problem formulated in this way required the use of specially adapted tools, available in the model for Deformed Geometry in COMSOL Multiphysics® software. When describing the rotational motion of the disc contours, taking into account the heating area, the following transformation equations were used for the components (b_x, b_y) and $(b_{x'}, b_{y'})$ of any vector $\vec{b}$ for the rotation of the Cartesian coordinate system $Ox'y'$ by the angle θ with respect to the stationary orthogonal system Oxy [23,24]:

$$b_{x'} = b_x \cos\theta + b_y \sin\theta,\ b_{y'} = -b_x \sin\theta + b_y \cos\theta, \tag{17}$$

or

$$b_x = b_{x'} \cos\theta - b_{y'} \sin\theta,\ b_y = b_{x'} \sin\theta + b_{y'} \cos\theta. \tag{18}$$

Then, in the system Oxy, the rotation of the disc contours with a separate heating area, corresponding to the rotation of the lining relative to the stationary disc, was described by a vector $\vec{d}$ with the following components:

$$d_x = b_x - b_{x'},\ d_y = b_y - b_{y'},\ d_z = 0. \tag{19}$$

During the calculations with the use of Equations (17)–(19), it should be remembered that the circumferential variable θ is not constant, but has a sense of the angular distance $\vartheta(t)$, which changes with time. According to Equation (2), the braking process consisted of two stages with linearly decreasing rotational velocity ω with constant, different at each stage, deceleration values. Therefore, in the COMSOL Multiphysics® program, first the rotational velocity values were calculated from Equation (2) with time steps $\Delta t = 0.001$ s, and then the appropriate values of the angular velocity $\omega(t)$ of the contact area Γ were determined on their basis:

$$\vartheta(t) = \int_0^t \omega(\tau)d\tau,\ 0 < t \le t_s. \tag{20}$$

The first calculations according to the above-described scheme showed that the temperature oscillations resulting from the displacement of the pad relative to the disc are so frequent that they require a minimum data reading step of no more than 0.002 s. Such a high concentration of data reading points for the braking process lasting several seconds leads to a significant increase in the computation time. In order to reduce this time, while maintaining the correct course of temperature evolution, it was decided to perform additional optimization calculations out of the COMSOL Multiphysics® environment. For this purpose, a code for determining the angular velocity $\omega(t)$, $0 < t \leq t_s$ was written in the Python programming and scripting language (Equation (2)). This made it possible to generate a table of values of $W_{p,d}^{diss}$ with a time step of 0.001 s. It should be noted that it can also be implemented on an ongoing basis in the program, using the tools for solving differential equations for Global ODEs (ordinary differential equations) and DAEs (differential-algebraic equations) (calculation of the angular distance) from the Mathematics module; however, after preliminary thermal analyses, it was found that the results obtained in this way were not accurate enough. Thus, in order to determine the angular distance ϑ (Equation (20)), the integration of the tabulated angular velocity ω was performed. The optimization of the temperature calculation time was performed using two time step values. A list of points was generated, such that the five time intervals before and after the left (leading) edge as well as the right (exit) edge of the pad had a time step of $\Delta t = 0.002$ s, and the remaining ones had a step of $\Delta t = 0.01$ s. A very small time step of $\Delta t = 0.00001$ s was applied to create this list in the original, in-house Python code. In addition, the data format in COMSOL Multiphysics® required writing the range in the form (t_0,Δt,t_{end}), where t_0 and t_{end}(the beginning and the end value of the time interval, respectively) had to be taken into account.

5. Numerical Analysis

Theoretical calculations were performed for the tangential vane type brake disc (640 mm × 110 mm) with the reduced ventilation and two organic composite brake pads (Figure 1). This friction pair was modeled using the solid disc (models I and III) and the ventilated disc (model II) with the same cover angle of the pad $2\theta_0 = 86.9°$ at the ambient temperature $T_a = 20\,°C$ (Figure 2). The values of the thermophysical and mechanical constants of the materials as well as the dimensions of the pads and the disc shown in Table 1 were used. The dimensions of the systems with the solid and ventilated disc are also presented in Figure 4a,b, respectively. In order to verify the obtained theoretical temperature values, corresponding experimental data were used, obtained on a full-scale dynamometer at the Railway Research Institute in Warsaw, Poland (Table 2). The temperature of the ventilated disc was measured using thermocouples in accordance with the UIC (International Union of Railways) Leaflet 541-3. The rotational velocity of the disc was adjusted for the wheel of a rail vehicle with a diameter of $R_w = 890$ mm. The six thermocouples T_n, $n = 1, 2, \ldots, 6$ were situated symmetrically 1 mm below the friction surface on both sides of the ventilated disc (Figure 4c). The radial distances from the axis of rotation of the disc to the individual measurement points were equal to $r = 207$ mm for $\hat{T}_1$ and $\hat{T}_2$, $r = 247$ mm for $\hat{T}_3$ and $\hat{T}_4$, and $r = 287$ mm for $\hat{T}_5$ and $\hat{T}_6$. In the circumferential direction, the distance between successive measurement points was 120°.

Table 1. Input parameters for numerical simulation.

Parameter	Disc	Pad
thermal conductivity, K(W m^{-1} K^{-1})	51	1.59
specific heat at constant pressure, c(J kg^{-1} K^{-1})	500	770
density, ρ(kg m^{-3})	7100	2450
surface emissivity, ε	0.28	0.8
heat transfer coefficient, h	100	100
initial temperature, T_0	28	28
thickness solid/ventilated, δ (mm)	55	25/35
outer radius, $R_{p,d}$ (mm)	320	303
inner radius, $r_{p,d}$ (mm)	175	178
inner radius of the hub solid/ventilated, $r_{p,d}$ (mm)	127.5/168.25	178
equivalent (friction) radius, r_{eq} (mm)	-	247
surface area of contact of the pad on one side of the disc, A_a(m^2)	-	0.034
surface area of the pad (m^2)	-	0.241
surface area the solid/ventilated disc (m^2)	1.287/2.002	-

Table 2. Experimental data.

No.	W_0 (kJ)	M_b (kg)	V_0 (km h^{-1})	ω_0 (rad s^{-1})	ω_1 (rad s^{-1})	ω_s (rad s^{-1})	F_1 (kN)	F_s (kN)	f (m)	l_1 (m)	l_s (m)	t_1 (s)	t_s (s)	$T_{6,exp}$ (°C)	$T^{I}_{6,theor}$ (°C)	$T^{II}_{6,theor}$ (°C)	$T^{III}_{6,theor}$ (°C)
1	985.8	6066.7	64.9	40.5	31.9	0.6	47.1	49.3	0.31	71.7	143.4	4.2	14.6	84.5	96.5	96.7	96.5
2	2085.3	5951.5	95.3	59.5	50.9	0	47.5	49.5	0.31	179.8	286.5	4.2	19.9	153	145.8	147.2	145.8

The average values of the coefficient of friction f were calculated according to the card UIC 541-3.

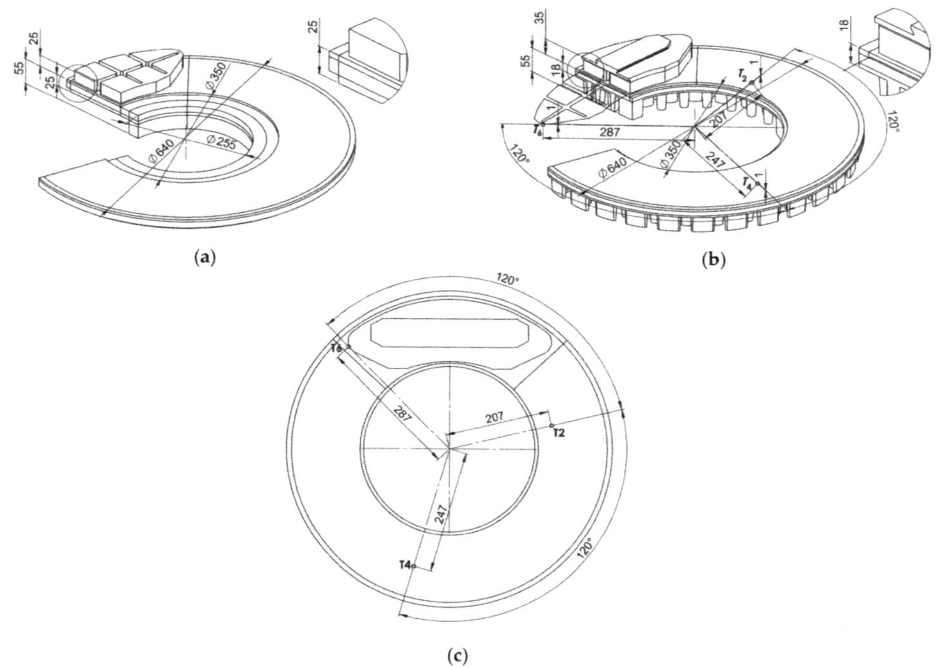

Figure 4. Dimensions of the solid (**a**) and the ventilated (**b**) disc brakes, with locations of the thermocouples (**c**).

Numerical simulations and experimental measurements with the use of thermocouples were performed for two (no. 1 and no. 2) brake applications with different, experimentally recorded operating parameters (Table 2). This table contains the values of the input parameters, such as simulated energy per braking system, brake cylinder filling time t_1 and braking times t_s, with their respective braking distances l_1 and l_s, forces F_1 and F_s, and angular velocities ω_0, ω_1 and ω_s. Table 2 also includes the temperature values at the point corresponding to the location of the thermocouple T_6, determined experimentally $T_{6,exp}$ and obtained numerically $T^{\mathrm{I}}_{6,\mathrm{theor}}$, $T^{\mathrm{II}}_{6,\mathrm{theor}}$ and $T^{\mathrm{III}}_{6,\mathrm{theor}}$ on the basis of the three above-mentioned FE calculation models. The point corresponding to the position of the thermocouple T_6 was chosen because of the highest temperature reached at this point compared to the temperature at the other five points. In the case of a solid disc, the relative percentage difference between the values $\left|T^{\mathrm{III}}_{6,\mathrm{theor}} - T^{\mathrm{II}}_{6,\mathrm{theor}}\right|$ is 0.2% and 1% for braking applications no. 1 and no. 2, respectively. The relative difference between the values of the temperature of the ventilated disc, determined theoretically $T^{\mathrm{III}}_{6,\mathrm{theor}}$ and experimentally $T_{6,\mathrm{exp}}$, is fully acceptable and is equal to 14.4% (no. 1) and 3.8% (no. 2).

The time profiles of the angular velocity $\omega(t)$ (Equation (2)) and the angular distance $\vartheta(t)$ (Equation (20)), obtained with the input parameters from Table 2 for the two variants of braking are presented in Figure 5. The corresponding temperature changes in time at a selected point $\mathrm{P}(-197\,\mathrm{mm}, 208\,\mathrm{mm})$ on the disc friction surface are shown in Figure 6. The temperature evolutions obtained on the basis of models III (solid disc) and II (ventilated disc) are marked in red and blue, respectively. These evolutions consist of a series of cycles (oscillations), each of which comprises two stages: first increase, and then decrease of the temperature. The first stage relates to the situation when the measuring point is still inside the nominal contact area during the movement of the pad, and the second—when the point P is already outside the contact area of the pad with the disc.

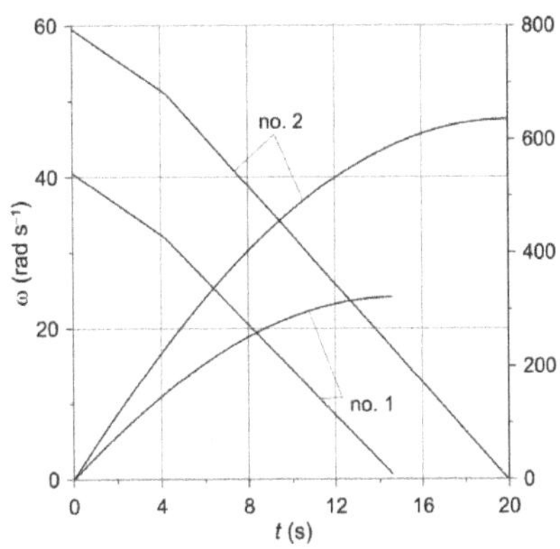

Figure 5. Changes in angular velocity $\omega(t)$ and angular distance $\vartheta(t)$ of the disc.

Due to the lower initial velocity ω_0 and shorter braking time t_s (Table 2), the number of cycles in the case of braking no. 1 (Figure 6a) is smaller than during braking no. 2 (Figure 6b). With each new cycle during the first stage $0 \leq t < t_1$ of the braking process, corresponding to a linear increase in pressure, the maximum temperature increases, suffers a slight decrease at $t_1 = 4.2\,\mathrm{s}$ as the pressure force increases abruptly from value F_1 to nominal F_s, and then increases again until the time corresponding to approximately half of the braking time ($t \cong 0.5t_s$). After this, a decrease in the maximum temperature is observed with each subsequent cycle, lasting until the end of the process $t = t_s$. It should be noted that such a time course of temperature during braking is determined by the time profile of the friction power density q (Equation (10)), including the stages increasing with the start of braking, reaching the maximum value, and the next reduction to the end time $t = t_s$. During braking no. 1, the maximum temperature of the ventilated disc (113.22 °C) is practically the same as the temperature of the solid disc (113.07 °C). For braking no. 2, the maximum temperature values were 170.47 °C and 169.61 °C, respectively. At the end of the braking process $t = t_s$, the temperature of the ventilated disc was 76.96 °C and for the solid disc 74.83 °C (braking no. 1), whereas for braking no. 2 it was 110.14 °C and 104.83 °C, respectively.

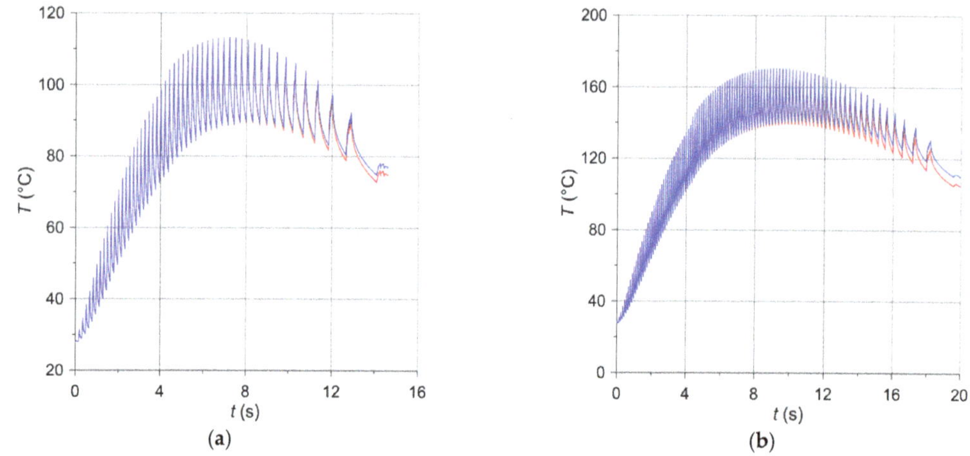

Figure 6. Temperature changes in time at location ($x = -197$ mm, $y = 208$ mm, $z = 0$ mm) for the solid (red lines) and the ventilated (blue lines) brake discs: (**a**) braking no. 1; (**b**) braking no. 2.

The oscillation amplitudes of the temperature time profiles decrease along with the distance from the disc friction surface, and at the depth above 5 mm they disappear altogether (Figure 7). During both brake applications, the temperature values determined with the use of the computational models of the solid and ventilated disc are very similar. Therefore, it can be concluded that the temperature distribution in the circumferential direction in both the solid and the ventilated disc at this depth is homogeneous. Therefore, the heating of the ventilated disc can be modeled using the simplification consisting of the fact that a layer with a thickness of about 5 mm rotates, and the remaining area of the disc towards its interior may be stationary. It would be far easier to simulate the rotation of the disc in relation to stationary pads, using for this purpose only the basic tools of the FEM programs, adapted in the COMSOL package. The available Translational Motion tool requires only the velocity field, with no additional disc deformation options. The obtained maximum temperatures at the point $P_6(-197\,\text{mm}, 208\,\text{mm}, -1\,\text{mm})$ corresponding to the location of the thermocouple T_6 for the ventilated and solid discs are, respectively, 96.7 °C and 96.5 °C during braking no. 1 and 147.2 °C and 145.8 °C during braking no. 2. These values are collated in Table 2 with the experimental data measured by the thermocouples.

The temperature distribution along the thickness of the disc for the selected time moments at point P are presented in Figure 8. For both brake applications, at the initial stage of the process, lasting approximately 6 s, the differences in temperature values of the solid and ventilated discs are insignificant and are observed at the depths greater than 10 mm. After 10 s, these differences appear on the friction surface and increase when moving away from it. The biggest temperature difference between the solid and ventilated discs occurred at the depth $|z| = 18$ mm at the time moment $t = 14$ s and was equal to 5.5 °C (Figure 8a) and 10.4 °C (Figure 8b) for braking no. 1 and no. 2, respectively.

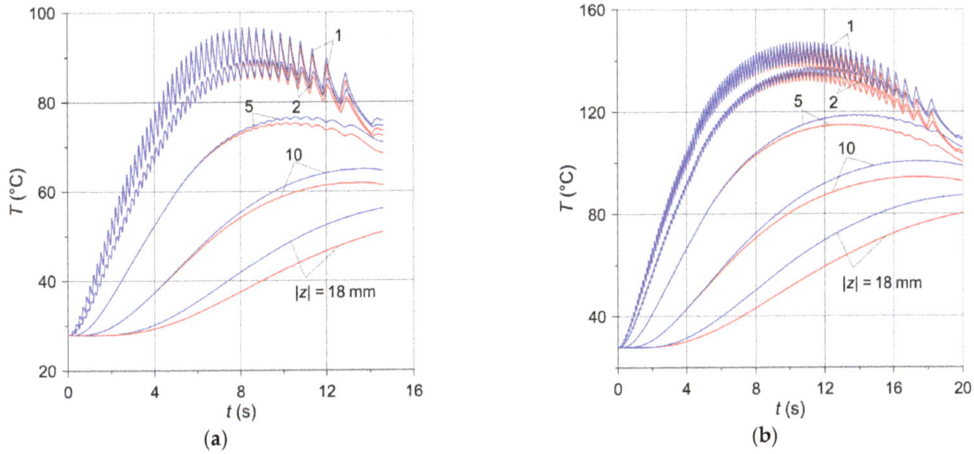

Figure 7. Temperature changes in time at location ($x = -197$ mm, $y = 208$ mm) under the friction surface of the solid (red lines) and ventilated (blue lines) brake discs: (**a**) braking no. 1; (**b**) braking no. 2.

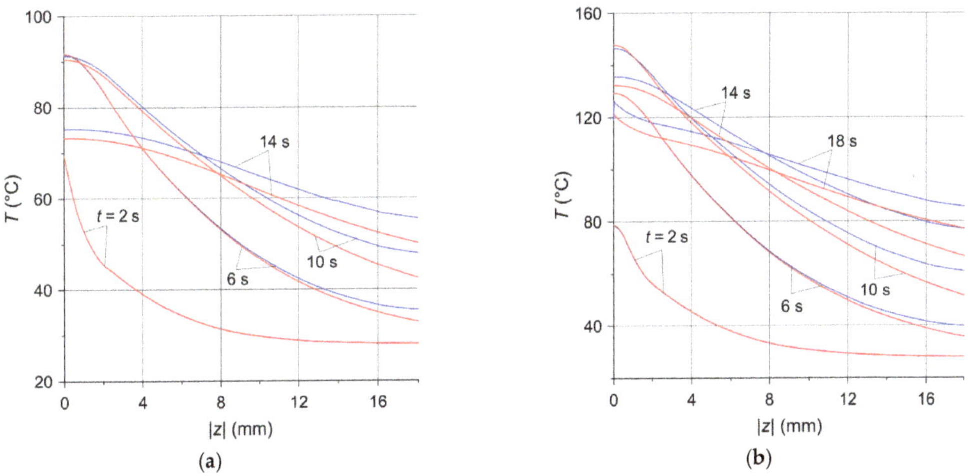

Figure 8. Temperature distributions in axial (z) direction of the solid (red lines) and ventilated (blue lines) disc at location ($x = -197$ mm, $y = 208$ mm) from (**a**) braking no. 1; (**b**) braking no. 2.

The evolutions of the work W (Equation (13)) done during braking are shown in Figure 9. Increasing monotonically with time, at the moment of stopping, the calculated values of the friction work are equal to 989.4 kJ (Figure 9a) and 2096 kJ (Figure 9b) for braking no. 1 and no. 2, respectively. These values agree well with the relevant experimental data contained in Table 2, concerning the initial kinetic energy of the system W_0 (985.8 kJ, 2085 kJ). Such compliance additionally confirms credibility of the obtained numerical results.

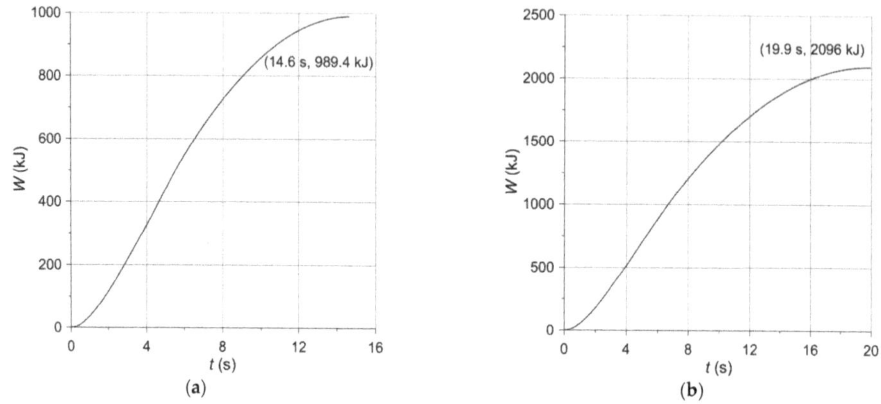

Figure 9. Kinetic energy W per one braking system converted into heat during (**a**) braking no. 1; (**b**) braking no. 2.

Based on Equations (16) and (17), the changes in the braking time of the heat $W_{p,d}^{diss}$ dissipated to the environment from the free surfaces of the pad and the disc according to two mechanisms—convection and radiation—are shown in Figures 10 and 11. For both brake applications, the heat $W_{p,d}^{rad}$ (Figures 10a and 11a) dispersed to the environment by radiation is small compared to the heat $W_{p,d}^{conv}$ (Figures 10b and 11b) transferred by convection, and even smaller compared to the work of friction W (Figure 9), converted into heat. Such small, almost negligible, amounts of energy dissipated due to thermal radiation were expected and are physically justifiable. The energy absorption from the surfaces of adjacent elements is of utmost importance. It should also be noted that the heat emitted as a result of radiation is reflected between the inner walls of the ventilated disc and is not dissipated to adjacent elements other than the disc. Therefore, significant amounts of thermal radiation energy are dissipated only through the exposed surfaces of the disc and partly internal and external cylindrical surfaces [11]. Despite the higher emissivity of the material of the pad than the disc material, the low thermal conductivity of the pad and its small free surface prevent it from playing a significant role in heat dissipation. In turn, the heat dissipated through convection from the free surfaces of the disc to the environment is small in relation to the total friction work during braking. The above shows that convection during a short-term single braking (from a few to several seconds) does not have any significant importance in the overall thermal balance of the brake. This explains, in particular, that the temperature of the ventilated disc is higher than that of the solid disc (Figure 6). As convection and radiation during short-term braking (from a few to several seconds) do not play a significant role in the overall thermal balance of the brake, thermal conductivity plays the main role in heat dissipation. The thickness of the material layer of the disc adjacent to the pad in a ventilated type of the disc is smaller than in a solid disc of the same mass. Hence, there is less heat dissipation by conduction in the ventilated disc.

The above conclusions are confirmed by the temperature distributions on the external surfaces of the ventilated disc and the solid disc of the same mass in selected time moments $t = 0.4$ s, 2 s, 10 s, and 14.6 s during braking no. 1, as shown in Figures 12–15. A slight difference of 0.1 °C between the maximum temperature value reached in the ventilated disc and the solid disc appears 2 s after the start of braking (Figure 13), and the greatest difference is 2.5 °C, i.e., about 3% of the maximum temperature value reached—at the end $t = 14.6$ s (Figure 15). In terms of quality, the temperature distributions in both types of discs (solid and ventilated) are the same. The absence of any displacements of the heating area in the ventilated and solid discs results from the adapted angular displacement in time, determined independently before the temperature calculations of the COMSOL Multiphysics® software. If this simulation step was omitted and all calculations were performed only in the COMSOL Multiphysics® environment with a standard time step, slight differences in the position of the heating area would be observed and thus would also be seen in the temperature distributions.

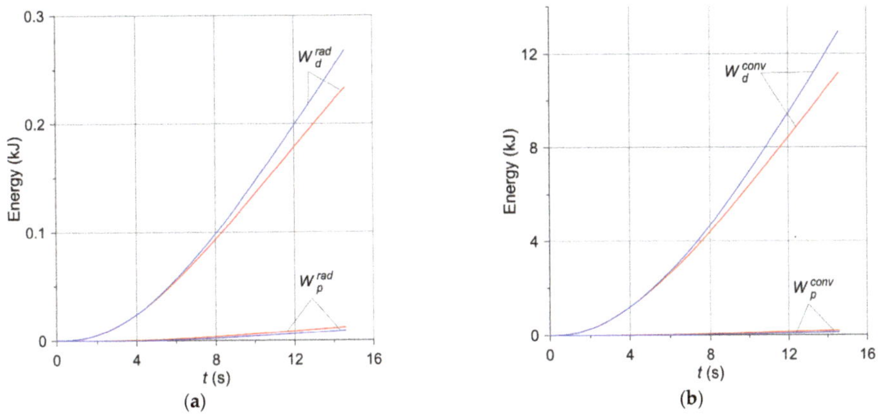

Figure 10. Changes in energy dissipated during braking no. 1 (solid disc—red lines, ventilated disc—blue lines) through (**a**) thermal radiation; (**b**) convection.

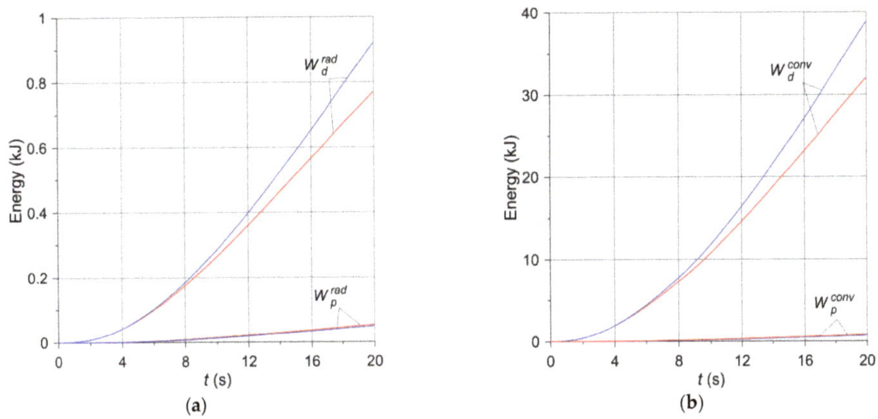

Figure 11. Changes in energy dissipated during braking no. 2 (solid disc—red lines, ventilated disc—blue lines) through (**a**) thermal radiation; (**b**) convection.

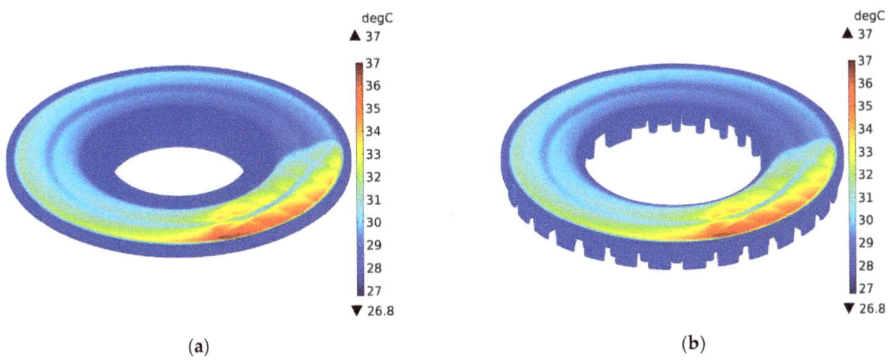

Figure 12. Temperature distribution in the solid (**a**) and ventilated (**b**) brake disc after $t = 0.4$ s of braking no. 1.

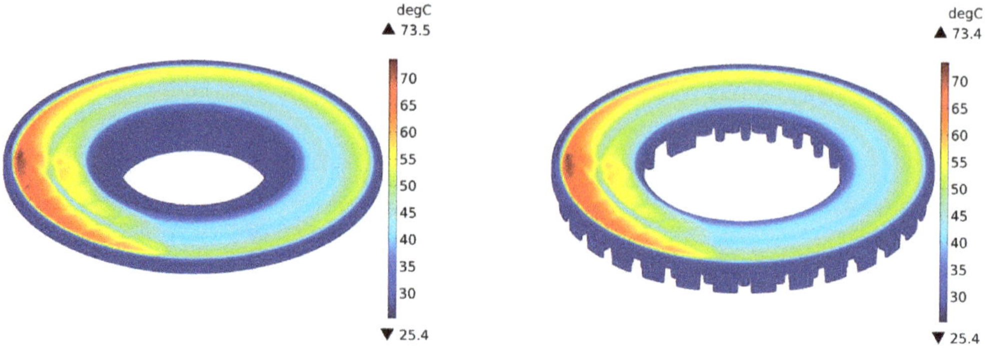

(**a**) (**b**)

Figure 13. Temperature distribution in the solid (**a**) and ventilated (**b**) brake disc after $t = 2$ s of braking no. 1.

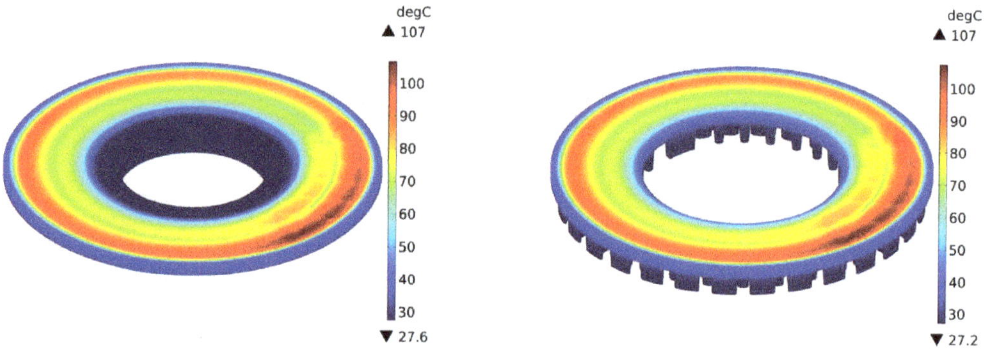

(**a**) (**b**)

Figure 14. Temperature distribution in the solid (**a**) and ventilated (**b**) brake disc after $t = 10$ s of braking no. 1.

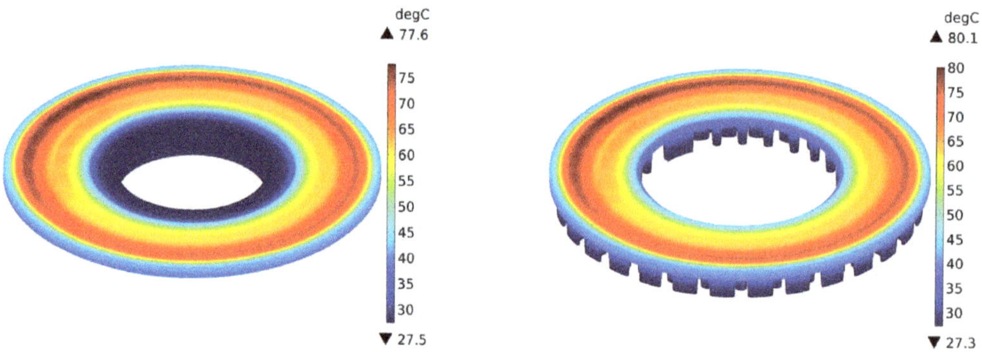

(**a**) (**b**)

Figure 15. Temperature distribution in the solid (**a**) and ventilated (**b**) brake disc after $t = 14.6$ s of braking no. 1.

6. Summary and Conclusions

Two calculation models of the ventilated disc temperature were developed, differing in their taking into account its rotational motion. For comparative purposes, in one of these models, the shape of the ventilated disc was simplified by replacing the pillars with an annular area, assuming that the new disc thus formed has the same mass as the ventilated disc. Based on the results of the simulation of the temperature mode during two single braking events (no. 1 and no. 2), differing in the initial kinetic energy, it was established that

- The maximum temperature values obtained with the use of both numerical models are consistent with the corresponding results of measurements with thermocouples on the full scale inertial test bench;
- The differences in the maximum temperature values of the ventilated disc (model II) and the solid disc (models I and III) are negligible and amount to about 0.15 °C (no. 1) and 0.86 °C (no. 2). At the end of the process, these differences are 2.13 °C (no. 1) and 5.31 °C (no. 2);
- The amplitude of temperature oscillations caused by the relative motion of the pad and the disc with a linear increase in pressure is the highest in the middle of the braking process, and the oscillations themselves decrease with the distance from the friction surfaces and appear in the disc to a depth of about 5 mm;
- Temperature distribution in both discs indicates the possibility of disregarding the rotation of the disc material at a depth exceeding 5 mm. In our opinion, the result is significant, since it allows for a simple modeling of the rotation of the axisymmetric ring area of the disc located at a distance from the friction surface and allows the area below this depth to remain stationary;
- Even after several seconds of braking, the disc is not heated evenly across its entire thickness. At the end of the process, only the layer reaching one-fourth of the thickness of the analyzed ventilated disc is heated;
- The heat dissipation due to convection and thermal radiation is not significant in the total heat balance during braking;
- Unforced convection heat exchange with the environment during short-term braking does not lead to significantly better cooling of the ventilated disc as compared to a solid disc. This effect may become significant only after a few or several dozen seconds after stopping [25,26];
- The performed calculations justify the use of the constant heat transfer coefficient with a single short-term braking.

The obtained results, apart from the differences between the ventilated and the solid disc, may suggest possible discrepancies in the temperature fields in the case when the ventilated disc, combined with the irregularly shaped pad, is replaced with an axially symmetric disc brake model. Such a brake model should actually be two models, with separate heating of the disc and the pad. In the case of the disc, the cross-section of the solid disc created in this work should be replicated; there is no other possibility of replacing the ribs (pillars) of the ventilated disc. On the other hand, it would be difficult to reproduce the cross-section of the pad, which would be a model of the axisymmetric heating of the disc.

Author Contributions: Conceptualization, A.Y. and P.G.; methodology, P.G.; software, P.G.; validation, M.K.; formal analysis, A.Y. and P.W.; investigation, P.G.; resources, M.K. and P.W.; data curation, P.G., P.W.; writing—original draft preparation, A.Y. and P.G.; writing—review and editing, A.Y., P.G. and P.W.; visualization, A.Y. and P.G.; supervision, M.K.; project administration, M.K.; funding acquisition, M.K. All authors have read and agreed to the published version of the manuscript.

Funding: This research was funded by National Science Centre, Poland, grant number 2017/27/B/ST8/01249, and project financing through the program of the Minister of Science and Higher Education of Poland, named "Regional Initiative of Excellence" in 2019–2022, project number 011/RID/2018/19, amount of financing 12,000,000 PLN.

Institutional Review Board Statement: Not applicable.

Informed Consent Statement: Not applicable.

Data Availability Statement: Not applicable.

Acknowledgments: We would like to thank the Railway Research Institute in Warsaw, Poland, for permission to use the photograph shown in Figure 1.

Conflicts of Interest: The authors declare no conflict of interest.

Nomenclature

A_a surface area of contact of the pad on one side of the disc (m^2)
$A_{p,d}$ half of the free surface areas $\Gamma_{p,d}$ of two pads and one disc (m^2)
c specific heat (J kg^{-1}K^{-1})
f coefficient of friction (dimensionless)
F_1 clamping force at time $t = t_1$ (N)
F_s nominal clamping force (N)
h heat transfer coefficient (W m^{-2}K^{-1})
k thermal diffusivity (m^{2}s^{-1})
K thermal conductivity (W m^{-1}K^{-1})
l_1 braking distance at time $t = t_1$ (m)
l_s total braking distance (m)
M_b mass per brake disc (m)
p contact pressure (MPa)
p_1 contact pressure at time $t = t_1$ (MPa)
p_s nominal contact pressure (MPa)
q specific friction power (W m^{-2})
r radial coordinate (m)
r, R inner and ouer radius, respectively (m)
r_{eq} equivalent (friction) radius of the contact region (m)
R_w outer (rolling) radius of the wheel (m)
t time (s)
t_1 time of increase of the clamping force to 95% of F_s (s)
t_s total braking time (s)
T temperature (°C)
T_a ambient temperature (°C)
T_0 initial temperature (°C)
$\hat{T}_1, \hat{T}_2$ temperatures at the radius $r = 207$ mm 1 mm under the firction surfaces of the disc (°C)
$\hat{T}_3, \hat{T}_4$ temperatures at the radius $r = 247$ mm 1 mm under the firction surfaces of the disc (°C)
$\hat{T}_5, \hat{T}_6$ temperatures at the radius $r = 287$ mm 1 mm under the firction surfaces of the disc (°C)
V_0 initial velocity of the vehicle
W_0 initial kinetic energy per one braking system (J)
$W_{p,d}^{conv}$ heat dissipated through convection from one pad or half of the disc free surfaces (J)
$W_{p,d}^{rad}$ heat dissipated through radiation from one pad or half of the disc free surfaces (J)
$W_{p,d}^{diss}$ heat dissipated from the free surfaces of two pads or the disc to the environment (J)
z axial coordinate (m)

Greek Symbols

$\Gamma_{p,d}$ free surface area of the pad and the disc (m^2)
δ thickness (m)
ε surface emissivity (m)
θ circumferential coordinate (rad)
θ_0 half of the coever angle of the pad (rad)
ϑ angular distance of the rotating pad to the stationary disc (rad)
ρ mass density (kg m^{-3})
σ Stefan-Boltzmann constant (W m^{-2}K^{-4})
ω angular velocity (rad s^{-1})
ω_0 initial angular velocity (rad s^{-1})
ω_1 angular velocity at time $t = t_1$ (rad s^{-1})
ω_s angular velocity at the end of braking (rad s^{-1})

Ω_d region within the half volume of the disc (m^{-3})
Ω_p region within the volume of the pad (m^{-3})

Subscripts

p, d pad, disc

References

1. Day, A.J. *Braking of Road Vehicles*, 1st ed.; Butterworth-Heinemann/Elsevier: Oxford, UK, 2014.
2. Yevtushenko, A.A.; Grzes, P.; Adamowicz, A. The temperature mode of the carbon-carbon multi-disc brake in the view of the interrelations of its operating characteristics. *Materials* **2020**, *13*, 1878. [CrossRef] [PubMed]
3. Baranowski, P.; Damziak, K.; Malachowski, J.; Mazurkiewicz, L.; Kastek, M.; Piatkowski, T.; Polakowski, H. Experimental and numerical tests of thermo-mechanical processes occurring on brake pad lining surfaces. In *Surface Effects and Contact Mechanics X, Computational Methods and Experiments*; De Hosson, J.T.M., Brebbia, C.A., Eds.; WIT Press: Southampton, UK, 2011; Volume 71, pp. 15–24.
4. Grzes, P. Finite element solution of the three-dimensional system of equations of heat dynamics of friction and wear during single braking. *Adv. Mech. Eng.* **2018**, *10*, 1–15. [CrossRef]
5. Adamowicz, A.; Grzes, P. Analysis of disc brake temperature distribution during single braking under non-axisymmetric load. *Appl. Therm. Eng.* **2011**, *31*, 1003–1012. [CrossRef]
6. Gao, C.H.; Lin, X.Z. Transient temperature field analysis of a brake in a non-axisymmetric three-dimensional model. *J. Mater. Process Tech.* **2002**, *129*, 513–517. [CrossRef]
7. Gao, C.H.; Huang, J.M.; Lin, X.Z.; Tang, X.S. Stress analysis of thermal fatigue fracture of brake disks based on thermomechanical coupling. *ASME J. Tribol.* **2007**, *129*, 536–543. [CrossRef]
8. Jian, Q.; Shui, Y. Numerical and experimental analysis of transient temperature field of ventilated disc brake under the condition of hard braking. *Int. J. Therm. Sci.* **2017**, *122*, 115–123. [CrossRef]
9. Yevtushenko, A.A.; Kuciej, M.; Grzes, P.; Wasilewski, P. Temperature in the railway disc brake at a repetitive short-term mode of braking. *Int. Commun. Heat Mass Transfer* **2017**, *84*, 102–109. [CrossRef]
10. Grzes, P.; Oliferuk, W.; Adamowicz, A.; Kochanowski, K.; Wasilewski, P.; Yevtushenko, A.A. The numerical–experimental scheme for the analysis of temperature field in a pad-disc braking system of a railway vehicle at single braking. *Int. Commun. Heat Mass Transfer* **2016**, *75*, 1–6. [CrossRef]
11. Tirovic, M. Energy thrift and improved performance achieved through novel railway brake discs. *Appl. Energ.* **2009**, *86*, 317–324. [CrossRef]
12. Wasilewski, P. Frictional heating in railway brakes: A review of numerical models. *Arch. Computat. Methods Eng.* **2020**, *27*, 45–58. [CrossRef]
13. Deressa, K.T.; Ambie, D.A. Thermal load simulations in railway disc brake: A systematic review of modelling temperature, stress and fatigue. *Arch. Computat. Methods Eng.* **2021**, 1–13. [CrossRef]
14. Zewang, Y.; Chun, T.; Mengling, W.; Guozhuang, W. A modified uniformly distributed heat source method for predicting braking temperature of railway brake disc. *Int. J. of Rail Transp.* **2021**, 1–14. [CrossRef]
15. Zewang, Y.; Chun, T.; Mengling, W.; Jiajun, Z.; Chao, C. Modeling and model validation of thermal behavior of railway disc during single braking. *ASME J. Thermal Sci. Eng. Appl.* **2021**, *13*, 051017. [CrossRef]
16. Grzes, P. Three-dimensional effect of heat generation in axisymmetric FE model of a brake disc. In Proceedings of the 15th International Symposium of Students and Young Mechanical Engineers "Advances in Chemical and Mechanical Engineering", Gdansk, Poland, 16–19 May 2012; EXPOL: Wloclawek, Poland, 2012; Volume 35, pp. 81–82.
17. Hwang, P.; Wu, X. Investigation of temperature and thermal stress in ventilated disc brake based on 3D thermo-mechanical coupling model. *J. Mech. Sci. Technol.* **2010**, *24*, 81–84. [CrossRef]
18. Ghadimi, B.; Kowsary, F.; Khorami, M. Thermal analysis of locomotive wheel-mounted brake disc. *Appl. Therm. Eng.* **2013**, *51*, 948–952. [CrossRef]
19. Ghadimi, B.; Sajedi, R.; Kowsary, F. 3D investigation of thermal stresses in a locomotive ventilated brake disc based on a conjugate thermo-fluid coupling boundary conditions. *Int. Commun. Heat Mass Transfer* **2013**, *49*, 104–109. [CrossRef]
20. Yevtushenko, A.A.; Kuciej, M. One-dimensional thermal problem of friction during braking: The history of development and actual state. *Int. J. Heat Mass Tran.* **2012**, *55*, 4148–4153. [CrossRef]
21. *COMSOL Multiphysics®*; v. 5.3. COMSOL AB: Stockholm, Sweden. Available online: www.comsol.com (accessed on 8 September 2021).
22. Adamowicz, A.; Grzes, P. Influence of convective cooling on a disc brake temperature distribution during repetitive braking. *Appl. Therm. Eng.* **2011**, *31*, 2177–2185. [CrossRef]
23. Kwela, J. *Physics Script, Classical and Relativistic Mechanics*; GUT Publishing House: Sopot, Poland, 2010. (In Polish)
24. Frei, W. COMSOL Blog, Deformed Mesh Interfaces: Rotations and Linear Translations. Available online: https://www.comsol.com/blogs/deformed-mesh-interfaces-rotations-and-linear-translations/ (accessed on 20 September 2021).

25. Adamowicz, A.; Grzes, P. Convective cooling of a disc brake during single braking. *Acta Mech. Autom.* **2012**, *6*, 5–10.
26. Adamowicz, A. Effect of convective cooling on temperature and thermal stresses in disk during repeated intermittent braking. *J. Frict. Wear* **2016**, *37*, 107–112. [CrossRef]

materials

MDPI

Article

An Effect of a Carbon-Containing Additive in the Structure of a Friction Material on Temperature of the Wet Clutch Disc

Aleksander Yevtushenko [1], Piotr Grzes [1,*], Aleksander Ilyushenko [2] and Andrey Liashok [2]

[1] Department of Mechanics and Applied Computer Science, Faculty of Mechanical Engineering, Bialystok University of Technology (BUT), 45C Wiejska Street, 15-351 Bialystok, Poland; a.yevtushenko@pb.edu.pl

[2] The State Scientific Institution "Powder Metallurgy Institute" (SSI PMI), National Academy of Sciences of Belarus, 41 Platonova Street, 220005 Minsk, Belarus; Alexil@mail.belpak.by (A.I.); sdilav@tut.by (A.L.)

* Correspondence: p.grzes@pb.edu.pl

Abstract: This paper consists of two parts. The first one contains a description and methodology of the composite material used as friction material in clutches. Four variants of such material, differing in the type of carbon additive (the elemental graphite, pencil graphite and foundry coke powder of various fractions) were considered. Thermal conductivity, thermal diffusivity as well as the specific heat all materials were determined experimentally. On the inertial IM-58 stand, a simulation of the braking process of the friction pair consisting of a steel disc with friction material and a counterpart in the form of a homogeneous steel disc was carried out. On this basis, averaged coefficients of friction, unchanging in the entire sliding process, were found for the four friction pairs. The experimental data obtained in the first stage were used in the second stage to develop two (2D and 3D) numerical models of the friction heating process of the friction pairs under consideration. For four variants of the friction material, a comparative spatial-temporal temperature analysis was performed using both models. It was found that a simplified axisymmetric (2D) model can be used to estimate the maximum temperature with high accuracy. The lowest maximum temperature (115.6 °C) obtained for the same total friction work was achieved on the friction surface of the material with the addition of GP-1.

Keywords: friction material; clutch; temperature; frictional heating; numerical model

Citation: Yevtushenko, A.; Grzes, P.; Ilyushenko, A.; Liashok, A. An Effect of a Carbon-Containing Additive in the Structure of a Friction Material on Temperature of the Wet Clutch Disc. *Materials* **2022**, *15*, 464. https://doi.org/10.3390/ma15020464

Academic Editor: Liaoliang Ke

Received: 2 December 2021
Accepted: 6 January 2022
Published: 8 January 2022

1. Introduction

The friction materials are widely used in friction pairs of vehicles, tractors, motorcycles, machine tools and others. Such are the hydro mechanical gearboxes, the oil-cooled brakes, the clutches and etc. [1–3]. From the reliable operation of such units and mechanisms depends on the safety of both service personnel and surrounding people [4].

The main advantages of friction materials are high efficiency in torque transmission, low wear [5], smoothness and transparency of operation [4], absence of seizure and formation of burns [6], effective dissipation of thermal energy generated during a slip [7], possibility to minimize the force of switching on and switching off the friction device due to creation of the optimum kinematic scheme.

Known friction materials containing fibers [8], are based on a polymer matrix [9], however, for heavy-loaded friction systems, the most common are powder sintered friction materials (PSFM) [10]. The PSFMs are produced by powder metallurgy and have a complex composite structure that combines metal and non-metal components. As the basis of the friction material, iron and copper have become the most widespread. An iron is used in friction pairs without lubrication. For friction pairs with lubrication, copper-based materials are the most common. Copper allows for efficient material processing, heat dissipation, and low wear [11,12]. However, pure copper does not provide the required values of the coefficient of friction and wear intensity. To change these properties, graphite powder is used. For these purposes natural graphite powders with a size of 80 μm, and finely dispersed, with a size of 8–10 μm can be used [13]. Coarse graphite provides a significantly higher coefficient of friction. It has been shown that foundry coke powder can be used as an additive in the PSFM composition [14]. Its presence makes it possible to increase significantly the value of the coefficient of friction of the material [14].

A feature of the operation of friction materials is slipping, accompanied by both an increase and a decrease in pressure for a short time period, while the velocity can either decrease in case of a friction pair closure, and increase—when it is opened. Short process durations and high velocities lead to the release of a large amount of heat and hence temperature. This causes the destruction of the surface layer, determining wear of the material and, consequently, the service life of the entire machine or mechanism [15]. The studies on the stability, corrosion-resistance, and applicability of ternary mixed metal oxides/metallates (tungstates) and the new additive-manufactured hybrid high entropy alloys were carried out in the articles [16,17].

The development of PSFMs requires the manufacture of samples, carrying out both full-scale and bench tests. The duration of such a period can be either several months or several years. In the case of PSFMs, solving the problem of heat dissipation during frictional sliding by computational methods, would significantly reduce the development time of the friction material for given operating conditions. Modeling the process of temperature change in the friction zone, depending on the geometrical dimensions of the friction disc and steel disc, the amount of lubricant supplied, as well as the composition of the PSFM itself, would allow to predict its operation at a preliminary stage.

The transient temperature fields and quasi-static stresses of the components of a wet multi-plate clutch during a typical engaging process were determined numerically in one of the pioneering works [18]. The analyzed friction pair consisted of alternately arranged steel discs and discs with a steel core lined with layers of friction material. In the calculation model of the friction clutch, the directionality of the thermophysical properties of materials (steel and sintered bronze) was taken into account. The temperature fields were determined using the finite difference method, while the thermal stresses at selected time moments of the friction process by the finite element method. The convective heat transfer coefficient applied in the computational model of the multidisc clutch was in the range 130–1890 W m^{-2} K^{-1}. A computer simulation of the friction process taking into account the thermomechanical contact of the working elements of the multidisc clutch with thermoelastic instability effect was carried out in the article [19]. An attempt to identify the features of the hot spots generation process in a wet multi-disc clutch was made in the paper [20]. The results obtained on the basis of the thermoelastic contact model were compared with the corresponding experimental data.

Computational models of friction clutches were discussed in terms of their transmissibility and engaging control strategy, and the influence of temperature, velocity, contact pressure and wear on the generated torque was analyzed [21].

Significant progress has been observed in the development of the analytical models of the heat generation process due to friction in friction components of clutches [22]. In this paper, apart from the calculations using an exact solution to the thermal problem of friction related to the heating of the dry clutch, the results of the relevant bench tests with the use of a thermal imaging camera are also presented. Analytical differential and integral wear was developed to prepare detailed maps of changes in friction material wear during clutch operation [23]. The proposed approach also allows for the analysis of changes with the braking time of the torque transmitted by the clutch. Two different (with a priori given heat partition coefficient or conditions of perfect thermal contact of friction) analytical models for determining the temperature during the operation of the clutch were proposed [24]. The temperature values obtained with the use of both theoretical models are in good agreement with the analogous results found with the finite element (FE) axisymmetric (2D) model. The above-mentioned analytical models that allow to determine not only the temperature, but also thermal stresses at a specific point on the working surface and inside the friction material were developed [25].

However, the greatest advances in temperature modeling of friction systems such as disc brakes and clutches stem from the use of numerical methods. Without listing all publications on this subject in detail, one should refer the reader to relevant reviews [26–29]. The studies that were not included in these articles cover the 2D FE computational model, implemented in ANSYS software [30]. It was assumed that both working elements of the clutch are heated separately with heat flux proportional to the friction power density. The ratio of heat distribution in the form of the Charron's formula was used as the proportionality coefficient. Single and multiple clutch operation modes were considered.

The subject of temperature is closely related to other aspects of clutch operation, in particular, such as strength of materials and mechanical properties. The analysis of the temperature field and its gradients was performed during a single engaging of the carbon fabric wet clutch in order to predict damage to the composite material [31]. The latter were defined as the increase in surface roughness and the occurrence of cracks as observed on scanning electron microscopy (SEM) micrographs.

A review and evaluation of materials used for friction elements of clutches was presented in relation to specific design solutions developed from the end of the 19th century [32]. The design aspects of the complex hybrid mechanical material system of a woven fiber yarn (glass fiber with aromatic polyamide, copper, and poly-acrylic-nitrile) reinforced friction material was performed with the intention of creating separated component groups (matrix and fiber groups). It was noted that the obtained data can be used as input parameters for the development of thermomechanical models of clutches.

The purpose of this article was:

(1) development of effective 2D and 3D computational models with the use FEM to determine the temperature of the friction clutch;
(2) carrying out a comparative analysis of the temperature fields obtained by means of both models with the same friction work done;
(3) investigating the effect of a carbon-containing additive in the structure of the new four friction materials on the temperature mode of the clutch.
(4) When developing the computational models of frictional heating of the clutch, the authors used their previous experience, gained earlier in the modeling of the temperature mode of disc brakes [33,34].

2. Friction Materials

The basis of the composite friction material was bronze obtained from copper powder and 10% tin (BrO_{10}), which contained 20% of the additive in the form of elemental graphite GE-1, pencil graphite GP-1 and foundry coke powder of various fractions C-1 and C-2.

The charge of the material was obtained by mixing the initial copper powders of the PMS-1 grade with an average particle size of 80 μm (Figure 1a), tin of the PO-1 grade with an average particle size of 30 μm (Figure 1b), graphite of the element grade GE-1 with an average particle size of 100 μm (Figure 1c), pencil graphite GP-1 with an average particle size of 8 μm (Figure 1d), foundry coke powder with a size of less than 60 μm (C-1) (Figure 1e) and 160–200 μm (C-2). Mixing of the initial powders was carried out in a mixer of the "drunken" barrel type for 50 min.

It should be noted that the tribotechnical properties of the friction material are largely influenced by the amount and size of the used powder additives, which determine the contact area during friction. In addition, structure has an impact. Graphite of any grade has a layered crystalline structure, while coke is amorphous.

The used coke powder additive is obtained by crushing large pieces of coke, which is used as fuel, the cost of which is not high. Graphites of grades GE-1 and GP-1 are natural, the cost of which is 20–30% higher than the cost of foundry coke.

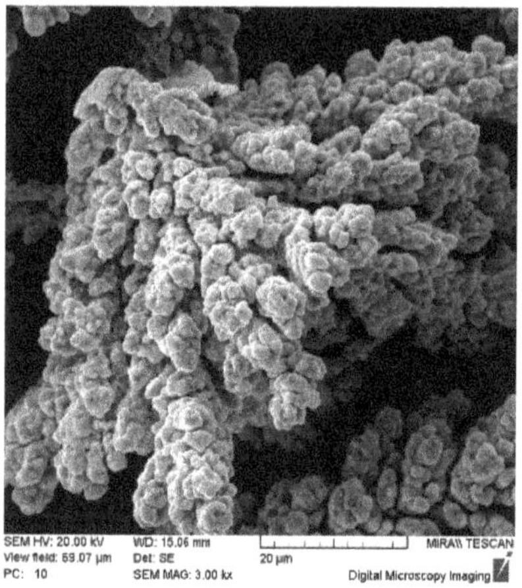

(**a**)

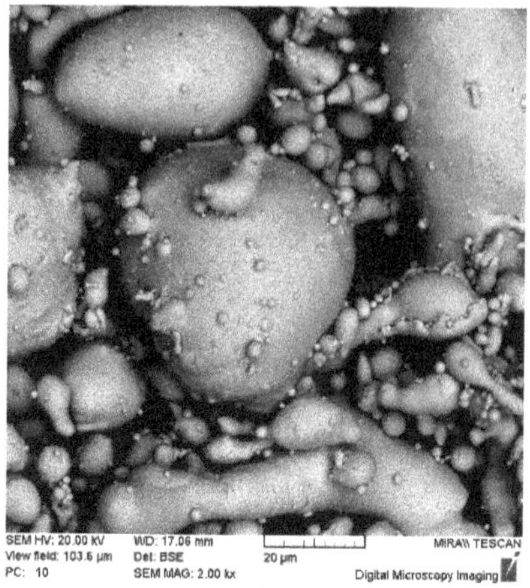

(**b**)

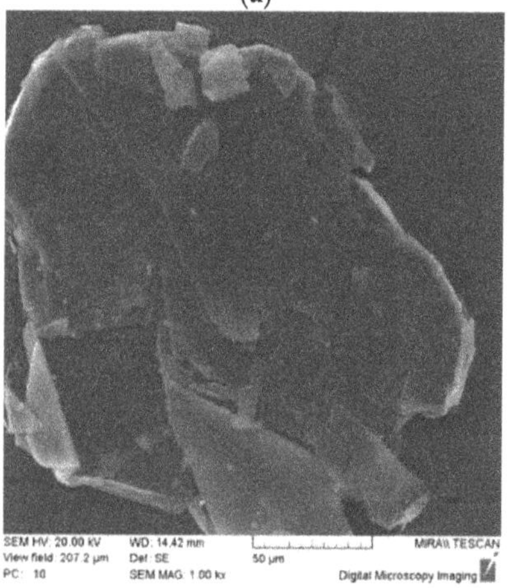

(**c**)

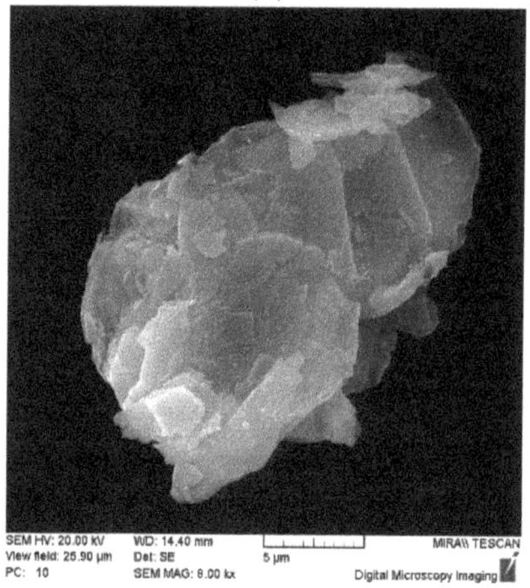

(**d**)

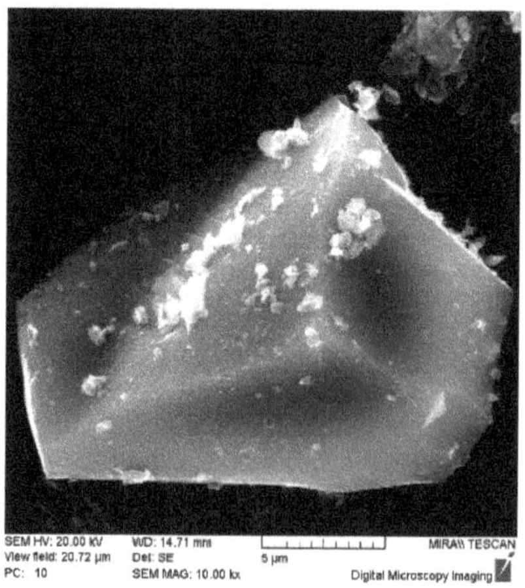

(**e**)

Figure 1. Shape of powder particles: (**a**) PMS-1; (**b**) PO-1; (**c**) GE-1; (**d**) GP-1; (**e**) C-1, C-2.

3. Experiment

The essence of the test method was to simulate the braking process on the inertial stand IM-58 and record the dependence of the friction moment on the velocity and time of braking. The scheme of the IM-58 stand is shown in Figure 2.

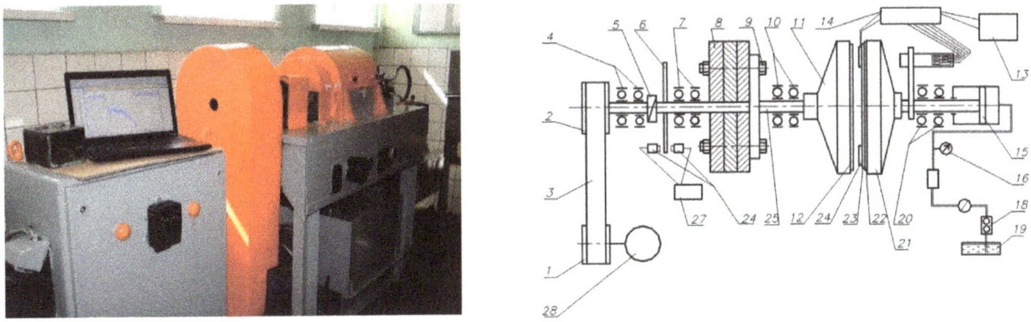

(**a**) (**b**)

Figure 2. Inertial stand IM-58: (**a**) physical appearance; (**b**) structural scheme.

The tests were carried out with friction by the ends of a rotating counterpart 12 and a stationary test specimen 24, which are mounted in the mandrels 11, 22 and 23 (Figure 2b). The axes of their rotation are located horizontally and mounted on bearings 4, 7, 10, 20. The rotating mandrel 11 is fixed on the spindle 25, which rotates from the electric motor 28 through the pulleys 1 and 2 and the V-belt transmission 3. The stationary mandrel 21 is fixed on the piston rod 15, mounted on two radial bearings and having the ability to rotate around its axis. The mandrel 21 and the rod 15 are held against rotation by a strain gauge beam. The load on the samples was created by a gear oil pump 18 powered from the oil tank 19 through the filter 17. The working oil pressure was regulated by the throttle and recorded by the pressure gauge 16. Fixation of readings of number of turns was carried out through the intermediate disk 6 and the speed sensor 27. The inertial masses 8 necessary to obtain the required kinetic energy are attached to the spindle flange 25 using bolts 9 and 26. After reaching the selected rotation velocity, the spindle 25 with the inertial masses 8 was disconnected from the electric motor using an overrunning cam clutch 5 and the stationary sample 24 was pressed with the required force against the rotating counterpart 12. With the electric motor turned off, the unique energy accumulator was the rotating inertial masses. Under the action of frictional forces between the counterpart and the test sample, the drive shaft along with the inertial masses gradually stopped, transmitting the torque to the amplifier 14 and personal computer 13.

The appearance of the test specimens is shown in Figure 3. The friction disc was a steel base in thickness 2.5 mm, made of 65G steel on which a layer of friction material with a thickness 0.5 mm was attached. The outer and inner diameters of the friction pad were 95 mm and 65 mm, respectively. The counterpart samples used for research were made of 65G steel (0.62–0.7% C, 0.17–0.37% Si, $\leq$0.9–1.2% Mn, $\leq$0.035% P, $\leq$0.035% S, $\leq$0.25% Cr, $\leq$0.2% Cu) with the thickness of 10.0 mm, the Brinell hardness of 180–200 MPa and the roughness of the surface of friction Ra = 0.7–0.8 μm. Hydraulic oil of grade A (TU 38.1011282-89) intended for operation in torque converters and automatic transmissions was used as a coolant. Oil was supplied from the inner part of the disc at a rate of 1.01 min^{-1}.

The results of the influence of the type of carbon-containing additive in the composition of the friction material on the thermal properties and the coefficient of friction are shown in Table 1. The coefficient of friction was recorded after 300 cycles of each test, i.e., in a steady state wear mode.

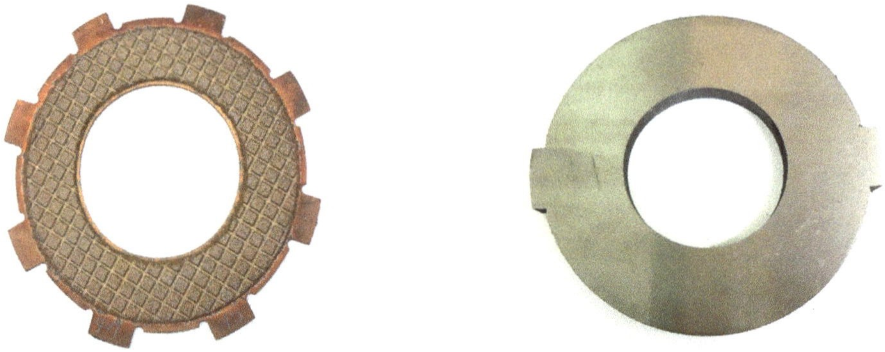

(a) (b)

Figure 3. Appearance of test samples: (**a**) the friction disc; (**b**) the steel counterpart.

Table 1. Influence of the type of carbon-containing additive on thermophysical properties and coefficient of friction.

No.	Additive	Thermal Conductivity $W\ m^{-1}\ K^{-1}$	Specific Heat $J\ kg^{-1}\ K^{-1}$	Mass Density $kg\ m^{-3}$	Thermal Diffusivity $mm^2\ s^{-1}$	Specific Heat Capacity $MJ\ m^{-3}\ K^{-1}$	Coefficient of Friction Dimensionless
1	GP-1	28.1	1189.2	6059	3.9	7.2	0.035
2	GE-1	44	2514.9	5832	3	15	0.045
3	C-1	13.6	438.3	5171	6	2.3	0.05
4	C-2	18.8	701.2	5362	5	3.7	0.06

The obtained data revealed that the highest value of the thermal conductivity of 44 $W\ m^{-1}\ K^{-1}$ was obtained for friction material GE-1, containing coarse graphite. In the case of fine graphite GK-1, the thermal conductivity was only 28 $W\ m^{-1}\ K^{-1}$. This may be due to the fact that the area of the metal contact of the formed tin bronze, in the case of GE-1, is significantly larger than that of GP-1 (Figure 4a). The fine powder of graphite GP-1 in the process of mixing is able to be located both between copper particles and inside them, leading to a decrease in the area of the metal contact (Figure 4b).

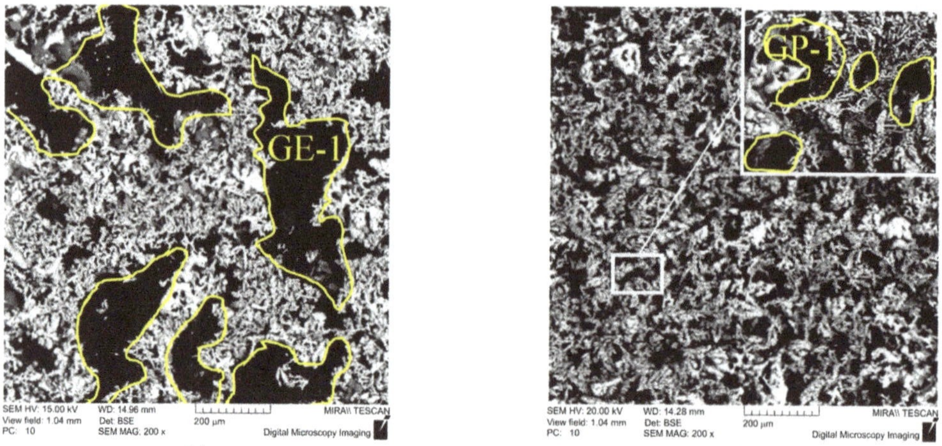

(a) (b)

Figure 4. Structure of a frictional material with various mark of graphite: (**a**) GE-1; (**b**) GP-1.

An even greater decrease in the thermal conductivity of the friction materials C-1 and C-2 containing foundry coke powder may be due to the significantly lower value of the thermal conductivity of coke in comparison with graphite. So in the monographs [35,36] it is shown that the thermal conductivity of coke powder of various fractions is 0.249–0.542 W m^{-1} K^{-1}, while after graphitization is equal to 0.58–1.34 W m^{-1} K^{-1}. For example, the thermal conductivity of graphite determined by the spatial orientation of the layers can be 233 W m^{-1} K^{-1}, while pure copper is equal to 400 W m^{-1} K^{-1} [37].

The highest value 0.06 of the friction coefficient was obtained for the friction material C-2 with the addition of foundry coke powder, while for graphite's GE-1 and GP-1 0.45 and 0.35, respectively. This may be due to both the structure of the crystal lattice and spatial orientation. Thus, graphite's GE-1 and GP-1 have a hexagonal crystal lattice, while coke is amorphous.

4. Numerical Model

Numerical simulation of the friction heating process clutch during a single deceleration from the initial speed to the stop was performed on the inertial stand of the IM-58 (Figure 2). The analyzed clutch consists of two working elements: a steel disc with friction material applied to one face, and a steel counterpart (Figure 3). The calculations were performed for four (no. 1, 2, 3, 4) variants of the friction material with thermophysical properties and friction coefficients from Table 1. The values of the other input parameters are shown in Table 2. The given values of contact pressure, velocity, moments of inertia are characterized as parameters of heavily loaded friction units, clutches, mechanisms of automotive vehicles, special purpose vehicles, devices, machine tools.

Table 2. Operating parameters from experimental tests.

Parameter	Value
contact pressure, $p_0^{(2D)}$ MPa	4
initial angular velocity, ω_0 rad s^{-1}	235.6
braking torque of the rotating masses, I_0 kg m^2	0.7
inner radius of the friction path, r_p mm	30
outer radius of the friction path, R_p mm	47.5
initial/ambient temperature, T_a °C	20
heat transfer coefficient, h W m^{-2} K^{-1}	600

The most important dimensions of both elements are shown in Figure 5.

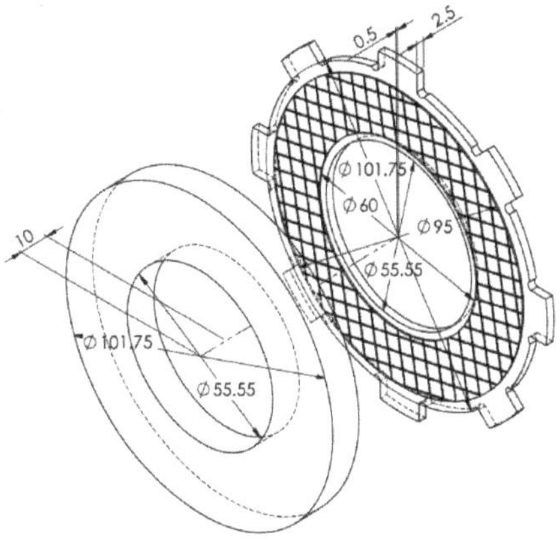

Figure 5. Geometrical 3D CAD model of the clutch.

Two numerical models were developed using FEM: axisymmetric (2D) and spatial (3D). The differences in these models were in the material of the lining and the disc connected to it, which had fasteners on the outer cylindrical surface limiting its rotation. In the case of the lining from the 3D model, cuts (grooves) were taken into. In the 2D model, these cuts were not included due to axial symmetry. Thus, nominal contact surfaces in 2D and 3D models were different. The final FE (quadratic Lagrange elements) meshes created for both models are shown in Figure 6, and the corresponding numbers of elements are included in Table 3.

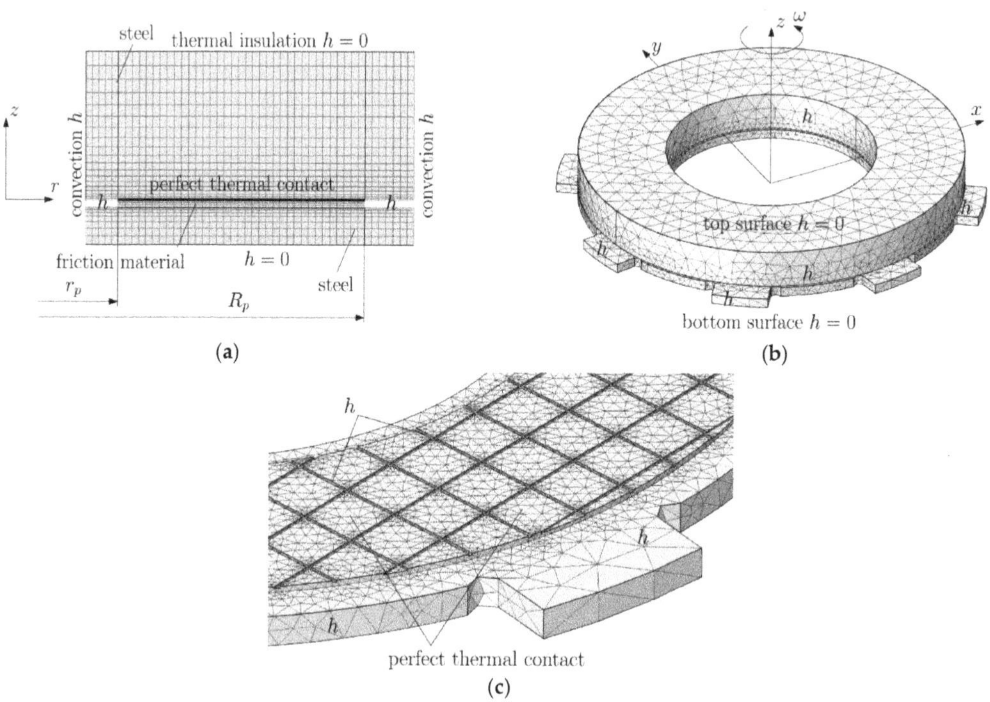

Figure 6. FE meshes of the clutch: (**a**) axisymmetric (2D) and (**b**,**c**) spatial (3D).

Table 3. Numbers of the 2D and 3D finite elements of the clutch models.

Type of Quadratic Lagrange Elements	Friction Material	Steel Plate	Steel Disc	Assembly
quadrilateral elements	175	376	940	1491
tetrahedral elements	69,014	69,946	129,267	268,227

Taking into account the data contained in Table 2, the nominal contact surface area A_a in the 2D model, determined by the values of the inner r_p and outer R_p radius of the disc containing the friction material, is equal to:

$$A_a^{(2D)} = \pi(R_p^2 - r_p^2) = 4261 \cdot 10^{-6}\,\text{m}^2, \tag{1}$$

and the clamping force P of the clutch elements during braking was:

$$P = p^{(2D)} A_a^{(2D)} = 17.04\,\text{kN}. \tag{2}$$

Because the friction surface area in the 3D model was smaller than in the 2D model, the contact pressure was higher. On the basis of the dimensions (Figure 5), an appropriate spatial CAD geometric model of two elements of the friction pair was developed and the friction surface area was determined $A_a^{(3D)} = 3759 \cdot 10^{-6}$ m^2. Then, taking into account the Formula (2), the contact pressure for the 3D model was determined:

$$p^{(3D)} = \frac{P}{A_a^{(3D)}} = 4.53 \text{ MPa}. \quad (3)$$

With the constant contact force P and the friction coefficients unchanged in the braking process f_i (Table 1), the reduction of the angular velocity ω in time t from the initial value ω_0 to zero in the moments of stopping $t = t_{s,i}$ is linear:

$$\omega(t) = \omega_0\omega^*(t),\ \omega^*(t) = 1 - \frac{t}{t_{s,i}},\ 0 \le t \le t_{s,i}, \quad (4)$$

$$t_{s,i} = \frac{\omega_0 I_0}{f_i P\, r_{eq}},\ i = 1,2,3,4, \quad (5)$$

where the equivalent radius r_{eq} was calculated from the formula [33]:

$$r_{eq} = \frac{2(R_p^3 - r_p^3)}{3(R_p^2 - r_p^2)} = 0.0394 \text{ m} \quad (6)$$

Taking into account the Formulas (3) and (6) and the values of the coefficients of friction from the Formula (5) contained in Table 1, the next stopping times were determined for each of the four variants of the friction material: $t_{s,1} = 7.02$ s, $t_{s,2} = 5.46$ s, $t_{s,3} = 4.91$ s and $t_{s,4} = 4.09$ s. Due to the constant contact force of the discs P (2) in both calculation models, also the stopping times for 2D and 3D models with the use of the corresponding friction materials were the same. The change in the process of braking power Q_i and work of friction W_i, $i = 1,2,3,4$:

$$Q_i(t) = Q_{0,i}\omega^*(t),\ 0 \le t \le t_{s,i},\ Q_{0,i} = f_i P\omega_0\, r_{eq}, \quad (7)$$

$$W_i(t) = \int_0^t Q_i(\tau)d\tau = W_{0,i}W_i^*(t),\ W_{0,i} = Q_{0,i}t_{s,i}, \quad (8)$$

$$W_i^*(t) = \frac{t}{t_{s,i}}\left(1 - \frac{t}{2t_{s,i}}\right), 0 \le t \le t_{s,i}, \quad (9)$$

are shown in Figure 7. Based on the Formulas (8) and (9), it was established that in the moments of stopping $t = t_{s,i}$, the friction work for all four calculation variants was the same and amounted to: $W_{s,i} \equiv W_i(t_{s,i}) = 0.5W_{0,i} = 19.43$ kJ, $i = 1,2,3,4$. This value is equal to the initial kinetic energy of the system, calculated from the formula:

$$W_0 = 0.5 I_0 \omega_0^2 = 19.43 \text{ kJ}. \quad (10)$$

The simulated process concerned frictional heating during braking in time $0 \le t \le t_{s,i}$, which made it possible to stop the rotating masses at a given moment of inertia, and then cooling the friction system during $t_{s,i} \le t \le 90$ s, $i = 1, 2, 3, 4$. During the slip of the clutch elements resulting in generation of heat in the time interval $0 \le t \le t_{s,i}$, in the contact area, the thermal contact of friction was perfect, i.e., the sum of heat fluxes directed perpendicularly from the contact surface of the friction material and the steel disc to the inside was equal to the power density of the frictional forces, and the temperature values of the opposite surfaces were the same. At $t > t_{s,i}$, $i = 1, 2, 3, 4$ the friction surfaces of the elements were adiabatic. Throughout the analyzed process ($0 \le t \le 90$ s), convection cooling on the friction-free surfaces with the constant heat transfer coefficient took place. The exception was the two outer frontal surfaces of the sliding elements, which were thermally insulated ($h = 0$) both during heating and cooling (Figure 6).

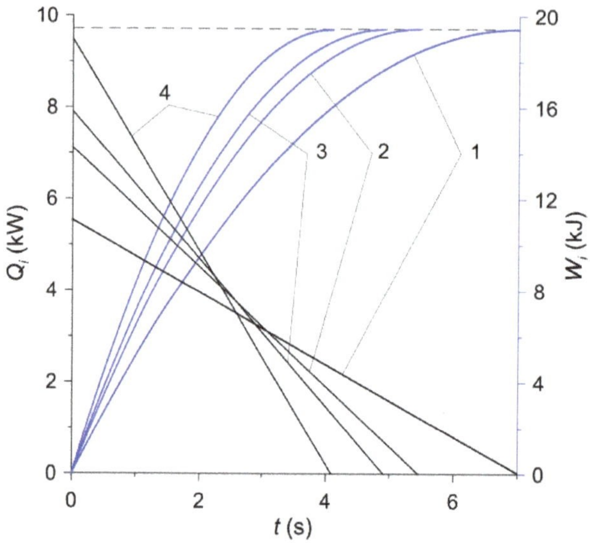

Figure 7. Time profiles of the power Q_i and the work W_i, $i = 1, 2, 3, 4$ of friction during braking.

In accordance with the experimental tests carried out, the relative rotational movement of the clutch components was simulated. The calculations were conducted in COMSOL Multiphysics® software by using the Heat Transfer Module [38]. Additionally, in order to determine the friction work done W_i (Figure 7), special tools, namely Global ODEs (ordinary differential equations) and DAEs (differential-algebraic equations) from Mathematics module were incorporated.

5. Numerical Analysis

The numerical simulation concerned the comparative analysis of the temperature fields in four friction pairs (no. 1, 2, 3, 4) obtained on the basis of the axisymmetric (2D) and spatial (3D) numerical models.

The temperature changes on the friction surface of the disc are shown in Figure 8. The assumptions made in the analysis regarding the constant contact pressure and the linearly decreasing angular velocity of the rotating elements lead to a typical change in the temperature of the contact surface of the friction clutch elements—its value increases at the beginning of the process to the maximum value within approximately 2–4 s and then decreases slightly until it stops (Figure 8a). The differences in the time courses of temperature on the friction surface, obtained by means of two FE models during braking, are insignificant and occur only at the stage of cooling, after stopping (Figure 8b). The difference of the maximum temperature values obtained at the braking stage with the use of both calculation models for all friction materials does not exceed 2% (Table 4). In the cooling phase after stopping, the temperature of the friction surface of the disc drops gradually. The temperature values at a fixed point in time obtained by means of the 2D or 3D model for all four friction materials differ negligibly. It is noticeable that the temperature found when the axisymmetric model is each time higher than that obtained with the use of the spatial model.

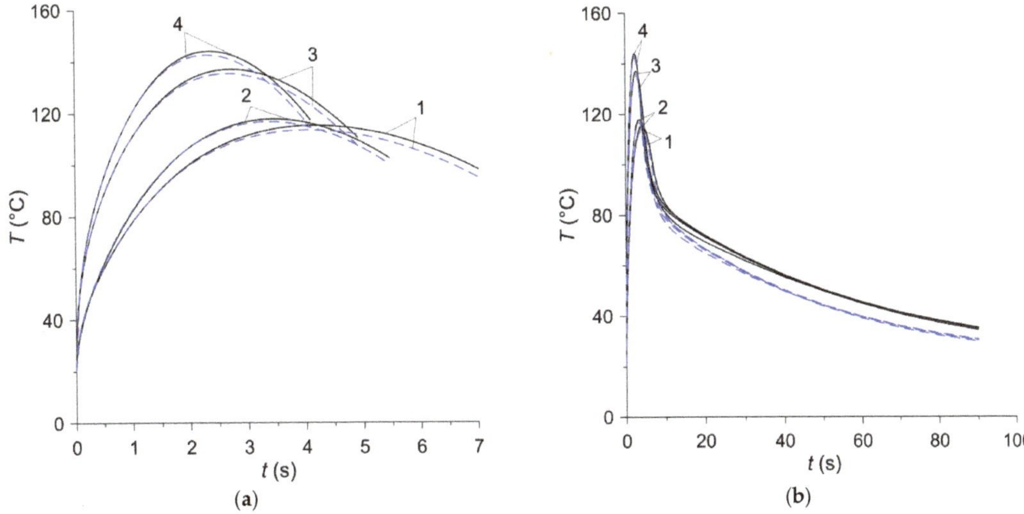

Figure 8. Temperature evolutions for four frictional material on the contact surface $z = 0$ at equivalent radius $r = r_{eq}$: (**a**) during braking; (**b**) during braking and subsequent cooling. Solid curves—2D model; dashed curves—3D model.

Table 4. Maximum temperature obtained using the 2D and 3D FE clutch models.

Model	T_{max} 1	T_{max} 2	T_{max} 3	T_{max} 4
2D	115.6 °C	118.4 °C	137.9 °C	145.1 °C
3D	113.7 °C	117.2 °C	136.1 °C	143.5 °C

The temperature distributions on the friction surface in the radial direction are presented in Figure 9. In the case of the axisymmetric model, when the temperature does not depend on the angular variable, the choice of the radial straight line along which the calculations were performed was arbitrary, while when using the 3D model, a line along the axis x was chosen (Figure 6b). In the contact area $\{30 \text{ mm} \leq r \leq 47.5 \text{ mm}, z = 0\}$, marked by vertical solid lines in Figure 9, the temperature distribution is shaped by a directly proportional linear dependence of the friction power density on the radial variable r (Figure 9a). The linear temperature rise in the contact area along the radius is the best seen in the initial braking stage at $t = 0.5$ s. Outside the area of heating, a clear drop in temperature is visible. The lack of heating after stopping results in a gradual leveling of temperature (Figure 9b). The highest temperature is achieved in the area of heating in the vicinity of the equivalent radius, which, according to Formula (6), is equal to $r_{eq} = 39.4$ mm. The difference between the maximum temperature values obtained with the 2D and 3D models is about 3 °C. Over time, the differences in temperature values obtained with both models increase and at the stopping time they are about 5 °C. At the time of the stop, the maximum temperature of 117.5 °C (2D model) was reached for the material no. 4 on the radius $r = 39.5$ mm close to the equivalent radius $r_{eq} = 39.4$ mm (Figure 9a). At the end of the cooling process of the clutch friction elements, the above-mentioned temperature differences are more noticeable, but they relate to a relatively much lower temperature than during the braking stage. At the time $t = 90$ s, the most heated material was that denoted no. 2, and the least—material no. 3, but the difference between the temperatures of these materials was insignificant (≈ 0.9 °C).

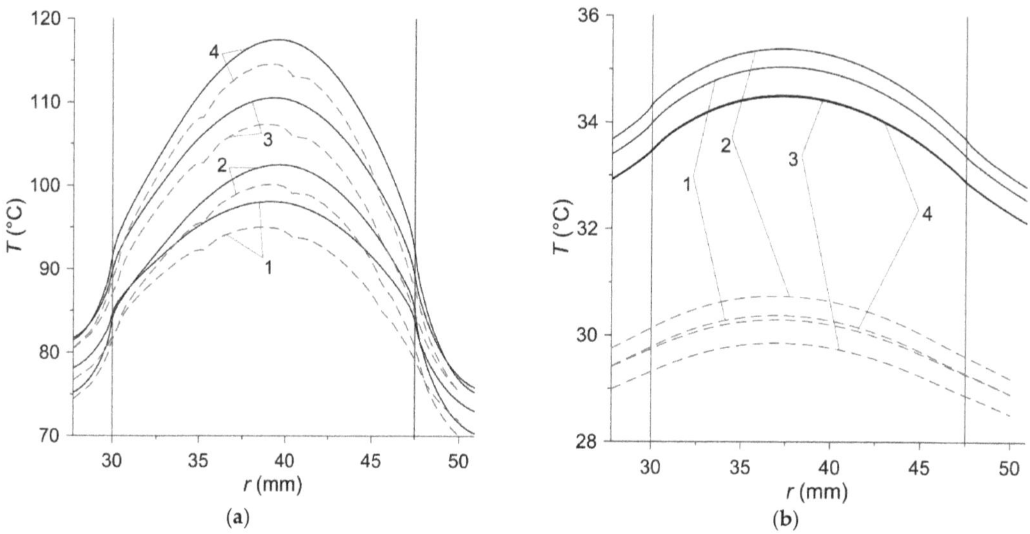

Figure 9. Temperature distributions for four materials along the radial variable r on the contact surface $z = 0$ in: (**a**) stop time moments $t = t_{s,i}$, $i = 1, 2, 3, 4$; (**b**) last moment of time $t_{end} = 90$ s. Solid curves—2D model; dashed curves—3D model.

Temperature changes in the axial z direction for a stationary disc with the friction material (-3 mm $\leq z \leq 0$) and a rotating steel disc (counterpart $0 \leq z \leq 10$ mm) are shown in Figure 10. In the stop times $t = t_{s,i}$, $i = 1, 2, 3, 4$ the calculation results obtained on the basis of 2D and 3D models agree well (Figure 10a). The temperature of the stationary steel disc decreases with the distance from the contact surface $z = 0$. The slight increase in temperature at a deflection from the contact surface in the stationary disc is associated with the known lagging effect—the decrease in temperature inside the body is slower than on its heated surface. At the end of the process $t = t_{end}$ of unforced convection cooling of the clutch after stopping, the temperature distribution along the thickness of the friction pair elements is even (Figure 10b). However, as in the radial direction (Figure 9b), the differences between the temperature values obtained with the two calculation models are noticeable (≈ 4 °C), but the temperature itself at that moment is not high.

The temperature distributions on the friction surface obtained by means of the 2D and 3D models at the time moments $t_{\max,i}$, $i = 1, 2, 3, 4$; when the maximum values of temperature $T_{\max}$ were reached (Table 4) are shown in Figure 11. They show the differences between the results obtained with the use of two calculation models on one hand, and on the other they show the explicitly temperature effect of using different friction materials. For the same total friction work, the temperature of the friction surface increases with the braking time reduction ($t_{s,1} = 7.02$ s, $t_{s,2} = 5.46$ s, $t_{s,3} = 4.91$ s, $t_{s,4} = 4.09$ s). The location of the area with the increased temperature in the vicinity of the equivalent radius is visible r_{eq} (6).

The isotherms in the cross-section of the clutch in the plane rz at the moments of reaching the maximum temperature in each of the four friction materials are shown in Figure 12. The highest value of the maximum temperature $T_{\max} = 145.1$ °C was observed in the case of the material no. 4, and the lowest $T_{\max} = 115.6$ °C was determined when using the material no. 1. The maximum temperature was reached on the radii of 40.4 mm, 40.7 mm, 40.75 mm, 40.85 mm, for the material no. 1, 2, 3, 4 respectively. It can be seen that the disc with the friction material is heated in the entire thickness of 3 mm, while the counterpart is heated only to about half of its thickness ($\approx$5 mm).

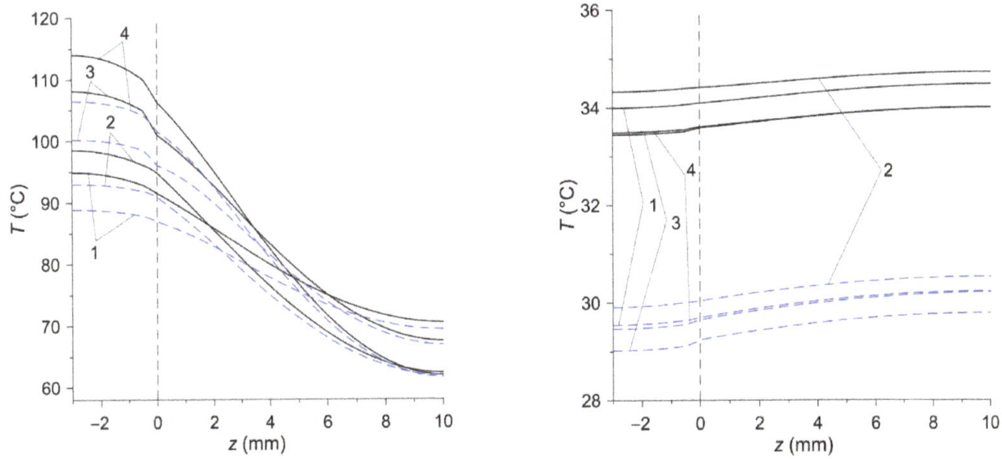

(**a**) (**b**)

Figure 10. Temperature distributions for four materials in axial direction r at radius $r = 45$ mm in: (**a**) stop time moments $t = t_{s,i}$, $i = 1, 2, 3, 4$; (**b**) last moment of time $t_{end} = 90$ s. Solid curves—2D model; dashed curves—3D model.

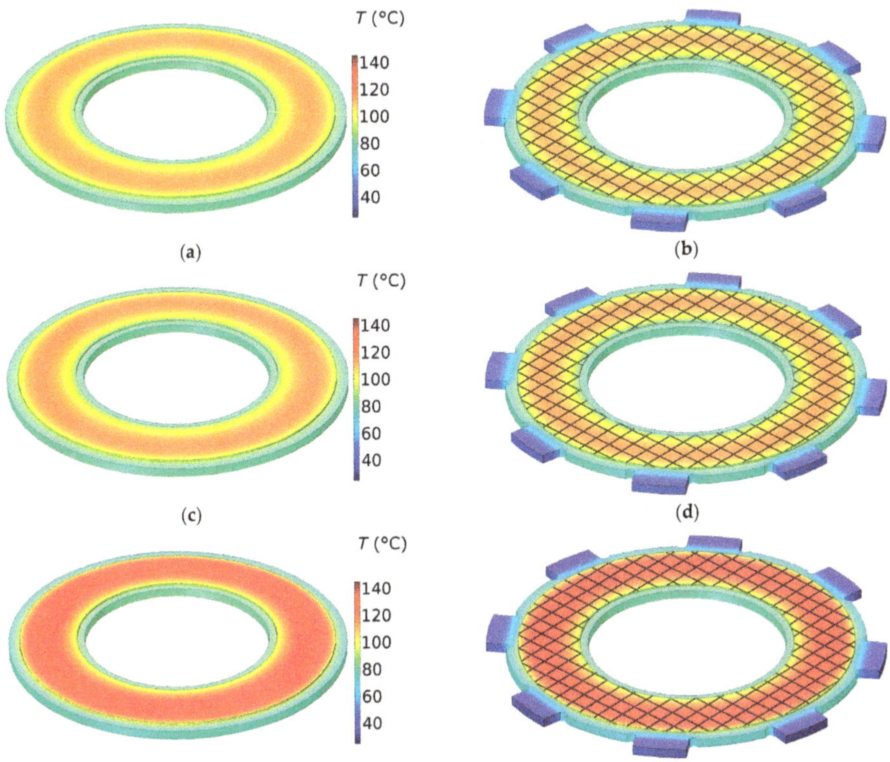

(**a**) (**b**)

(**c**) (**d**)

(**e**) (**f**)

Figure 11. *Cont.*

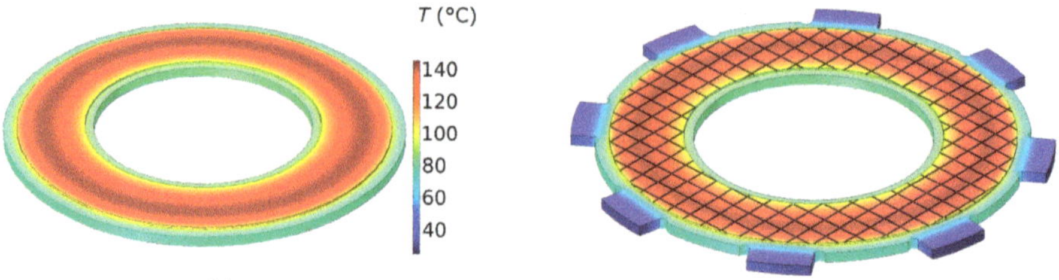

(g) (h)

Figure 11. Temperature distributions on the surface of friction at the time $t_{max,i}$, $i = 1, 2, 3, 4$ of reaching the maximum temperature for four frictional materials: (**a**,**b**) $t_{max,1} = 4.16$ s; (**c**,**d**) $t_{max,2} = 3.45$ s; (**e**,**f**) $t_{max,3} = 2.72$ s; (**g**,**h**) $t_{max,4} = 2.35$ s. Model 2D—(**a**,**c**,**e**,**g**); model 3D—(**b**,**d**,**f**,**h**).

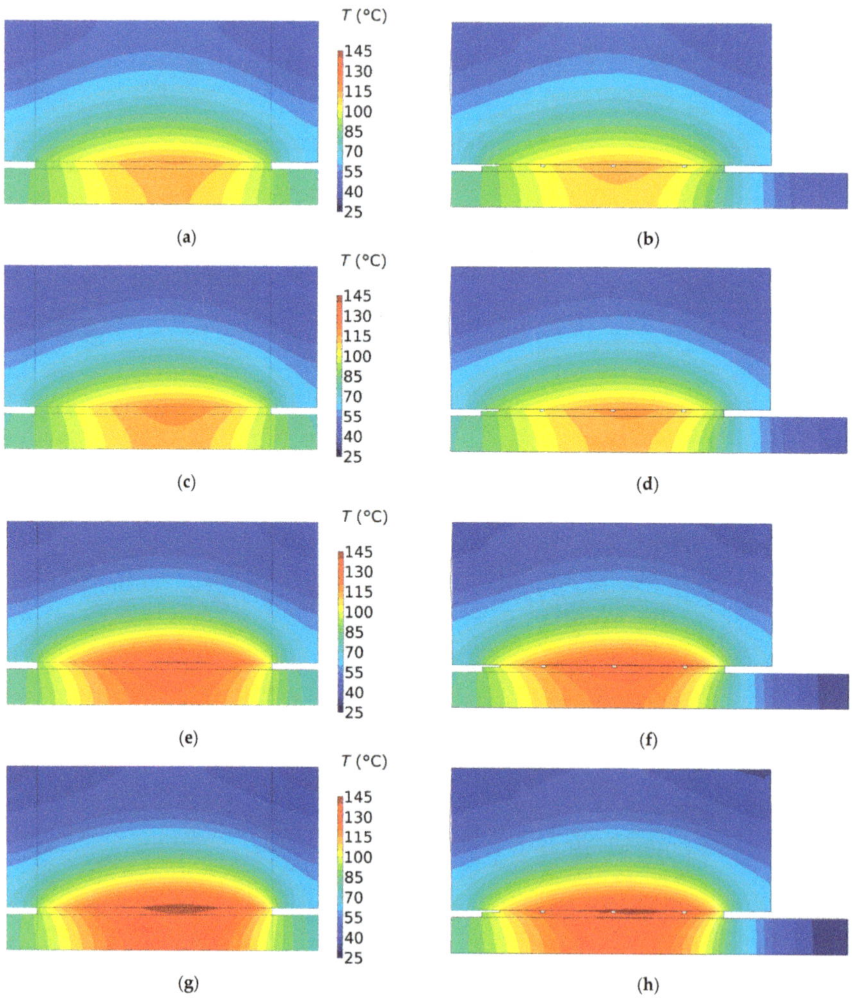

Figure 12. Isotherms in a plane rz at the time $t_{max,i}$, $i = 1, 2, 3, 4$ of reaching the maximum temperature for four frictional materials: (**a**,**b**) $t_{max,1} = 4.16$ s; (**c**,**d**) $t_{max,2} = 3.45$ s; (**e**,**f**) $t_{max,3} = 2.72$ s; (**g**,**h**) $t_{max,4} = 2.35$ s. Model 2D—(**a**,**c**,**e**,**g**); model 3D—(**b**,**d**,**f**,**h**).

6. Conclusions

This paper presents the results of experimental studies on the thermal and physical properties and the coefficient of friction as well as numerical simulations of the temperature mode of the new composite friction material used in the clutches. The data obtained for the four variants (no. 1, 2, 3, 4) of such material, differing in the type of carbon additive, were analyzed. The axisymmetric (2D) and the spatial (3D) numerical models of the wet clutch in order to determine the transient temperature fields of the friction discs during a single engagement and after the stop were developed. Based on the obtained results, it was found that:

(1) the highest thermal conductivity and specific heat has the material no. 2, while the lowest values of the quantities has the material no. 3;
(2) the highest coefficient of friction appears for the steel disc combined with the friction material no. 4, and the lowest for steel disc and friction material no. 1;
(3) the estimation of the maximum clutch temperature can be carried out with sufficient accuracy using a 2D model, which allows for the reduction of labor losses at the preparatory stage and computational time. The single simulation case carried out on the workstation with CPU Intel® Xeon® E5-2698 v4 @ 2.20GHz; RAM 64 GB (DDR4) lasted approximately 40 s, and 2200 s, when using 2D and 3D models, respectively. On the other hand, the determination of the temperature field in the elements of the friction clutch is better to carry out with the use of the 3D model;
(4) At the same total friction work done during a single clutch engagement, the lowest temperature was achieved when using friction material no. 1, and the highest for material no. 4. Also, the use of these materials resulted in the longest and shortest periods of frictional sliding, respectively.

It should be noted that in the proposed numerical models with the use of FEM, the averaged, experimentally obtained, thermophysical properties of the composite material were used. The development of the computational models that take into account the structure of the composite and the properties of individual components belongs to the future.

Subsequent studies will aimed at using the developed 2D and 3D models for assessing the effect of additives of powders (ceramic, metallic, intermetallic compounds), their particle size, distribution and the amount in the composition of the friction material on the thermal effects during friction. One of the important aspects would be to include realistic properties of a coolant. This will make it possible to assess the efficiency of their use without expensive studies on the full scale inertial test bench.

Author Contributions: Conceptualization, A.Y. and A.I.; methodology, A.Y. and A.I.; software, P.G. and A.L.; validation, P.G. and A.L.; formal analysis, A.Y. and A.I.; investigation, P.G. and A.L.; resources, A.Y., A.I., P.G. and A.L.; data curation, P.G. and A.L.; writing—original draft preparation, P.G. and A.L.; writing—review and editing, A.Y. and A.I.; visualization, P.G. and A.L.; supervision, A.Y. and A.I.; project administration, P.G.; funding acquisition, A.Y. and P.G. All authors have read and agreed to the published version of the manuscript.

Funding: Project financing through the program of the Minister of Science and Higher Education of Poland named "Regional Initiative of Excellence" in 2019–2022, project number 011/RID/2018/19, amount of financing 12,000,000 PLN.

Institutional Review Board Statement: Not applicable.

Informed Consent Statement: Not applicable.

Data Availability Statement: Not applicable.

Conflicts of Interest: The authors declare no conflict of interest.

Nomenclature

$A_a^{(2D)}$	nominal contact surface area in the 2D model (m^2)
$A_a^{(3D)}$	nominal contact surface area in the 3D model (m^2)
f_i	coefficient of friction of the $i - th = 1,2,3,4$ friction material (dimensionless)
I_0	braking torque of the rotating masses (kg m^2)
h	heat transfer coefficient (W $m^{-2}K^{-1}$)
$p_0^{(2D)}$	contact pressure in the 2D model (MPa)
$p^{(3D)}$	contact pressure in the 3D model (MPa)
P	clamping force of the clutch elements (N)
Q_i	friction power of the $i - th = 1,2,3,4$ friction material (W)
r	radial coordinate (m)
r_p, R_p	inner and ouer radius of the friction path, respectively (m)
r_{eq}	equivalent radius of the contact region (m)
t	time (s)
t_{end}	total time of the process equal to 90 s (s)
$t_{s,i}$	braking time of the $i - th = 1,2,3,4$ friction material (s)
T	temperature (°C)
T_a	initial/ambient temperature (°C)
W_0	kinetic energy of the system (J)
W_i	work of friction of the $i - th = 1,2,3,4$ friction material (J)
z	axial coordinate (m)
Greek Symbols	
ω	angular velocity (rad s^{-1})
ω_0	initial angular velocity (rad s^{-1})

References

1. Yu, L.; Ma, B.; Chen, M.; Li, H.Y.; Liu, J. Experimental study on the friction stability of paper-based clutches concerning groove patterns. *Ind. Lubr. Tribol.* **2020**, *72*, 541–548. [CrossRef]
2. Yu, L.; Ma, B.; Chen, M.; Li, H.; Ma, C.; Liu, J. Comparison of the friction and wear characteristics between copper and paper based friction materials. *Materials* **2019**, *12*, 2988. [CrossRef] [PubMed]
3. Wang, X.; Ru, H. Effect of Lubricating Phase on Microstructure and Properties of Cu–Fe Friction Materials. *Materials* **2019**, *12*, 313. [CrossRef] [PubMed]
4. Liu, J.; Ma, B.; Li, H.; Chen, M.; Li, G. Control strategy optimization for a dual-clutch transmission downshift with a single slipping clutch during the torque phase. *Proc. Inst. Mech. Eng. Part D J. Automob. Eng.* **2018**, *232*, 651–666. [CrossRef]
5. Marklund, P.; Larsson, R. Wet clutch friction characteristics obtained from simplified pin on disc test. *Tribol. Int.* **2008**, *41*, 824–830. [CrossRef]
6. Biryukov, V.P.; Il'yushenko, A.F.; Leshok, A.V.; Pinchuk, T.I. Influence of carbon-containing additives in the composition of copper-based friction materials on boundary friction in mineral and synthetic oils. *J. Frict. Wear* **2020**, *41*, 377–382. [CrossRef]
7. Anderson, A.E. Friction and wear of automotive brakes. In *ASM Handbook, Volume 7: Powder Metallurgy*; Samal, P.K., Newkirk, J.W., Eds.; ASM International: Almere, The Netherlands, 1984; pp. 569–577.
8. Xie, F.; Hu, W.; Ning, D.; Zhuo, L.; Deng, J.; Lu, Z. ZnO nanowires decoration on carbon fiber via hydrothermal synthesis for paper-based friction materials with improved friction and wear properties. *Ceram. Int.* **2018**, *44*, 4204–4210. [CrossRef]
9. Bijwe, J.; Majumdar, N.; Satapathy, B.K. Influence of modified phenolic resins on the fade and recovery behavior of friction materials. *Wear* **2005**, *259*, 1068–1078. [CrossRef]
10. Ilyushchanka, A.P.; Leshok, A.V.; Dyachkova, L.N.; Alekseenko, N.A. Formation of the friction surface of a friction material based on copper depending on the amount of tin under lubrication condition. In Proceedings of the 10th international scientific conference Balttrib' 2019, Vytautas Magnus University, Agriculture Academy, Kaunas, Lithuania, 14–16 November 2019; pp. 20–26. [CrossRef]
11. Prapai, J.; Morakotjinda, M.; Yotkaew, T.; Vetayanukul, B. Tribological properties of PM Cu-based dry friction clutch. *Key Eng. Mater.* **2013**, *545*, 163–170. [CrossRef]
12. Tosangthum, N.; Krataitong, R.; Tongsri, R. Sintered frictional materials based on Cu powders. *Adv. Mat. Res.* **2013**, *747*, 55–58. [CrossRef]
13. Liashok, A.; Dyachkova, L.; Feldshtein, E. On the effect of the grade and content of graphite on the structure, strength and tribological bechavior of friction materials based on tin bronze. In *Inżynieria Produkcji, Badania w Inżynierii Mechanicznej*; Jakubowski, J., Patalas-Maliszewska, J., Eds.; Wydawnictwo Naukowe Instytutu Inżynierii Mechanicznej; Uniwersytet Zielonogórski: Zielona Góra, Poland, 2020; Volume 17, pp. 7–18. (In Polish)

14. Leshok, A.V. Influence of graphite-containing additives of cast and shungite coke on tribotechnical properties of iron-based friction material. In Proceedings of the International Scientific and Technical Conference "Polymer Composites and Tribology" "Polikomtrib-2019", Gomel, Belarus, 25–28 June 2019.
15. Abdullah, O.I.; Schlattmann, J. Contact Analysis of a Dry Friction Clutch System. *Int. Sch. Res. Not.* **2013**, *2013*, 495918. [CrossRef]
16. Sarkar, S.; Sarswat, P.K.; Free, M.L. Metal oxides and novel metallates coated stable engineered steel for corrosion resistance applications. *Appl. Surf. Sci.* **2018**, *456*, 328–341. [CrossRef]
17. Sarswat, P.K.; Sarkar, S.; Murali, A.; Huang, W.; Tan, W.; Free, M.L. Additive manufactured new hybrid high entropy alloys derived from the AlCoFeNiSmTiVZr system. *Appl. Surf. Sci.* **2019**, *476*, 242–258. [CrossRef]
18. Zagrodzki, P. Numerical analysis of temperature fields and thermal stresses in the friction discs of a multidisc wet clutch. *Wear* **1985**, *101*, 255–271. [CrossRef]
19. Zagrodzki, P. Analysis of thermomechanical phenomena in multidisc clutches and brakes. *Wear* **1990**, *140*, 291–308. [CrossRef]
20. Zagrodzki, P.; Truncone, S.A. Generation of hot spots in a wet multidisk clutch during short-term engagement. *Wear* **2003**, *254*, 474–491. [CrossRef]
21. Della Gatta, A.; Iannelli, L.; Pisaturo, M.; Senatore, A.; Vasca, F. A survey on modeling and engagement control for automotive dry clutch. *Mechatronics* **2018**, *55*, 63–75. [CrossRef]
22. Awrejcewicz, J.; Grzelczyk, D. Modeling and analysis of thermal processes in mechanical friction clutch—Numerical and experimental investigations. *Int. J. Struct. Stab. Dy.* **2013**, *13*, 1340004. [CrossRef]
23. Grzelczyk, D.; Awrejcewicz, J. Wear Processes in a Mechanical Friction Clutch: Theoretical, Numerical, and Experimental Studies. *Math. Probl. Eng.* **2015**, *2015*, 725685. [CrossRef]
24. Abdullah, O.I.; Schlattmann, J.; Majeed, M.H.; Sabri, L.A. The temperatures distributions of a single-disc clutches using heat partitioning and total heat generated approaches. *Case Stud. Therm. Eng.* **2018**, *11*, 43–54. [CrossRef]
25. Topczewska, K.; Schlattmann, J.; Abdullah, O.I. Temperature and thermal stresses distributions in a dry friction clutch. *J. Theor. Appl. Mech.* **2020**, *58*, 351–360. [CrossRef]
26. Yevtushenko, A.A.; Grzes, P. The FEM-modeling of the frictional heating phenomenon in the pad/disc tribosystem (a review). *Numer. Heat Tr. A Appl.* **2010**, *58*, 207–226. [CrossRef]
27. Yevtushenko, A.A.; Grzes, P.; Adamowicz, A. Numerical analysis of thermal stresses in disk brakes and clutches (a review). *Numer. Heat Tr. A Appl.* **2015**, *67*, 170–188. [CrossRef]
28. Wasilewski, P. Frictional heating in railway brakes: A review of numerical models. *Arch. Computat. Methods Eng.* **2020**, *27*, 45–58. [CrossRef]
29. Deressa, K.T.; Ambie, D.A. Thermal load simulations in railway disc brake: A systematic review of modelling temperature, stress and fatigue. *Arch. Computat. Methods Eng.* **2021**, 1–13. [CrossRef]
30. Pisaturo, M.; Senatore, A. Simulation of engagement control in automotive dry-clutch and temperature field analysis through finite element model. *Appl. Therm. Eng.* **2016**, *93*, 958–966. [CrossRef]
31. Wenbin, L.; Jianfeng, H.; Jie, F.; Liyun, C.; Chunyan, Y. Simulation and application of temperature field of carbon fabric wet clutch during engagement based on finite element analysis. *Int. Commun. Heat Mass* **2016**, *71*, 180–187. [CrossRef]
32. Biczó, R.; Kalácska, G.; Mankovits, T. Micromechanical model and thermal properties of dry-friction hybrid polymer composite clutch facings. *Materials* **2020**, *13*, 4508. [CrossRef] [PubMed]
33. Yevtushenko, A.A.; Grzes, P.; Adamowicz, A. The temperature mode of the carbon-carbon multi-disc brake in the view of the interrelations of its operating characteristics. *Materials* **2020**, *13*, 1878. [CrossRef] [PubMed]
34. Yevtushenko, A.A.; Grzes, P. Initial selection of disc brake pads material based on the temperature mode. *Materials* **2020**, *13*, 822. [CrossRef] [PubMed]
35. Agroskin, A.A. *Thermal Physics of Solid Fuel*; Nedra: Moscow, Russia, 1980; pp. 1–256. (In Russian)
36. Lutkov, A.I. *Thermal and Electrical Properties of Carbon Materials*; Metallurgy: Moscow, Russia, 1990; pp. 1–176. (In Russian)
37. Norley, J.; Tzeng, J.J.-W.; Getz, G.; Klug, J.; Fedor, B. The development of a natural graphite heat-spreader. In Proceedings of the Seventeenth Annual IEEE Semiconductor Thermal Measurement and Management Symposium, San Jose, CA, USA, 20–22 March 2001; pp. 107–110.
38. COMSOL. *Multiphysics®v. 5.3.*; COMSOL AB: Stockholm, Sweden; Available online: www.comsol.com (accessed on 8 September 2021).

materials

Article

Influence of the Additive of Ceramic and Intermetallic Powders on the Friction Properties and Temperature of the Wet Clutch Disc

Aleksander Yevtushenko [1], Michal Kuciej [1,*], Piotr Grzes [1], Aleksander Ilyushchanka [2] and Andrey Liashok [2]

[1] Department of Mechanics and Applied Computer Science, Faculty of Mechanical Engineering, Bialystok University of Technology (BUT), 45C Wiejska Street, 15-351 Bialystok, Poland

[2] The State Scientific Institution "Powder Metallurgy Institute" (SSI PMI) of the National Academy of Sciences of Belarus, 41 Platonova Street, 220005 Minsk, Belarus

* Correspondence: m.kuciej@pb.edu.pl; Tel.: +48-85-746-9200

Abstract: The basic function of friction clutches is to transfer the torque in the conditions of its smooth engagement without vibrations. Hard working conditions under high thermal and mechanical loads, leading to high temperature in the contact area, intense wear, and instability of the coefficient of friction impose restrictive criteria in the design of friction materials. In this paper, the results of experimental research of the effect of ceramic and intermetallic additives to the copper-based material of the friction disc of the clutch on the thermophysical and frictional properties were presented. Next, these properties were incorporated in the proposed contact 3D numerical model of the clutch to carry out computer simulations of the heating process and subsequent cooling. Based on the obtained experimental data and transient temperature changes of the friction and steel discs, the relations between the powder additives, thermophysical properties of the five friction materials, and coefficients of friction, wear, and temperature reached were discussed. Among these, it was found that when working with the lubrication, the largest values of the coefficient of friction 0.068 and wear 13.5 $\mu m\ km^{-1}$ were reached when using the 3 wt.% SiC additive.

Keywords: friction material; clutch; temperature; frictional heating; finite element analysis

Citation: Yevtushenko, A.; Kuciej, M.; Grzes, P.; Ilyushchanka, A.; Liashok, A. Influence of the Additive of Ceramic and Intermetallic Powders on the Friction Properties and Temperature of the Wet Clutch Disc. *Materials* **2022**, *15*, 5384. https://doi.org/10.3390/ma15155384

Academic Editor: Shengqiang Ma

Received: 28 June 2022
Accepted: 31 July 2022
Published: 4 August 2022

1. Introduction

Powder sintered friction materials (PSFM) have become widely used in the friction units of the automotive vehicles, motorcycles, tractors, airplanes, boats, machine tools, etc. Such friction units include, in particular, hydro mechanical gearboxes, oil-cooled brakes, clutches, etc. [1,2]. Powder metallurgy allows obtaining composite materials using powders of different types and at various chemical compositions.

The PSFM should comply with the following requirements: stable value of the coefficient of friction, high wear resistance, effective adaptation, and high thermal conductivity [3]. As a rule, PSFM on the basis of copper are used to operate under lubrication conditions, while materials based on iron are used at dry friction.

The achievement of the given level of tribotechnical and operating properties of the PSFM has been achieved by the use of additives of various kinds, and granulometric composition, the content of which is within the range 0.5–15 vol.%. The additives are able to interact with the metal base, and to localize in the form of individual inclusions. The main additive used in the composition of PSFM are graphites of various types, as well as carbon-containing additives with an amorphous structure, such as coke and anthracite [4,5].

It has been shown that graphite with a size of 80 μm provides a high value of the coefficient of friction, compared with graphite with a size of 8–10 μm. However, a much greater increase in the coefficient of friction was achieved with the use of coke powder [6]. An additive in the form of graphite allows increasing the operating properties of the friction unit material [7–9]. For a material containing 5 masses % of graphite, an increase in the sliding velocity leads to a sharp increase in the coefficient of friction, while the material itself is characterized by the increased antifriction and anti-pressure properties, and performance under increased load-velocity conditions. For the tin bronze containing 10% of graphite, the wear resistance is significantly higher than bronze with the graphite contents [10].

However, the use of graphite by itself does not provide specific tribological properties. Their further improvement is achieved through the use of additives in the PSFM structure of solid ceramic powders and their compounds.

The composition of friction material (FM) for clutches and brake units, which has a high coefficient of friction and a small difference between dynamic and static coefficients of friction, was presented in the patent [11]. It was noted that this effect is achieved using 8–15% of Al_2O_3. By means of a method of powder metallurgy, a metal-based FM class with high coefficient of friction and wear resistance, as well as reduced acoustic characteristics was created [12–17]. These materials contain in their composition 2–30% of the component of solid particles selected from metal oxides (composites), metal nitrides (carbonitrides), metal carbides, metal borides, intermetalides, and minerals with a Mohs hardness of 3.5 or more.

Titanium dioxide is widely used in industry as an addition of a small cost, characterized by stable properties and non-toxicity. This additive is characterized by a very high specific surface area, up to 600 $m^2\ g^{-1}$, and a low thermal conductivity [18].

The carbides of transition metals are characterized by high hardness and, at the same time, brittleness. The most commonly used is silicon carbide, which has high hardness and thermal conductivity at low density. Its main disadvantage is the low ($2-3$ MPa $m^{1/2}$) viscosity of destruction [19].

Currently, the use of the intermetallic powder additives in tribotechnical materials is of great interest. The intermetalides Ni_3Al and NiAl appear in difficult operating conditions due to the set of unique properties, such as increased value of impact toughness, resistance to oxidation at elevated temperatures and thermal resistance. The above-mentioned intermetalides have a density 7.3 and 5.9 g cm^{-3}, respectively, are characterized by a high Young's elasticity, and may be used in products for constructional and tribotechnical purposes [20]. An effect of the addition of intermetallic powder Ti-46Al-8Cr, obtained by the method of mechanoactivable self-propagating high-temperature synthesis, on the tribological properties of the copper-based antifriction material was investigated [21]. It was shown that an increase in the content of aluminite from 0.5% to 1% leads to a decrease in the intensity of wear of the material by more than 3 times. The inclusion of the additive of $NiAl/Al_2O_3$ powder system in the frictional material based on copper in the range of 0.5–2.5% revealed an increase in the dynamic coefficient of friction from 0.040 to 0.051, while wear ranged from 4.2 to 5.7 $\mu m\ km^{-1}$ [22].

Newly designed PSFM materials used for friction elements of clutches, before implementation to production process, undergo a series of restrictive tests, both in full scale and laboratory tests. Even at the stage of their preselection and elimination of the least promising ones, this may be an expensive and time-consuming process. Numerical models are effective (time, costs) in this first phase of designing a new friction pair, allowing for a preliminary analysis of the level and temperature distribution of the friction pair. The novelty presented in this article is the proposed 3D numerical models to analyze the transient temperature fields of the new designed PSFM friction materials. The study presented an extensive numerical finite element (FE) model (from ref. [6]) of friction heating for the estimation of temperature distributions in a wet clutch. Unlike the numerical model from the article [6], this model takes into account the change of thermal properties of the steel disc under the influence of temperature and two different phases of the clutch operation, i.e., after its engagement (heat generation) and disconnection (cooling). The analysis of temperature distribution, based on the structural and tribological tests of newly developed PSFM materials, allowed selecting the most effective friction pair for the use in the wet clutch disc.

2. Materials and Experimental Methods

As a basis for the friction material, a mixture of copper powders (81 wt.%), tin (11 wt.%), and elemental graphite GE-1 (8 wt.%) was used. The initial mixture was prepared in a blade mixer by mixing, within 45 min, the copper powders obtained by electrolysis with a mean particle size of 100 μm (Figure 1a), the tin obtained by spraying the melt, with a mean particle size of 20 μm (Figure 1b), the elemental graphite GE-1, the natural origin obtained by extraction, grinding, and processing in acid solution and having a scaly shape with a mean size of 100 μm (Figure 1c). As test additives, the powders of silicon carbide with a size 4–9 μm (Figure 1d) and the titanium dioxide, which is a conglomerate with a size of 100–150 μm, were used. The conglomerate consisted of the ultra-disperse powders of predominantly spherical shape up to 0.2 μm in size (Figure 1e) and the intermetallic powder Ti-46Al-8Cr. Particles of the powder Ti-46Al-8Cr in the size of 50–500 nm formed agglomerates in the size 5–20 μm with a high specific surface area (Figure 1f). The micro hardness of the powder particles was 4000–5140 MPa. The powders were supplied by manufacturers.

Figure 1. Morphology of the surface of the particles of powders: (**a**) the copper; (**b**) the tin; (**c**) the elemental graphite GE-1; (**d**) the silicon carbide; (**e**) the titanium dioxide; (**f**) the intermetalide Ti-46Al-8Cr.

The TI-46AL-8CR system was obtained by the method of the mechanoactivated self-propagating high-temperature synthesis (MASHS) [23]. The preliminary mechanical processing of the reaction mixture of the powders of the titanium, aluminum, and chromium was carried out in a mill A-4.5 with the following parameters: rotational speed of the impeller shaft 360 r min^{-1}, the ratio of the mass of spheres and powder 10:1, and the duration of processing 3 h. The subsequent self-propagating high-temperature synthesis was carried out in the experimental reactor for MASHS in argon environment. The mixture of powders was ignited with a tungsten spiral heated by the passage of an electric current. After cooling, the resulting sinter was milled in the planetary mill Pulverisette 6 (Fritsch, Germany) in an alcohol medium at the following parameters: the diameter of the spheres 5 mm, the mass ratio of the spheres and the powder 20:1, rotational speed of the impeller drive shaft 400 r min^{-1}, and the grinding time 30 min.

The synthesized material Ti-46Al-8Cr, according to X-ray diffraction, consisted of the basis in the form of intermetalide $\gamma-$ TiAl (the spatial group P4/mmm) (Figure 2a,d phase 1), doped by chromium, containing 64–68 at. % Ti, 30–34 at. % Al and up to 5 at. % Cr, the inclusions of double intermetalides Ti_3Al and $AlCr_2$, and the triple intermetalide $Al_{0.67}Cr_{0.08}Ti_{0.25}$ (Figure 2a,c). Thin secondary τ-phase $Al_{0.67}Cr_{0.08}Ti_{0.25}$ smaller than 0.5 μm (Figure 2b, phase 3), falling out in grains of titanium monoaluminide, contained about 68–71 at. % Al, 20–25 at. % Ti, 7–12 at. % Cr and had a cubic grate of Pm-3m type, which provided coherence of boundaries with the γ-phase.

(a) (b)

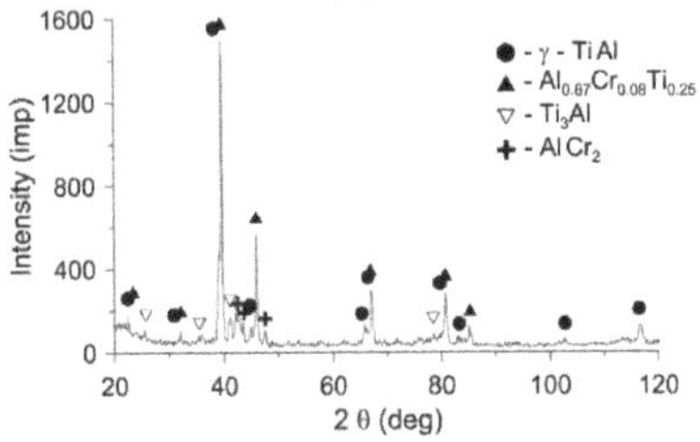

(c)

Variant, No.	Composition, at. %		
	Ti	Al	Cr
1	64–68	30–34	1.4–5
2	72.5–78	20–26	1.2–3
3	20–25	68–71	7–12
4	3–5	46–53	45–60

(d)

Figure 2. Microstructure and phase composition of the synthesized SHS powder of the Ti-46Al-8Cr system before grinding: (**a**,**b**) structure; (**c**) radiograph; (**d**) results of the micro X-ray spectral analysis (MRSA).

The $\alpha_2 -$ Ti_3Al phase (the spatial group P63/mmc) was localized mainly along the boundaries of grains and contained about 2 at. % Cr (Figure 2b, the phase 2). In addition, at the grain boundaries of titanium monoaluminide, there were also inclusions of excessive phases of chromium compounds with aluminum and titanium containing 46–53 at. % Al, 45–60 at. % Cr, and 3–5 at. % Ti (Figure 2b, phase 4), the formation of which was probably due to the problems of diffusion redistribution of components in the SHS process under conditions of predominantly solid-phase interaction.

The samples of the friction discs for testing were made as follows: obtained charge from the initial powders was applied by free filling to the surface of the steel base using special technological equipment, and then preliminary sintering was carried out in dissociated ammonia at a temperature of 840 °C within 50 min. For forming a system of oil-removing channels and grooves on the surface of the sintered material, as well as obtaining a porosity of 12–18%, the sintered workpiece of the friction disc was subjected to plastic deformation (embossing) with a punch having a profile in the form of a "grid" on the surface. Then, the final sintering was carried out at a pressure of 0.1 MPa in a medium of dissociated ammonia, which contains 75% H_2 and 25% N_2 at a temperature of 840 °C within 3 h. The friction and steel discs are shown in Figure 3.

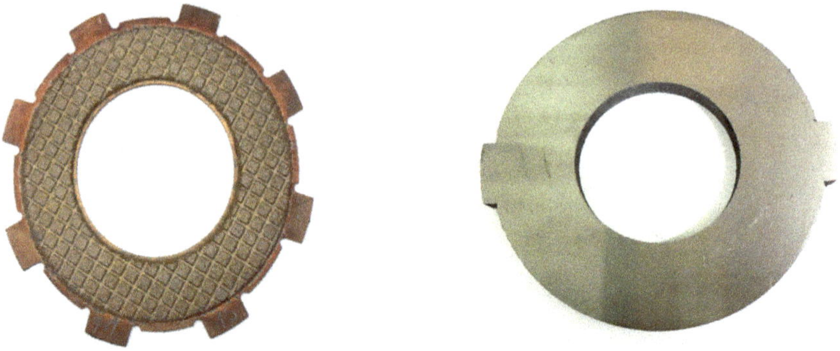

(a) (b)

Figure 3. Clutch elements: (**a**) friction disc; (**b**) steel disc-counter body [6].

The study of the tribotechnical properties of the friction material was carried out on a friction machine IM-58 according to the scheme friction disc-counter body at the following input parameters. The initial velocity of braking was 10 m s^{-1}, the contact pressure was 4 MPa, the moment of inertia of rotating masses was 0.56 N m s^2, and the work of friction was 27.5 kJ. As a counterpart, a disc made of 65H steel with a hardness of 260–320 HB and a surface roughness of Ra = 0.7–0.8 were used [6]. The bedding-in (burnishing) of the working surfaces by 300 engaging cycles was carried out. Then, 10 measurements of the values of the coefficients of friction and wear were made. From these data, mean values were determined.

The investigation of structure was carried out by means of the optical microscope MEF-3 (Austria). The morphology of the surface of the friction disc and its microstructure were studied on a high-resolution scanning electron microscope MIRA (Czech Republic) with a micro-X-ray spectral console INCA 350 of the Oxford Instruments (UK) company. The phase composition was examined on an X-ray diffractometer Ultima IV (Rigaky) in Cu Kα-radiation at an X-ray tube voltage of 40 kV and the anode current of 40 mA. The parameters of the crystal grate of the alloys were determined by diffraction lines located at the large scattering angles. For a phase analysis, a standard PDF card files was used. The thermophysical properties of investigated compositions of friction materials were carried out on the analyzer of thermal properties Hot Disk TPS2500S. As a sensor, a spiral, being a source of heat, was used. The sensor was located between the sample under the study and the sample, with the known thermophysical properties. Ten measurements were made after a given period of time, and the mean values of the thermophysical properties were established. The tested samples had a diameter of 50 mm, a thickness of 10 mm, and were obtained by compressing at the pressure of 2.5 t cm^{-2} and sintering at 840 °C for 3 h.

3. Results of Experimental Investigations

The results of the study of the physical and frictional properties of five compositions of friction materials are given in Table 1.

Table 1. Influence of the type of carbon-containing additive on the thermophysical properties and coefficient of friction.

Variant, No.	Material	Thermal Conductivity, W m^{-1} K^{-1}	Specific Heat, J kg^{-1} K^{-1}	Density, kg m^{-3}	Coefficient of Friction, Dimensionless	Wear, μm km^{-1}
0	Basic	7.9	5.4	4340	0.036	3.1
1	SiC 3%	6.5	5.4	4110	0.068	13.5
2	Ti-46Al-8Cr 2%	6.75	5.6	4220	0.055	5.1
3	TiO_2 2%	5.9	4.8	4210	0.043	3.6
4	TiO_2 5%	4.83	3.9	3980	0.052	4.3

The data obtained showed that use of additive SiC obtains the greatest value of the coefficient of friction. The solid inclusions of SiC in the process of friction are crumbled, displacing coarsely dispersed graphite from the surface of the friction material. The change in the morphology of the surface layer, the closure of pores, and increase in the area of the metal phase were fixed (Figure 4b).

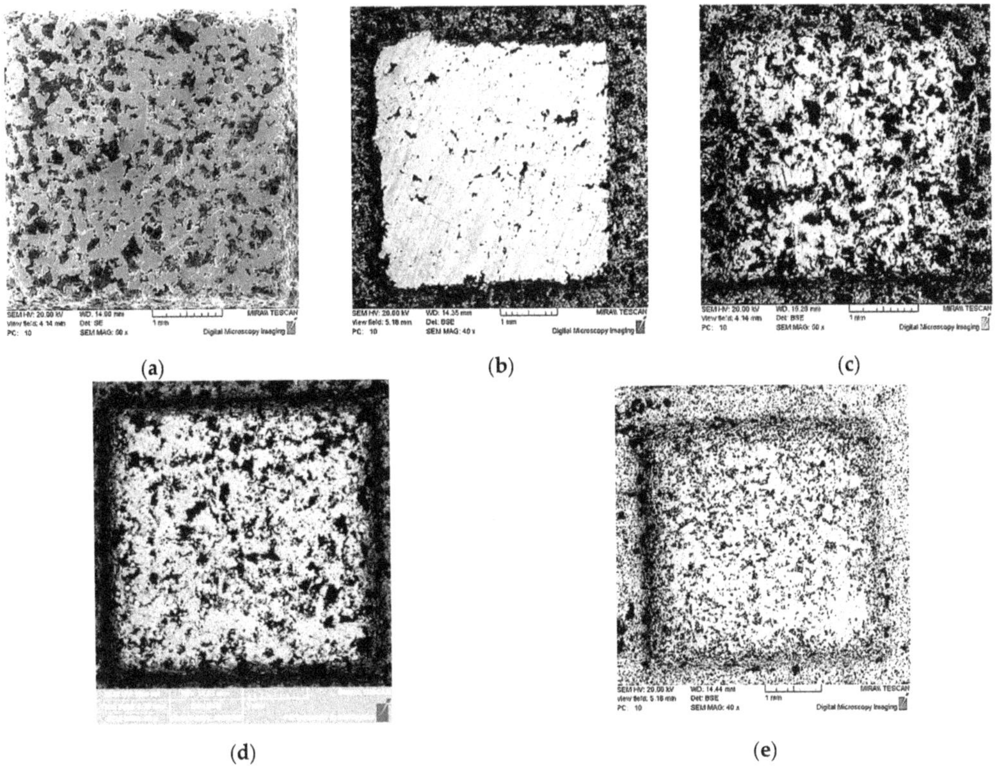

Figure 4. Morphology of the friction surface of the: (**a**) basic material; with powder additives (**b**) SiC; (**c**) Ti-46Al-8Cr; (**d**) 2 wt.% TiO_2; (**e**) 5 wt.% TiO_2.

The introduction of the additive of the intermetallic powder of the Ti-46Al-8Cr in an amount of 2 wt.% showed an increase in the coefficient of friction to 0.055, whereas for the basic composition, without additives of powders, it was 0.036. An analysis of the morphology of the surface layer showed that the initial porosity of the friction material was preserved, and there is no replacement of graphite particles (Figure 4c), which is characteristic of the basic composition of the friction material (Figure 4a).

The use of TiO_2 powder additive in an amount of 2 wt.% and 5 wt.% led to an increase in the coefficient of friction to 0.043 and 0.052, respectively. An increase in the addition of TiO_2 powder from 2 to 5 wt.% showed a change in the morphology of the friction surface of the friction material with a slight increase in the area of the metal phase (Figure 4d,e).

4. Numerical Simulation of the Temperature Mode of the Clutch

Operating Parameters

The aim of the numerical simulations was to investigate an effect of the above-mentioned powder additives, namely one ceramic (SiC) denoted as variant 1 and three intermetalides (2–Ti-46Al-8Cr, 3–2 wt.% TiO_2 and 4–5 wt.% TiO_2) to 0—the friction base material, on the clutch temperature, presented in Figure 3. The analyzed friction pair consisted of two discs—a fixed one with a steel substrate (65H) and a friction material applied to it—and a steel (65H) disc rotating against the specimen. The thermophysical properties of the materials at the ambient (initial) temperature $T_0 = 20$ °C are presented in Table 1. The changes in the properties of 65H steel with temperature increasing from 20 °C to 800 °C are shown in Table 2.

Table 2. Temperature-dependent properties of the steel 65H [24].

Temperature, °C	Thermal Conductivity, W m^{-1} K^{-1}	Density, kg m^{-3}	Specific Heat, J kg^{-1} K^{-1}
20	37	7850	490
100	36	7830	490
200	35	7800	510
300	34	7800	525
400	32	7730	560
500	31	7730	575
600	30	7730	590
700	29	7730	625
800	28	7730	705

The dimensions of the clutch components and the initial kinetic energy of the system were the same as in the article [6] (Table 2). The calculations were carried out for five variants of friction materials: 0—basic, 1—TiC, 2—Ti-46Al-8Cr, 3—2 wt.% TiO_2, 4—5 wt.% TiO_2 with the corresponding values of the coefficients of friction listed in Table 1.

5. Heating Taking into Account the Thermal Sensitivity of 65H Steel (First Calculation Model)

Two 3D numerical models were developed using the finite element method (FEM) adapted in the Heat Transfer Module of the COMSOL Multiphysics® programme. The first model was a generalization of the linear (with material properties unchanged) model from the article [6] for the case of thermally sensitive materials (with temperature-varying properties of 65H steel). The finite element analysis was limited only to the friction heating stage during braking. The results of the calculations are presented in Figure 5 and in Table 3.

The evolutions of temperature of the friction surfaces at the equivalent radius $r_{eq} = 39.4$ mm shown in Figure 5 for thermosensitive (dashed lines) and constant (solid lines) properties of materials revealed typical changes for braking at constant deceleration. Namely, temperature increased rapidly at the beginning, reached maximum value, and decreased until the stop. The obtained maximum temperature values did not exceed 165 °C (higher values appeared when taking into account thermosensitivity of the steel), hence the omission of the thermal sensitivity of the friction materials hardly influencing the simulation results. It should also be noted that in this temperature range (from 20 °C to 165 °C) the changes in the properties of steel 65H are negligible (Table 2).

Comparison of the results for 5 materials analyzed shows how braking time affects the maximum value of the temperature. Since frictional sliding lasts only a few seconds, generated heat cannot be absorbed by the components of the clutch and convection. Therefore, differences in braking times have a strong effect on the maximum temperature reached. The highest value (at thermosensitive material) is equal to 161.8 °C, whereas the lowest is equal to 123.1 °C.

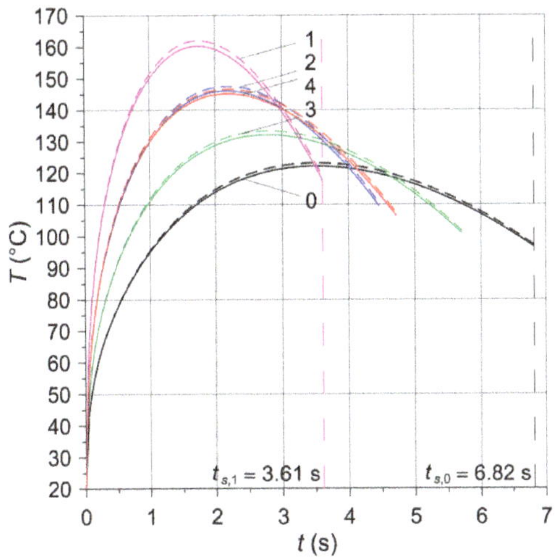

Figure 5. Evolution of the temperature of the friction surfaces of the clutch on the equivalent radius r_{eq} = 39.4 mm with constant (solid lines) and temperature-dependent (dashed lines) properties of 65H steel for five friction materials. Numbers 0–4 denote friction materials given in Table 1.

Table 3. Calculated parameters of the braking process for five friction materials.

Variant, No.	t_{max}, s	T_{max}, °C	t_s, s	T_s, °C
0	3.5	122.1	6.82	96.6
1	1.75	160.2	3.61	117.5
2	2.2	146.1	4.46	109.5
3	2.8	132.2	5.71	101.0
4	2.2	145.1	4.72	106.4

6. Heating with Subsequent Cooling of the Clutch Elements (Second Calculation Model)

The second computational model concerned both the clutch heating stage due to friction during operation as well as the next, after stopping, disengagement of the discs and their oil cooling. Due to the negligible influence of the thermal sensitivity of 65H steel on temperature (Figure 5), the calculations were performed with the constant, adapted to the initial temperature, material properties. This stemmed from the relatively short heating time of the clutch, less than 7 s, and thus limited ability to heat conduction to other neighboring parts of the assembly. On the other hand, the cooling step following the friction heating and lasting ≈90 s took place in the environment of the oil, which absorbed heat from the surface of the components intensely compared to air. The construction stages of the second model are presented below.

6.1. Boundary Conditions

As mentioned above, the analyzed friction pair consisted of three geometric objects representing the basic elements of the clutch. The calculations were divided into two stages:

1. Heating of the friction surfaces during sliding contact with convection cooling of the side free surfaces;

2. Exclusive convection cooling of the lateral surfaces and working faces of the discs, where frictional heating occurred in the first stage. The second stage simulated the state when the components were disconnected (no friction).

It should be noted that in both stages the surfaces of the discs parallel to the friction surfaces were adiabatic.

In the first stage, during the frictional heating with duration time denoted $t_{s,i}$, $i = 0, 1, 2, 3, 4$, the type of connection of geometric elements "create union" was used. This meant that the conditions of temperature continuity and heat flux intensity (perfect thermal contact) were required at the interface between the steel substrate and the clutch facing (friction material). On the other hand, on the contact surface of the friction material and the steel counterpart, there was a perfect thermal contact of friction, which consisted of meeting the following equality of:

1. Temperatures of friction surfaces;
2. The sum of the heat flux densities directed to each part and the boundary heat source power density.

On the lateral surfaces of both discs, heat exchange with the surrounding environment according to Newton's law of cooling at the constant heat transfer coefficient $h = 600$ W m^{-2} K^{-1} took place.

After stopping and disconnecting the clutch components, it was necessary to change the connection type of the parts in the "geometry" domain in COMSOL. Such a change affects almost all stages of the model creation (finite element mesh, selecting surfaces for heat transfer due to convection, etc.). Therefore, a new file was created, into which the temperature field from the last time step from the braking stage study was imported. Then, modifications were made to rebuild the geometry into an assembly. Creating an assembly, instead of the union formulation, allowed for the separation of the objects and the introduction of heat transfer due to convection also on the friction surfaces. The presence of such cooling better reflects the actual conditions in the clutch on the test bench. It was not possible when using the "create union" option in the computational model from the article [6].

6.2. Modeling Rotational Motion

As on the test stand, in the developed numerical models, it was assumed that the discs with the clutch facing are stationary, and the steel counterpart rotates at the angular velocity ω. The rotation of the counterpart in relation to the stationary disc was carried out using the well-known and verified approach of changing the velocity field at each point of the rotating part. The components of the linear velocity V vector were determined respectively from the dependence $V_x = -y\omega$ and $V_y = x\omega$ using a special tool available as the Translational Motion option of the Heat Transfer module of the COMSOL Multiphysics® software (Heat Transfer in Solids-Solid-Translational Motion).

6.3. Construction of a Finite Element Mesh of the Clutch

Apart from the counterpart (steel disc) characterized by geometrical axial symmetry (mounting elements were omitted) (Figure 3b), there were differences in the shape in the circumferential direction of the other parts (steel plate with the clutch facing) (Figure 3a). The spatial (3D) model of the clutch was selected for the thermal finite element analysis.

When dividing the 3D geometric objects of the clutch into finite elements, an automatic mesh generator with an option of tetrahedral elements (free tetrahedral) and the general default size appearing under the name "normal" was used. This method takes into account the type of the problem as well as the curvature and geometric details that change mesh (divide into smaller elements) only in critical areas. Initial attempts to manually create mapped or free quad elements and then building regular hexagonal finite elements on the basis of the sweep method showed a number of warnings and errors at the edges of the objects. This was due to the large difference in the size of the contacting edges of the two parts—the smallest edges in the case of a friction material with many cuts on the working surface, steel plate, and the counterpart.

The final mesh created from of tetrahedral elements is shown in Figure 6, and before the actual calculations, it was additionally verified in terms of distributions and maximum values of temperature in the braking process.

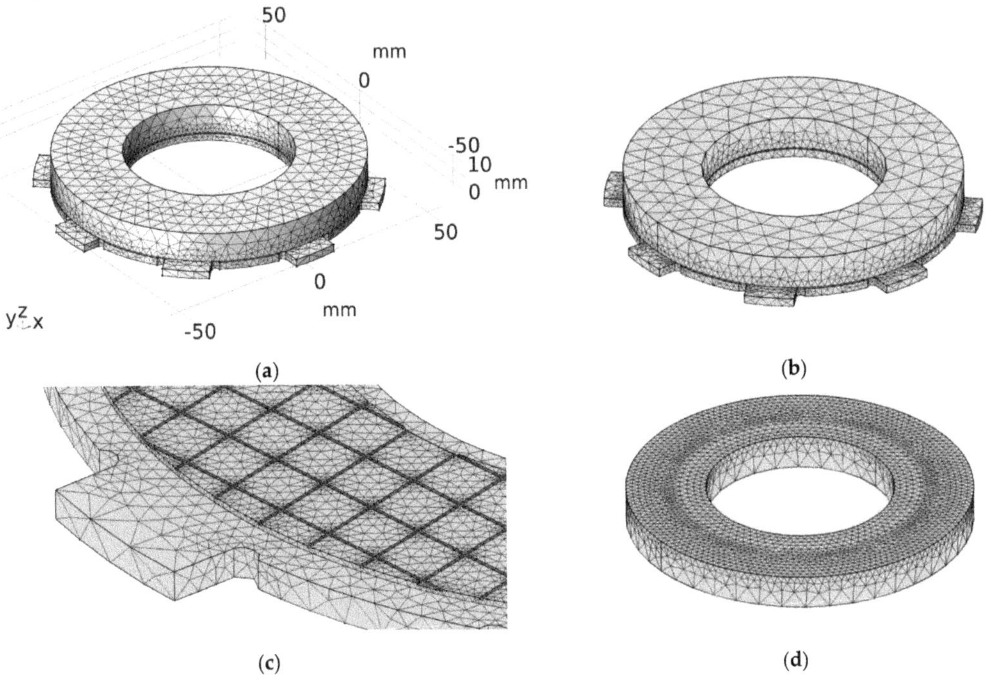

Figure 6. Finite element mesh used in finite element analysis of: (**a**) heating; (**b**–**d**) cooling.

6.4. Results of Computer Simulations

The second order shape function (quadratic Lagrange) of elements was used to calculate the temperature fields at both stages (heating and convection cooling). Such finite elements generated the most accurate results without the need to use an extremely fine mesh in the area of high temperature gradients. An experience in the construction of a finite element grid was obtained from previously conducted simulations of heat generation in disc brakes [25] and tread brakes (wheel-rail) of railway vehicles [26]. It was found that the linear finite elements significantly falsify the calculations (over 20%) even at many times higher than the default mesh density. The results of the calculations of the working surfaces temperature of the clutch are shown in Figure 7 and in Table 4.

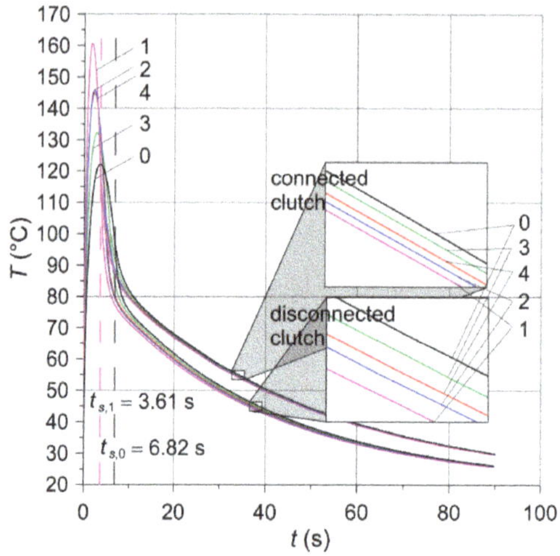

Figure 7. Evolutions of the temperature of the friction surfaces of the clutch on the equivalent radius $r_{eq} = 39.4$ mm obtained by means of the computational model: from the article [6]—connected clutch; developed in this paper—disconnected clutch. Numbers 0–4 denote friction materials given in Table 1.

Table 4. Calculated parameters for the cooling stage for five friction materials.

Variant, No.	t_s, s	t_c, s	$T_c^{(c)}$, °C Connected [6]	$T_c^{(d)}$, °C Disconnected
0	6.82	83.18	29.8	22.2
1	3.61	86.39	29.6	25.6
2	4.46	85.54	29.6	25.7
3	5.71	84.29	29.8	25.9
4	4.72	85.28	29.7	25.8

In order to investigate the effect of oil cooling in the contact area, the temperature distributions of the clutch in the cross-section (r, z) were compared under the condition of perfect thermal contact and with the disconnected parts after stopping time moments $t_s + 0$ s, $t_s + 15$ s, and $t_s + 55$ s (Figures 8–10). It should be noted that because of different braking durations for each of the five friction materials, the presented distributions occur at slightly different points in time from the beginning at $t = 0$ s.

The temperature distributions in Figure 8 show the stopping times $t_s + 0$ s. Slight differences in the distributions for variants a and b result from the fact that for connected clutch components (variant a) these are the values calculated and displayed from the model in which the perfect thermal contact condition was maintained all the time, while for disconnected components (variant b) the field is imported to the model with the separate cooling. The highest temperature is accumulated in the central part of the friction path near the friction radius.

Significant differences in temperature distributions resulting from the cooling method appeared after time $t_s + 15$ s (Figure 9). Due to the smaller total cooled area of the clutch, at this time moment, significantly higher temperature values were achieved for the model in which the friction pair remained connected (Figure 9a). Only for materials 1 and 2, for which the shortest cooling times take place, was the maximum temperature of the friction disc for variant b equally high.

The temperature evolutions are confirmed by the temperature distributions shown in Figure 10. It can be clearly seen that the temperature field for each of the tested materials was similar. However, while at the time moment $t_s + 15$ s, a higher temperature was obtained for the friction disc, and at $t_s + 55$ s a higher average temperature occurred for the steel disc.

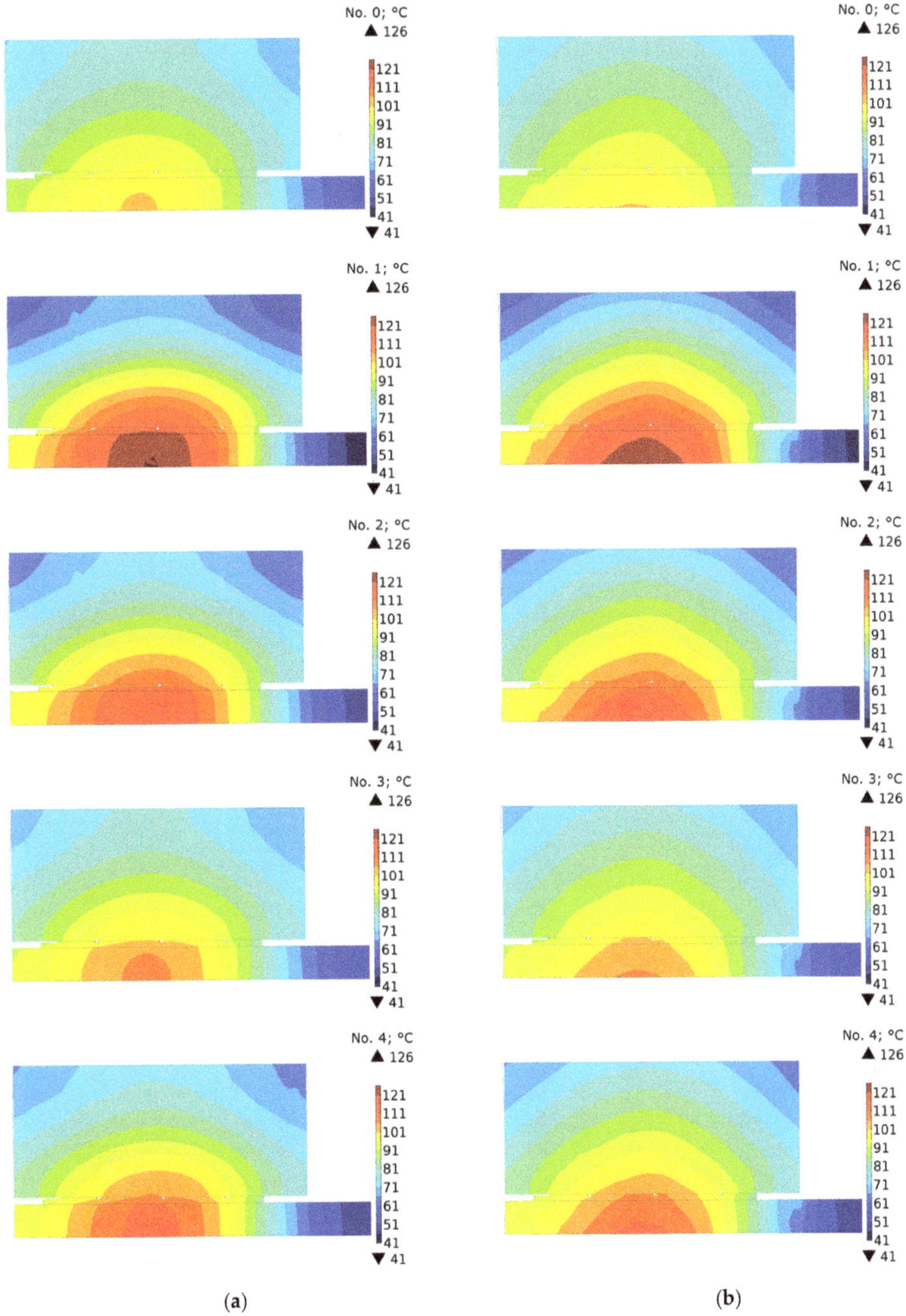

Figure 8. Temperature distribution at time $t_s + 0$ s obtained using: (**a**) connected [6]; and (**b**) disconnected parts of the clutch for materials no. 0, 1, 2, 3, 4.

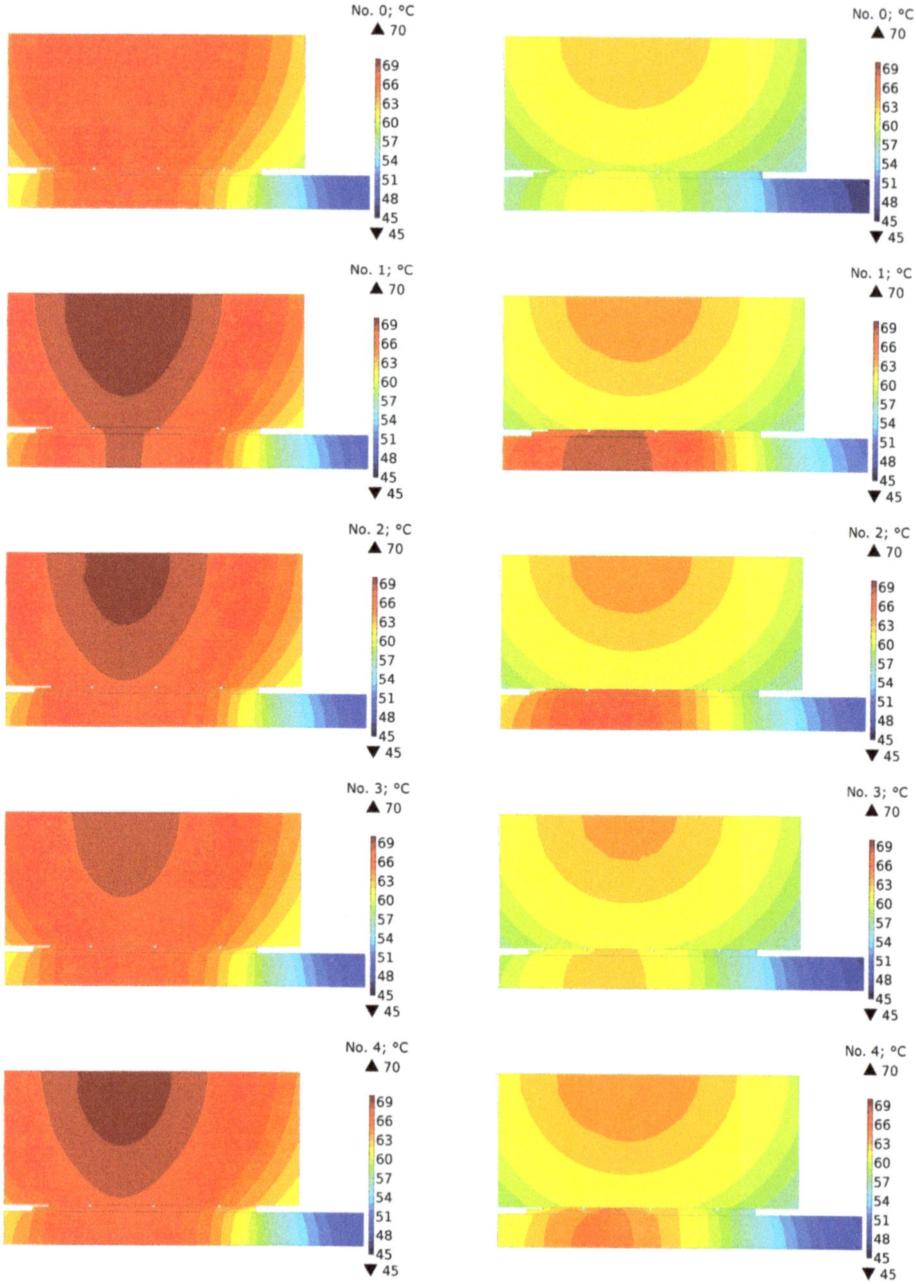

(a) (b)

Figure 9. Temperature distribution at time $t_s + 15$ s obtained using: (**a**) connected [6]; and (**b**) disconnected parts of the clutch for materials no. 0, 1, 2, 3, 4.

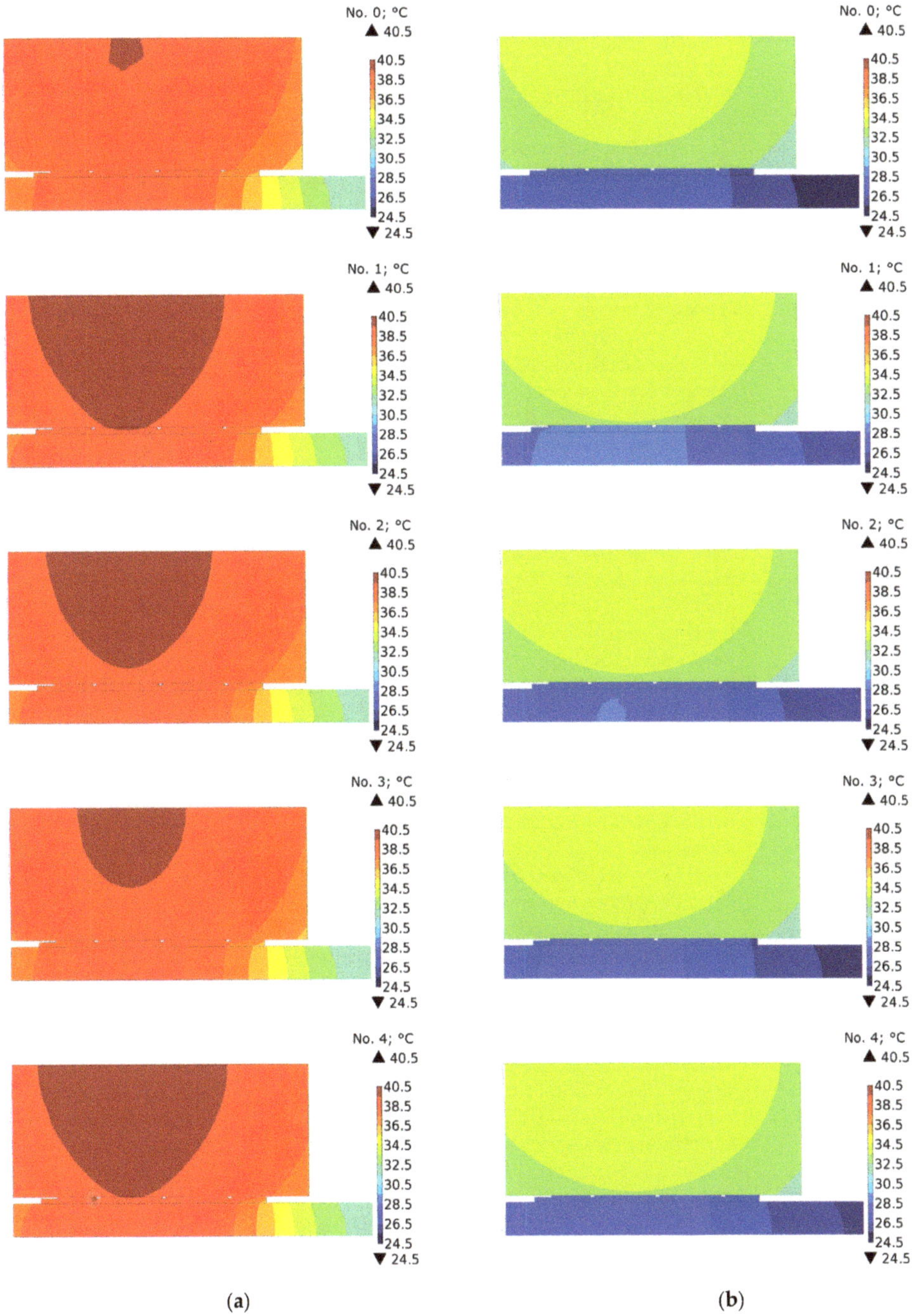

Figure 10. Temperature distribution at time $t_s + 55$ s obtained using: (**a**) connected [6]; and (**b**) disconnected parts of the clutch for materials no. 0, 1, 2, 3, 4.

7. Results and Discussion

The article presented an experimental analysis of material properties and thermal finite element analysis of friction heating for new PSFM materials used for clutch facing under lubricated conditions. Experimental tests were carried out on the IM-58 friction machine for four different PSFM materials with different additives (SiC, Ti-46Al-8Cr, and TiO_2) and one base material. The materials produced were formed into friction discs and combined with a steel 65H disc, determining the values of the friction coefficients for each pair. The thermophysical properties for the new materials were investigated using the Hot Disk TPS2500S analyzer of thermal properties. These properties and values of the friction coefficient, as well as the input parameters (initial velocity, contact pressure, moment of inertia of rotating masses) of the experiment were adapted to 3D numerical models of friction heating. Based on the computer simulations carried out for the heating stage only with five friction materials, the temperature distributions (its maximum value on the contact surface and the time to reach this value), taking into account the temperature changes of the material properties of the steel disc, were analyzed. In the second part of the numerical tests, both the friction heating stage and the cooling stage after the clutch was disengaged were taken into account.

One of the main results of the material, tribological, and numerical tests carried out is the selection of such additives that had the greatest impact on the operation of the friction pair, and thus also on the temperature level during clutch engagement. It was shown that the greatest change of the tribological properties was obtained using addition of 3 wt.% SiC in the composition of the friction material based on copper with 12% tin and 30 vol.% graphite GE-1, namely, the coefficient of friction increased from 0.036 to 0.068. At the same time, wear increased from 3.1 to 13.5 μm km^{-1}. The least influence on the tribological properties of the base material has 2 wt.% TiO_2 powder, i.e. the coefficient of friction was the smallest (0.043) at the greatest wear resistance.

The basic factors influencing the changes in temperature distribution in the friction pair components and the evolution of the maximum temperature in the contact zone include (1) the amount of mechanical energy converted into heat, and thus the initial angular velocity and the moment of inertia of rotating masses; (2) the velocity at which this energy is dissipated, i.e. the braking torque dependent on the clamping force, coefficient of friction, and the friction radius; (3) type and dimensions of the given friction pair (thickness, number of neighboring elements absorbing heat), (4) thermophysical properties, and (5) cooling conditions due to convection and thermal radiation.

Assuming that in the analyzed friction pairs, braking takes place at the same input parameters (initial angular velocity, moment of inertia of rotating masses and clamping force), and assuming that the process time is short enough to ignore the influence of cooling, the key factors that affect the maximum temperature are the thermophysical properties and the coefficient of friction. As shown in Table 1, the thermophysical properties were very similar, while the greatest difference in the values of the friction coefficients was 89% (SiC in relation to the base material). Therefore, it is the coefficient of friction and the resulting braking time that in this case play a key role in reaching the maximum temperature value. For the higher coefficient of friction, the braking time is shorter, and the maximum temperature higher since the time for heat dissipation from the contact area due to conduction being limited.

The shortest braking time $t_{s,1} = 3.61$ s and the highest temperature value equal to 160.2 °C among the five numerically tested materials was for the material with the addition of ceramic powder (SiC)—the greatest value of the coefficient of friction (Table 1). The longest braking time $t_{s,0} = 6.82$ s and lowest temperature on the working surfaces, equal to 122.1 °C, was reached for the base material—the least value of the coefficient of friction.

Taking into account the disconnection of the clutch elements after stopping and convection cooling of the working surface with oil at the heat transfer coefficient h influences the value of the maximum contact surface temperature. The difference in the average temperature value for the five materials with the clutch components disconnected and the average temperature value obtained while maintaining the condition of perfect thermal contact was about 13 °C (30%) in the middle of the cooling stage ($t \approx 40$ s) and 5 °C (17%) at the end ($t \approx 90$ s).

Based on the presented research, we can conclude that the most promising from the point of view of achieving the shortest braking time with the same total friction work is the friction material with a 3% addition of SiC ceramics.

As a part of the future research, it is planned to determine the mechanical properties of the considered friction materials and to carry out numerical calculations of thermal stresses. In addition, attempts will be made to take into account the thermal contact resistance instead of using the perfect thermal contact condition.

Author Contributions: Conceptualization, A.Y., M.K. and A.I.; methodology, A.Y., M.K. and A.I.; software, P.G. and A.L.; validation, M.K. and A.L.; formal analysis, A.Y., M.K. and A.I.; investigation, P.G. and A.L.; resources, A.Y., A.I., P.G. and A.L.; data curation, P.G. and A.L.; writing—original draft preparation, M.K., P.G. and A.L.; writing—review and editing, A.Y., M.K. and A.I.; visualization, P.G. and A.L.; supervision, A.Y., M.K. and A.I.; project administration, M.K.; funding acquisition, A.Y., M.K. and P.G. All authors have read and agreed to the published version of the manuscript.

Funding: Project financing through the program of the Minister of Science and Higher Education of Poland named "Regional Initiative of Excellence" in 2019–2022, project number 011/RID/2018/19, amount of financing 12,000,000 PLN.

Institutional Review Board Statement: Not applicable.

Informed Consent Statement: Not applicable.

Data Availability Statement: Not applicable.

Conflicts of Interest: The authors declare no conflict of interest.

Nomenclature

h	heat transfer coefficient (W m^{-2} K^{-1})
r	radial coordinate (m)
r_{eq}	equivalent radius of the contact region (m)
t	time (s)
t_c	cooling time equal to $t_c = t_{end} - t_s$ (s)
t_{end}	total time of the braking and cooling process equal to 90 s (s)
$t_{\max}$	time, at which maximum temperature on the equivalent radius was reached (s)
$t_{s,i}$	braking time of the $i = 0, 1, 2, 3, 4$ friction material (s)
T	temperature (°C)
$T_c^{(c)}$	temperature of the friction surfaces at $r = r_{eq}$, obtained using numerical model with connected clutch components at the end of cooling $t = t_{end}$
$T_c^{(d)}$	temperature of the friction surfaces at $r = r_{eq}$, obtained using numerical model with disconnected clutch components at the end of cooling $t = t_{end}$ (°C)
$T_{\max}$	maximum temperature obtained during braking at time $t_{\max}$ (°C)
T_s	temperature obtained at the end of braking process at time t_s (°C)
T_0	initial/ambient temperature (°C)
V	linear velocity vector (m s^{-1})
x, y	Cartesian coordinates (m)
z	axial coordinate (m)
Greek Symbols	
ω	angular velocity (rad s^{-1})

References

1. Yu, L.; Ma, B.; Chen, M.; Li, H.Y.; Liu, J. Experimental Study on the Friction Stability of Paper-Based Clutches Concerning Groove Patterns. *Ind. Lubr. Tribol.* **2019**, *72*, 541–548. [CrossRef]
2. Yu, L.; Ma, B.; Chen, M.; Li, H.; Ma, C.; Liu, J. Comparison of the Friction and Wear Characteristics between Copper and Paper Based Friction Materials. *Materials* **2019**, *12*, 2988. [CrossRef] [PubMed]

3. Marklund, P.; Larsson, R. Wet Clutch Friction Characteristics Obtained from Simplified Pin on Disc Test. *Tribol. Int.* **2008**, *41*, 824–830. [CrossRef]
4. Leshok, A.; Dykha, A. Regularities of the Influence of Submicron Ceramic Powders TiO2, AlN, Cr2O3 on the Tribological Properties of a Friction Material. *Probl. Tribol.* **2021**, *99*, 20–27. [CrossRef]
5. Shibata, K.; Goto, A.; Yoshida, S.; Azuma, Y.; Nakamura, K. *Development of Brake Friction Material*; SAE Technical Paper 930806; SAE International: Warrendale, PA, USA, 1993.
6. Yevtushenko, A.; Grzes, P.; Ilyushenko, A.; Liashok, A. An Effect of a Carbon-Containing Additive in the Structure of a Friction Material on Temperature of the Wet Clutch Disc. *Materials* **2022**, *15*, 464. [CrossRef] [PubMed]
7. Ma, W.; Lu, J. Effect of Sliding Speed on Surface Modification and Tribological Behavior of Copper–Graphite Composite. *Tribol. Lett.* **2011**, *41*, 363–370. [CrossRef]
8. Kato, H.; Takama, M.; Iwai, Y.; Washida, K.; Sasaki, Y. Wear and Mechanical Properties of Sintered Copper–Tin Composites Containing Graphite or Molybdenum Disulfide. *Wear* **2003**, *255*, 573–578. [CrossRef]
9. Ghorbani, M.; Mazaheri, M.; Afshar, A. Wear and Friction Characteristics of Electrodeposited Graphite–Bronze Composite Coatings. *Surf. Coat. Technol.* **2005**, *190*, 32–38. [CrossRef]
10. Kestursatya, M.; Kim, J.K.; Rohatgi, P.K. Wear Performance of Copper–Graphite Composite and a Leaded Copper Alloy. *Mater. Sci. Eng. A* **2003**, *339*, 150–158. [CrossRef]
11. Kondoh, K.; Takano, Y. Sintered Friction Material, Composite Copper Alloy Powder Used Therefor and Manufacturing Method Thereof. JPS5879073A, 1 May 1996.
12. Matsumoto, M.; Nakazawa, S.; Okazaki, T. Dry-Process Sintered Friction Material. JPH0415285A, 28 May 1998.
13. Kano, G.; Karikome, T. Sintered Alloy as Friction Material. JPS60116751A, 24 June 1985.
14. Naohiro, H.; Susumu, U. Dry Sintered Friction Material. JPS58151444A, 8 September 1983.
15. Senda, Y. Sintered Friction Material. JPS6167737A, 7 April 1986.
16. Kawakami, Y.; Takahashi, H. Frictional Material for Brake. JP2010222635A, 7 October 2010.
17. Oyabu, H.; Saito, K.; Tomiyama, Y. Friction Material of Sintered Alloy. JPS63109131A, 12 July 1995.
18. Tarasov, A.B. Synthesis, Structure and Functional Properties of Nanostructured Titanium Dioxide Obtained by Heterogeneous Hydrolysis of Titanium Tetrachloride in Aerosol Systems. Ph.D. Thesis, Candidate of Chemical Sciences: 02.00.04. Chernogolovka, Lomonosov Moscow State University, Moscow, Russia, 2016.
19. Perevislov, S.N. Materials Based on Silicon Carbide and Silicon Nitride with Oxide Activating Additives for Structural Products. Ph.D. Thesis, St. Petersburg State Institute of Technology, St. Petersburg, Russia, 2018.
20. Kablov, E.N.; Lukin, V.I. Intermetallic Compounds Based on Titanium and Nickel for Products of New Technology. *Autom. Weld.* **2008**, *11*, 76–82.
21. Kolobov, Y.R.; Kablov, E.N.; Kozlov, Y.V. *Structure and Properties of Intermetallic Materials with Nanophase Hardening*; Kablova, E.N., Kolobov, Y.R., Eds.; Center for Nanostructural Materials and Nanotechnologies of the Belgorod State University MISiS: Moscow, Poland, 2008.
22. Ilyushchenko, A.F. Investigation of the Influence of Additives of Ultrafine Powder of the NiAl/Al2O3 System on the Tribotechnical Properties of a Friction Material Based on Copper, Ilyushchenko, A.Ph.; Talako, T.L.; Leshok, A.V.; Letsko, A.I.; Pinchuk, T.I. *Rep. Natl. Acad. Sci. Belarus* **2022**, *66*, 113–122.
23. Dyachkova, L.N.; Fel'dshtein, E.E.; Letsko, A.I.; Kelek, P.; Kelek, T. Tribotechnical Properties of Sintered Bronze Strengthened by Aluminide of Ti–46Al–8Cr. *J. Frict. Wear* **2017**, *38*, 98–103. [CrossRef]
24. Handbook of the Properties of Substances and Materials. Available online: http://thermalinfo.ru/svojstva-materialov/metally-i-splavy/teploprovodnost-stali-i-chuguna-teplofizicheskie-svojstva-stali (accessed on 23 June 2022).
25. Yevtushenko, A.; Kuciej, M.; Grzes, P.; Wasilewski, P. Comparative Analysis of Temperature Fields in Railway Solid and Ventilated Brake Discs. *Materials* **2021**, *14*, 7804. [CrossRef] [PubMed]
26. Kuciej, M.; Grzes, P.; Wasilewski, P. A Comparison of 3D and 2D FE Frictional Heating Models for Long and Variable Applications of Railway Tread Brake. *Materials* **2020**, *13*, 4846. [CrossRef] [PubMed]

materials

MDPI

Article

Inverse Thermoelastic Analysis of a Cylindrical Tribo-Couple

Roman Kushnir [1,2], Anatoliy Yasinskyy [1], Yuriy Tokovyy [1,2,*] and Eteri Hart [3]

1 Pidstryhach Institute for Applied Problems of Mechanics and Mathematics, National Academy of Sciences of Ukraine, 79060 Lviv, Ukraine; dyrector@iapmm.lviv.ua (R.K.); yasinskyy.anatoliy@gmail.com (A.Y.)
2 Department of Applied Mathematics, Institute of Applied Mathematics and Fundamental Sciences, National University Lviv Polytechnic, 79013 Lviv, Ukraine
3 Department of Theoretical and Computer Mechanics, Oles Honchar Dnipro National University, 49010 Dnipro, Ukraine; hart@ua.fm
* Correspondence: tokovyy@gmail.com

Abstract: Within the framework of the one-dimensional model for a tribo-couple consisting of two elastic cylinders accounting for the frictional heat generation on the interface due to the roughness of the contacting dissimilar materials, a problem on the identification of the unknown temperature on one of the limiting surfaces of either inner or outer cylindrical layers is formulated and reduced to an inverse thermoelasticity problem via the use of the circumferential strain given on the other surface. To solve the latter problem, a semi-analytical algorithm is suggested, and its stability with respect to the small errors in the input data is analyzed. The efficiency of the proposed solution algorithm is validated numerically by comparing its results with the solution of a corresponding direct problem. The temperature and thermal stresses in the tribo-couple are analyzed.

Keywords: tribo-couple; cylindrical layers; frictional heating; unknown thermal loading; inverse thermoelasticity problem; Volterra integral equation; stable algorithm

Citation: Kushnir, R.; Yasinskyy, A.; Tokovyy, Y.; Hart, E. Inverse Thermoelastic Analysis of a Cylindrical Tribo-Couple. *Materials* **2021**, *14*, 2657. https://doi.org/10.3390/ma14102657

Academic Editor: Aleksander Yevtushenko

Received: 8 April 2021
Accepted: 3 May 2021
Published: 19 May 2021

1. Introduction

Construction and improvement of the elements of present-day techniques, along with the development and implementation of new materials with advanced properties, necessitate the comprehensive analysis of the heat transfer and the stress–strain state in composite materials under the simultaneous action of force and thermal fields while accounting for a wide range of the operational and constructional features, as well as the interaction of the structural elements of different geometry [1]. The importance of such analysis for both mechanical engineering and material science is also motivated by the prioritized implementation of non-destructive testing, which is important for ensuring the safety and durability of the operational performance of the heat and power equipment [2–4].

The comprehensive thermoelastic analysis is extremely important for the structural elements, some surface parts of which appear to be inaccessible (due to specific structural, technological, operational, or environmental reasons) for the direct reading of the thermal and mechanical signatures that are to be in use as the boundary conditions for the corresponding direct heat-transfer and thermoelasticity problems. As a result, the corresponding heat-transfer and thermoelasticity problems for such structural elements become ill-posed and require some supplementary information about the thermal or mechanical process, collected, preferably, on the accessible segments of the surface. It is worth noting that the type of additional information can be regarded as a critical point of the methodologies for solving the ill-posed problems of this kind.

If, for example, the original problem is supplemented with the information about some parameters of the thermal process (e.g., temperature or heat flux) at some points of a solid or its surface, the problem of the identification of the unknown thermal loading can be reduced to solving an inverse heat conduction problem [5,6]. The inverse problems obtained in this case are substantially ill-posed and their solution is concerned with the application of the corresponding regularizing algorithms.

There are numerous practical cases, however, when the reproduction of all the components of thermal loading appears to be impossible within the framework of an inverse heat-transfer problem alone. While accounting for coupling between the temperature and strain fields, the original mathematical models in these cases can be extended to non-classical thermoelastic problems by implementing the additional information on the thermomechanical parameters (displacements, strains, or stresses) on the accessible segment of the surface. The problem of determining the temperature and thermostressed fields in a solid under the above conditions can thereby be reduced to an inverse thermoelasticity problem [7]. The inverse problems of this kind are conditionally well-posed; i.e., they may become well-posed under certain supplementary conditions. This can be explained by the fact that the components of the stress–strain state have the form of integral dependences on the temperature at all points of a solid, including its boundary [1,8,9]. For non-stationary processes, these conditions usually express the fitting between the input data at the initial moment of time or interrelation of the mechanical components on the surfaces of the solid [1,10]. Some methods for solving one- and two-dimensional inverse thermoelasticity problems have been addressed in [11–16].

The inverse analysis of the temperature on inaccessible surfaces is an important issue in the analysis of tribo-systems. Such analysis is a key point in evaluating the characteristics of frictional interaction and material properties and is vital for both mechanical engineering and material science. Therefore, many practical cases of thermoelasticity problems, those that focus on the coupling between the thermal and mechanical fields, are concerned with frictional heating induced by the roughness of the contacting surfaces of interacting solids (see, e.g., [17–19]).

In [20–22], a technique for solving the inverse thermoelasticity problems was presented based on the reduction to integral equations. Particularly in [21,22], one-dimensional thermoelasticity problems were considered for interacting layers with friction. In engineering practice and material science experiments, numerous tribo-systems involve elements of cylindrical shape. In this paper, we extend the technique for solving inverse one-dimensional thermoelasticity problems, which are obtained for the identification of the time-dependence of a temperature of one of the circumferences of a cylindrical tribo-couple by making use of the additionally known circumferential strain on the surface where the thermal loading is known.

2. Formulation of the Problem

Consider a one-dimensional model of a cylindrical tribo-couple consisting of two cylindrical elements made of dissimilar materials generating heat due to mutual interfacial friction [17]. The model is schematized in Figure 1 and presented by a circular hollow cylinder "1" of the inner and outer radii $r = R_1$ and $r = R_0$ encapsulated without tension and gap into another cylinder "2" of the same shape with inner and outer radii $r = R_0$ and $r = R_2$, respectively. Assume the inner, $r = R_1$, and outer, $r = R_2$, circumferences of the tribo-couple to be kept under the given transient temperatures $t_1^*(\tau_*)$ and $t_2^*(\tau_*)$ while being subjected to the compressive forces $P_1(\tau_*)$ and $P_2(\tau_*)$. Here, r is the radial coordinate and τ_* is time. The mechanical and thermal contact of the cylindrical layers occurring on the interface $r = R_0$ is assumed to be imperfect in view of the roughness of the material on contacting surfaces so that the linear relationship

$$\widetilde{u}^{(i)}(\tau_*) = (-1)^i n_i P(\tau_*),\ i = 1,2 \tag{1}$$

obtains between the radial displacements on the interface $\widetilde{u}^{(1)}(\tau_*)$ and $\widetilde{u}^{(2)}(\tau_*)$ induced by the deformation of micro-roughness and the contact pressure $P(\tau_*)$. Here, n_i are the coefficients characterizing the deformative features of the contacting surfaces.

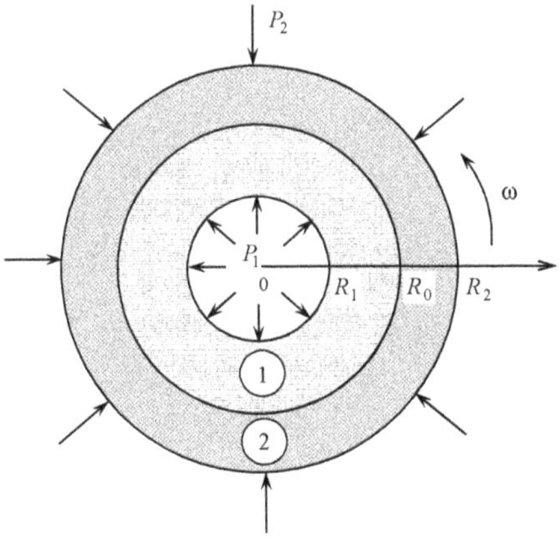

Figure 1. The scheme of the considered tribo-couple, where the inner and outer cylindrical layers are denoted by "1" and "2", respectively; the thermal and force loadings are imposed on the inner and outer surfaces R_1 and R_2 and the frictional heating occurs on the interface R_0.

Assume that one of the cylinders (let it be the outer one) rotates against the other cylinder. Let us restrict our attention to the case when the rotation reaches a steady-state condition at a constant angular velocity $\omega = const$. Due to the frictional forces according to Amonton's law, the interface $r = R_0$ is subjected to the non-stationary heat generation, and the specific power of the frictional heating sources equals the specific work of the friction forces. The mechanical and thermo-physical properties of cylinders 1 and 2 are constant and indicated with upper indices accordingly. Within the framework of the formulated problem, the transient temperature field in the considered tribo-couple varies along the radial coordinate r only, and in view of the plane strain condition, $u_z^{(1)} = u_z^{(2)} = 0$, where $u_z^{(i)}$ is the axial displacement of the ith layer of the cylinder.

In view of the foregoing model, the one-dimensional thermoelasticity problem for the tribo-couple is governed by the following system of equations, including:

(i) the heat-transfer equation

$$\frac{\partial^2 T_i^*(\rho,\tau)}{\partial\rho^2} + \frac{1}{\rho}\frac{\partial T_i^*(\rho,\tau)}{\partial\rho} = \frac{1}{b_i}\frac{\partial T_i^*(\rho,\tau)}{\partial\tau} \tag{2}$$

(ii) and the Lamé equations

$$\frac{\partial^2 u_r^{(i)}(\rho,\tau)}{\partial\rho^2} + \frac{1}{\rho}\frac{\partial u_r^{(i)}(\rho,\tau)}{\partial\rho} - \frac{u_r^{(i)}(\rho,\tau)}{\rho^2} = \beta_i R_0 \frac{\partial\, T_i^*(\rho,\tau)}{\partial\rho} \tag{3}$$

$$\frac{\partial^2 u_\varphi^{(i)}(\rho,\tau)}{\partial\rho^2} + \frac{1}{\rho}\frac{\partial u_\varphi^{(i)}(\rho,\tau)}{\partial\rho} - \frac{u_\varphi^{(i)}(\rho,\tau)}{\rho^2} = 0 \tag{4}$$

under the set of complementary conditions consisting of:

(i) the mechanical boundary conditions

$$\sigma_{rr}^{(i)}(k_i,\tau) = -P_i(\tau),\ u_\varphi^{(i)}(k_i,\tau) = 0 \tag{5}$$

(ii) the mechanical interface conditions

$$\begin{gathered}\sigma_{rr}^{(1)}(1,\tau) = \sigma_{rr}^{(2)}(1,\tau) = -P(\tau),\ \sigma_{r\varphi}^{(1)}(1,\tau) = \sigma_{r\varphi}^{(2)}(1,\tau) = -fP(\tau)\\ u_r^{(1)}(1,\tau) + \widetilde{u}^{(1)}(\tau) = u_r^{(2)}(1,\tau) + \widetilde{u}^{(2)}(\tau),\ u_z^{(i)}(\rho,\tau) = 0\end{gathered} \tag{6}$$

(*iii*) the thermal boundary conditions

$$T_i^*(k_i,\tau) = t_i^*(\tau) \tag{7}$$

(*iv*) the thermal interface conditions

$$\begin{aligned} \lambda_1 \frac{\partial T_1^*(1,\tau)}{\partial \rho} - \lambda_2 \frac{\partial T_2^*(1,\tau)}{\partial \rho} &= \omega R_0^2 f P(\tau) \\ \lambda_1 \frac{\partial T_1^*(1,\tau)}{\partial \rho} + \lambda_2 \frac{\partial T_2^*(1,\tau)}{\partial \rho} &= \frac{R_0}{R}\left(T_2^*(1,\tau) - T_1^*(1,\tau)\right) \end{aligned} \tag{8}$$

(*v*) and the initial condition

$$T_i^*(\rho,0) = T_0 = \text{const} \neq 0 \tag{9}$$

where, $i = 1,2$ ($i = 1$ corresponds to the range $\rho \in [k_1, 1)$ and $i = 2$ corresponds to the range $\rho \in (1, k_2]$), $\rho = r/R_0$ is the dimensionless radial coordinate $\rho \in [k_1, k_2]$, $k_i = R_i\ /\ R_0$, $\tau = a_2\tau_*/R_0^2$ is the Fourier criterion, $\tau \in (0, \tau_m]$, τ_m is a constant parameter, $b_1 = a_1\ /a_2$, $b_2 = 1$, a_i is the coefficient of thermal diffusivity, $\beta_i = \alpha_T^{(i)}(1+\nu_i)/(1-\nu_i)$, $\alpha_T^{(i)}$ is the coefficient of linear thermal expansion, ν_i is the Poisson ratio, λ_i denotes the heat-conduction coefficient, T_i^* is the temperature, $u_r^{(i)}$ and $u_\varphi^{(i)}$ are the radial and circumferential displacements, $\sigma_{rr}^{(i)}$ and $\sigma_{r\varphi}^{(i)}$ are the radial and tangential stress-tensor components, f is the coefficient of friction, and R is the coefficient of contact thermal resistance.

It is well known that in the case of plane strain, the thermoelasticity problem (3) and (4) can be represented by two independent problems [23], when (i) $u_r^{(i)} \neq 0$, $\varepsilon_{rr}^{(i)} \neq 0$, $\varepsilon_{\varphi\varphi}^{(i)} \neq 0$, $u_\varphi^{(i)} = \varepsilon_{r\varphi}^{(i)} = 0$ and (ii) $u_\varphi^{(i)} \neq 0$, $\varepsilon_{r\varphi}^{(i)} \neq 0$, $u_r^{(i)} = \varepsilon_{rr}^{(i)} = \varepsilon_{\varphi\varphi}^{(i)} = 0$, $i = 1,2$. Here, $\varepsilon_{rr}^{(i)}$, $\varepsilon_{\varphi\varphi}^{(i)}$, and $\varepsilon_{r\varphi}^{(i)}$ are, respectively, the radial, circumferential, and tangential strains of the ith cylindrical layer.

If all the input functions and coefficients in Equations (2)–(9) are properly imposed, then the formulated problem appears to be a well-posed direct thermoelasticity problem. Assuming, however, the transient temperature $t_1^*(\tau)$, $\tau \in [0,\ \tau_m]$, on the inner surface $\rho = k_1$ to be unknown (a typical situation due to the inaccessibility of the inner surface for the direct measurement) necessitates the determination of this function prior to solving the direct problem.

In order to identify this function appearing in the boundary condition (7), we use the supplementary information about the thermo-mechanical state of the compound cylinder, i.e., the condition

$$\varepsilon_{\varphi\varphi}^{(2)}(k_2,\tau) = \varphi_*(\tau),\ \tau \in [0,\ \tau_m] \tag{10}$$

imposing the circumferential strain measured on the accessible outer surface $\rho = k_2$. Here, $\varphi_*(\tau)$ is a given function of time.

Let us determine the temperature field and thermal stresses in the considered tribo-couple by making use of condition (10) in order to identify the unknown temperature distribution $t_1^*(\tau)$ on the inner circumference of the cylinder.

3. Solution Technique

By implementing the technique suggested in [17], a solution to the formulated thermoelastic problem (1), (3)–(6) can be given in the form expressing the circumferential strain in the cylindrical tribo-couple explicitly through the force loadings and thermal field as follows:

$$\begin{array}{c}\varepsilon_{\varphi\varphi}^{(i)}(\rho,\tau)=\alpha_T^{(i)}(1+\nu_i)T_0+\left((1-\overline{\nu}_i)+\frac{1+\overline{\nu}_i}{\rho^2}\right)\frac{c_ip_i(\tau)}{2}\\ -\left((1-\overline{\nu}_i)+k_i^2\frac{1+\overline{\nu}_i}{\rho^2}\right)\frac{c_ip(\tau)}{2k_i^2}+(-1)^{i+1}\frac{1-\overline{\nu}_i}{1+\overline{\nu}_i}\frac{\beta_iT_0}{1-k_i^2}\int\limits_{v_1^{(i)}}^{v_2^{(i)}}\xi T_i(\xi,\tau)\mathrm{d}\xi\\ +\frac{\beta_iT_0}{2\rho^2}\int\limits_{v_1^{(i)}}^{v_2^{(i)}}\xi\left((-1)^{i+1}\frac{1+k_i^2}{1-k_i^2}+\mathrm{sgn}(\rho-\xi)\right)T_i(\xi,\tau)\mathrm{d}\xi\end{array} \quad (11)$$

and

$$c_3p(\tau)=c_1p_1(\tau)-c_2p_2(\tau)+\ell_1\int\limits_{k_1}^{1}\xi T_1(\xi,\tau)\mathrm{d}\xi+\ell_2\int\limits_{1}^{k_2}\xi T_2(\xi,\tau)\mathrm{d}\xi+(\overline{\alpha}_1-\overline{\alpha}_2)T_0 \quad (12)$$

where $i=1,2$, $p(\tau)=P(\tau)/\sigma_*$ and $p_i(\tau)=P_i(\tau)/\sigma_*$ are the dimensionless contact pressure and compressive pressures on the inner and outer surfaces, σ_* is a constant in the dimension of stresses, $\overline{\nu}_i=\nu_i/(1-\nu_i)$, $\overline{E}_i=E_i/(1-\nu_i^2)$, $T_i=(T_i^*-T_0)/T_0$, $v_i^{(i)}=k_i$, $v_1^{(2)}=v_2^{(1)}=1$, $\ell_i=2\alpha_T^{(i)}(1+\nu_i)T_0/(1-k_i^2)$, $\overline{\alpha}_i=\alpha_T^{(i)}(1+\nu_i)$, $c_i=2k_i^2\sigma_*/((1-k_i^2)\overline{E}_i)$, E_i denotes the Young modulus of the ith cylindrical layer, and

$$c_3=\sum_{i=1}^{2}(-1)^{i+1}\frac{1-\overline{\nu}_i+(1+\overline{\nu}_i)k_i^2}{1-k_i^2}\frac{\sigma_*}{\overline{E}_i}+\frac{(n_1+n_2)\sigma_*}{R_0}$$

A general solution to Equation (4) for the circumferential strain $u_\varphi^{(i)}$ can be given [23] as

$$u_\varphi^{(i)}(\rho,\tau)=\frac{A_i(\tau)R_0\rho}{2}+\frac{B_i(\tau)}{R_0\rho}$$

where $A_i(\tau)$ and $B_i(\tau)$ are arbitrary and yet unknown functions of time, $i=1,2$. By making use of conditions (5) and (6) for the displacement $u_\varphi^{(i)}$ and stress $\sigma_{r\varphi}^{(i)}$, we can finally derive

$$\begin{array}{c}u_\varphi^{(i)}(\rho,\tau)=\frac{f(1+\overline{\nu}_i)R_0P(\tau)}{\overline{E}_i\rho}\left(1-\frac{\rho^2}{k_i^2}\right)\\ \varepsilon_{r\varphi}^{(i)}(\rho,\tau)=-\frac{f(1+\overline{\nu}_i)P(\tau)}{\overline{E}_i\rho^2},\ \sigma_{r\varphi}^{(i)}(\rho,\tau)=-\frac{fP(\tau)}{\rho^2}\end{array} \quad (13)$$

Equation (13) allow for expressing the thermal stresses and displacements in the two-layer cylindrical tribo-couple through the contact pressure found by formula (12).

Assuming the function $t_1^*(\tau)$ to be known for $\tau\geq 0$ and making use of the integral Laplace transform [24] by the time-variable τ yields a solution to the heat-conduction problem (2), (7), (8) and (9) in the form as follows

$$T_i(\rho,\tau)=\sum_{j=1}^{2}\int\limits_0^\tau G_j^{(i)}(\rho,\tau-\xi)t_j(\xi)\mathrm{d}\xi+\Omega\int\limits_0^\tau G_3^{(i)}(\rho,\tau-\xi)p(\xi)\mathrm{d}\xi \quad (14)$$

where $i = 1, 2,$

$$
\begin{gathered}
G_1^{(1)}(\rho,\tau) = \sum_{n=1}^{\infty} \frac{\exp(-\mu_n^2\tau)}{\partial_s(\Delta(s_n))}(2\lambda Z_{10}^{(1)}(1,\rho,s_n)\, Z_{10}^{(2)}(1,k_2,s_n) \\
+\vartheta(\lambda Z_{00}^{(2)}(k_2,1,s_n) Z_{10}^{(1)}(1,\rho,s_n) + Z_{10}^{(2)}(1,k_2,s_n) Z_{00}^{(1)}(1,\rho,s_n)\,)\,) \\
G_2^{(1)}(\rho,\tau) = \vartheta \sum_{n=1}^{\infty} \frac{\exp(-\mu_n^2\tau)}{\partial_s(\Delta(s_n))}\, Z_{00}^{(1)}(\rho,k_1,s_n) \\
G_3^{(1)}(\rho,\tau) = \sum_{n=1}^{\infty} \frac{\exp(-\mu_n^2\tau)}{\partial_s(\Delta(s_n))}\left(Z_{10}^{(2)}(1,k_2,s_n) + \vartheta Z_{00}^{(2)}(k_2,1,s_n)\right) Z_{00}^{(1)}(\rho,k_1,s_n) \\
G_1^{(2)}(\rho,\tau) = -\lambda\vartheta \sum_{n=1}^{\infty} \frac{\exp(-\mu_n^2\tau)}{\partial_s(\Delta(s_n))}\, Z_{00}^{(2)}(\rho,k_2,s_n) \\
G_2^{(2)}(\rho,\tau) = \sum_{n=1}^{\infty} \frac{\exp(-\mu_n^2\tau)}{\partial_s(\Delta(s_n))}(2\lambda Z_{10}^{(1)}(1,k_1,s_n)\, Z_{10}^{(2)}(1,\rho,s_n) \\
+\vartheta(\lambda Z_{10}^{(1)}(1,k_1,s_n) Z_{00}^{(2)}(\rho,1,s_n) + Z_{00}^{(1)}(1,k_1,s_n) Z_{10}^{(2)}(1,\rho,s_n))) \\
G_3^{(2)}(\rho,\tau) = -\sum_{n=1}^{\infty} \frac{\exp(-\mu_n^2\tau)}{\partial_s(\Delta(s_n))}\,(\lambda Z_{10}^{(1)}(1,k_1,s_n) + \vartheta Z_{00}^{(1)}(1,k_1,s_n))\, Z_{00}^{(2)}(\rho,k_2,s_n) \\
\Delta(s) = 2\lambda Z_{10}^{(2)}(1,k_2,s) Z_{10}^{(1)}(1,k_1,s) \\
+\vartheta\left(\lambda Z_{10}^{(1)}(1,k_1,s) Z_{00}^{(2)}(k_2,1,s) + Z_{10}^{(2)}(1,k_2,s) Z_{00}^{(1)}(1,k_1,s)\right) \\
Z_{10}^{(j)}(x,y,s) = q_j x (I_1(q_j x) K_0(q_j y) + I_0(q_j y) K_1(q_j x)) \\
Z_{kk}^{(j)}(x,y,s) = I_k(q_j x) K_k(q_j y) + I_k(q_j y) K_k(q_j x)
\end{gathered}
$$

$j = 1, 2$; $k = 0, 1$, $\Omega = \omega R_0^2 f \sigma_* / (\lambda_2 T_0)$ is the dimensionless angular velocity, $\lambda = \lambda_1/\lambda_2$, $\vartheta = R_s/\overline{R}$, $R_s = R_0/(R_*\lambda_2)$, $q_1^2 = s/b_1$, $q_2^2 = s$, $\overline{R} = R/R_*$ is the dimensionless interfacial thermal resistance, R_* is a constant in the dimension of thermal resistance, $I_k(s)$ and $K_k(s)$ are the modified Bessel functions of the first and second kind, $k = 0, 1$, s stands for the parameter of the Laplace transform, ∂_s denotes the partial derivative by s, and $s_n = -\mu_n^2$ are the roots of the characteristic equation $\Delta(s) = 0$, $\mu_n > 0$, $n = 1, 2, \ldots$

Formula (14) expresses the dependence of the temperature field within the tribo-couple on the contact pressure, while formula (12) shows the dependence of the contact pressure on the temperature. By making use of these two formulas along with expression (11) for the circumferential strain, the condition for the radial displacement in (5) and (6) yields the following formula for the contact pressure on the interface:

$$
\begin{gathered}
p(\tau) = \int_0^{\tau} M(\tau-\eta)(c_1 p_1(\eta) - c_2 p_2(\eta))\mathrm{d}\eta + \sum_{i=1}^{2}\int_0^{\tau} N_i(\tau-\eta) t_i(\eta)\mathrm{d}\eta \\
+\left((1+\nu_1)\alpha_T^{(1)} - (1+\nu_2)\alpha_T^{(2)}\right) T_0 \int_0^{\tau} M(\eta)\mathrm{d}\eta
\end{gathered}
\tag{15}
$$

where

$$
\begin{gathered}
M(\tau) = \sum_{n=1}^{\infty} \frac{\Delta(s_n^*)\exp(s_n^*\tau)}{\partial_s(\Delta_*(s_n^*))},\; N_i(\tau) = \sum_{n=1}^{\infty} \frac{V_i(s_n^*)\exp(s_n^*\tau)}{\partial_s(\Delta_*(s_n^*))} \\
V_1(s) = \ell_1\Bigg(2\lambda k_1 Z_{11}^{(1)}(1,k_1,s) Z_{10}^{(2)}(1,k_2,s) + \vartheta\Bigg(\lambda k_1 Z_{11}^{(1)}(1,k_1,s) Z_{00}^{(2)}(k_2,1,s) \\
+ Z_{10}^{(2)}(1,k_2,s)\frac{Z_{10}^{(1)}(k_1,1,s)-1}{q_1^2}\Bigg)\Bigg) - \ell_2\lambda\vartheta\frac{1-Z_{10}^{(2)}(1,k_2,s)}{q_2^2} \\
V_2(s) = \ell_1\vartheta\frac{Z_{10}^{(1)}(1,k_1,s)-1}{q_1^2} + \ell_2\Bigg(2\lambda k_2 Z_{10}^{(1)}(1,k_1,s) Z_{11}^{(2)}(k_2,1,s) \\
+\vartheta\Bigg(\lambda Z_{10}^{(1)}(1,k_1,s)\frac{Z_{10}^{(2)}(k_2,1,s)-1}{q_2^2} + k_2 Z_{00}^{(1)}(1,k_1,s) Z_{11}^{(2)}(k_2,1,s)\Bigg)\Bigg) \\
\Delta_*(s) = c_3\Delta(s) - \Omega\Bigg(\ell_1\left(Z_{10}^{(2)}(1,k_2,s) + \vartheta Z_{00}^{(2)}(k_2,1,s)\right)\frac{Z_{10}^{(1)}(1,k_1,s)-1}{q_1^2} \\
-\ell_2\left(\lambda Z_{10}^{(1)}(1,k_1,s) + \vartheta Z_{00}^{(1)}(1,k_1,s)\right)\frac{1-Z_{10}^{(2)}(1,k_2,s)}{q_2^2}\Bigg)
\end{gathered}
$$

s_n^* are the roots of the characteristic equation $\Delta_*(s) = 0, n = 1, 2, \ldots$

By putting (14) and (15) into formula (11) at $i = 2$ and $\rho = k_2$ within the context of the supplementary condition (10) for the circumferential strain, we arrive at the convolution-type Volterra integral equation of the first kind [25] for the determination of function $t_1(\tau)$ in the following form:

$$\begin{gathered}\int_0^\tau K_1(\tau-\eta)\, t_1(\eta)\mathrm{d}\eta = \varphi_*(\tau) - \int_0^\tau K_2(\tau-\eta)\, t_2(\eta)\mathrm{d}\eta \\ -\int_0^\tau L(\tau-\eta)\,(c_1 p_1(\eta) - c_2 p_2(\eta))\mathrm{d}\eta - \tfrac{c_2}{2k_2}(1 + k_2^2 + (1-k_2^2)\overline{\nu}_2)p_2(\tau) \\ -\Big((1+\nu_1)\alpha_T^{(1)} - (1+\nu_2)\alpha_T^{(2)}\Big)T_0 \int_0^\tau L(\eta)\mathrm{d}\eta - (1+\nu_2)k_2\alpha_T^{(2)}T_0\end{gathered} \tag{16}$$

where $\tau \in [0, \tau_m]$ and

$$\begin{gathered}K_i(\tau) = \sum_{k=1}^{\infty} \tfrac{U_i(s_k)\exp(s_k\tau)}{\partial_s(\Delta(s_k)\Delta_*(s_k))},\ L(\tau) = \sum_{n=1}^{\infty} \tfrac{V_3(s_n^*)\exp(s_n^*\tau)}{\partial_s(\Delta_*(s_n^*))} \\ U_1(s) = V_1(s)V_3(s) + k_2\ell_2\lambda\vartheta\tfrac{1-Z_{10}^{(2)}(1,k_2,s)}{q_2^2}\Delta_*(s) \\ U_2(s) = V_2(s)V_3(s) - k_2\ell_2(2\lambda k_2 Z_{10}^{(1)}(1,k_1,s)Z_{11}^{(2)}(k_2,1,s) \\ +\vartheta\Big(\lambda Z_{10}^{(1)}(1,k_1,s)\tfrac{Z_{10}^{(2)}(k_2,1,s)-1}{q_2^2} + k_2 Z_{00}^{(1)}(1,k_1,s)Z_{11}^{(2)}(k_2,1,s)\Big)\Big)\Delta_*(s) \\ V_3(s) = -\tfrac{c_2}{k_2}\Delta(s) + \Omega\, k_2\ell_2\Big(\lambda Z_{10}^{(1)}(1,k_1,s) + \vartheta\, Z_{00}^{(1)}(1,k_1,s)\Big)\tfrac{1-Z_{10}^{(2)}(1,k_2,s)}{q_2^2}\end{gathered}$$

s_k are roots of equations $\Delta(s) = 0$ and $\Delta_*(s) = 0$ combined, which are negative real numbers $s_k = -\gamma_k^2, \gamma_k > 0, k = 1, 2, \ldots$, when the angular velocity does not exceed a critical value [17].

By setting $\tau = 0$ in (16) and allowing $t_i(0) = 0, i = 1, 2$, we derive the fitting condition for the initial temperature, the circumferential strain imposed on the outer surface $\rho = k_2$, and the dimensionless pressures on the inner and outer circumferences of the tribo-couple at the initial moment of time in the form as follows:

$$\begin{gathered}\tfrac{c_2}{2k_2}(1 + k_2^2 + (1-k_2^2)\overline{\nu}_2)p_2(0) - \tfrac{c_2}{k_2 c_3}(c_1 p_1(0) - c_2 p_2(0)) \\ -\tfrac{c_2}{k_2 c_3}\Big((1+\nu_1)\alpha_T^{(1)} - (1+\nu_2)\alpha_T^{(2)}\Big)T_0 + (1+\nu_2)k_2\alpha_T^{(2)}T_0 = \varphi_*(0)\end{gathered}$$

The latter condition ensures the continuity of the solution of integral Equation (16).

In such a manner, the original heat-conduction problem for the considered cylindrical tribo-couple with frictional hating is reduced to an inverse thermoelasticity problem, which is verbalized by the integral Equation (16) and implies the determination of the temperature on the inner surface via the temperature and circumferential strain given on the outer surface.

It can be shown that the kernel $K_1(\tau - \eta)$ of Equation (16) is always positive for $\eta \in [0, \tau]$, increases monotonically and suffers the root singularity at $\eta = \tau$. This means that Equation (16) is the Abel integral equation [25]. The fact that the kernel $K_1(\tau - \eta)$ has the integrable singularity at $\eta = \tau$ implies the absence of the time delay in the maximum response of the thermal constituent of the circumferential strain $\varphi_*(\tau)$ to the variation of temperature $t_1(\tau)$.

Assume the unknown temperature $t_1(\eta)$ to be a continuous function on the interval $[0, \tau]$, i.e., $t_1(\eta) \in C_{[0,\tau]}$, to construct a solution to Equation (16). Let us represent the time interval $[0, \tau_m]$ by the mesh consisting of m intervals of the length $h = \tau_m/m$ and represent the sought-out function on each of these intervals by a linear spline

$S_j^{(1)}(\tau) = ((\tau_j - \tau)t_1^{(j-1)} + (\tau - \tau_{j-1})t_1^{(j)})h^{-1}$, $\tau \in [\tau_{j-1}, \tau_j]$, $\tau_j = hj$, $t_1^{(j)} = t_1(\tau_j)$, $j = 1, \ldots, m$. As a result, Equation (16) yields the following system of linear algebraic equations:

$$t_1^{(1)} = \frac{\Phi_1}{c_0}, \sum_{j=1}^{l-1} \Theta_{l\,j} t_1^{(j)} + t_1^{(l)} = \frac{\Phi_l}{c_0}, \; l = 2, \ldots, m \tag{17}$$

Here, $\Phi_l = \Phi(\tau_l)$ is the values of the right-hand side of Equation (16) at the knots of the mesh $\tau = \tau_l$ and

$$\Theta_{lj} \equiv q_{l-j} = \frac{1}{c_0} \sum_{k=1}^{\infty} \frac{U_1(s_k)}{\partial_s(\Delta(s_k)\Delta_*(s_k))} \frac{(1-\exp(-\gamma_k^2 h))^2}{\gamma_k^4 h^2} \exp(-\gamma_k^2 h(l-j-1)), \; j < l$$

$$c_0 = \sum_{k=1}^{\infty} \frac{U_1(s_k)}{\partial_s(\Delta(s_k)\Delta_*(s_k))} \frac{1}{\gamma_k^2 h} \left(1 - \frac{1-\exp(-\gamma_k^2 h)}{\gamma_k^2 h}\right)$$

The matrix of system (17) is the lower diagonal matrix with equal elements on each diagonal below the main one:

$$\mathbf{Q}_1 = \begin{pmatrix} 1 & 0 & 0 & \cdots & 0 & 0 \\ q_1 & 1 & 0 & \cdots & 0 & 0 \\ q_2 & q_1 & 1 & \cdots & 0 & 0 \\ \vdots & \vdots & \vdots & \ddots & \vdots & \vdots \\ q_{m-1} & q_{m-2} & q_{m-3} & \cdots & q_1 & 1 \end{pmatrix}, \; 0 < q_1 < 1, \; q_{i+1} < q_i, \; i = 1, \ldots, m-1$$

It can be shown that for $h > 0$, the norm $\|\mathbf{Q}_1\| = \max_j \left(\sum_i |\Theta_{i\,j}| \right) < \infty$.

System (17) can be represented in the following form:

$$\mathbf{T} = \mathbf{Q}_2 \mathbf{T} + \mathbf{F} \tag{18}$$

where

$$\mathbf{Q}_2 = \begin{pmatrix} 0 & 0 & 0 & \cdots & 0 & 0 \\ q_1^* & 0 & 0 & \cdots & 0 & 0 \\ q_2^* & q_1^* & 0 & \cdots & 0 & 0 \\ \vdots & \vdots & \vdots & \ddots & \vdots & \vdots \\ q_{m-1}^* & q_{m-2}^* & q_{m-3}^* & \cdots & q_1^* & 0 \end{pmatrix}, \; \mathbf{T} = \begin{pmatrix} t_1^{(1)} \\ t_1^{(2)} \\ t_1^{(3)} \\ \vdots \\ t_1^{(m)} \end{pmatrix}, \; \mathbf{F} = \frac{1}{c_0} \begin{pmatrix} \Phi_1 \\ \Phi_2 - \Phi_1 \\ \Phi_3 - \Phi_2 \\ \vdots \\ \Phi_m - \Phi_{m-1} \end{pmatrix}$$

$q_1^* = 1 - q_1$, $q_i^* = q_{i-1} - q_i$, $q_i^* > 0$, $i = 2, \ldots, m-1$. Due to the fact that $\|\mathbf{Q}_2\| = \sum_{i=1}^{m-1} q_i^* = 1 - q_m < 1$ for $h > 0$, the simple iteration routine [26,27] implies that the problem on solving the system of Equation (18), and, consequently, (17) is well-posed. Based on this fact, system (17) allows for deriving a recursive formula for determination of $t_1^{(l)}, l = 1, \ldots, m$.

Having derived the temperature $t_1(\tau)$, $\tau \in [0, \tau_m]$, by means of the foregoing routine, we can use Formula (14) to determine the temperature field within the tribo-couple. The thermal stresses and displacements can be computed accordingly by making use of the Formulae (11)–(13) and (15), along with the basic thermoelasticity equations [23].

4. Numerical Example and Discussion

In order to verify the proposed solution technique, consider a solution to the formulated identification problem for the tribo-couple, whose inner layer 1 is made of steel ($\lambda_1 = 21$ [Wt/(m × K)], $a_1 = 5.9 \times 10^{-6}$ [m^2/s], $\alpha_T^{(1)} = 14 \times 10^{-6}$ [1/K], $E_1 = 190$ [GPa], and $\nu_1 = 0.3$) and the outer one 2 is made of copper ($\lambda_2 = 381$ [Wt/(m × K)], $a_2 = 101.9 \times 10^{-6}$ [m^2/s], $\alpha_T^{(2)} = 17 \times 10^{-6}$ [1/K], $E_2 = 121$ [GPa], and $\nu_2 = 0.33$).

Herein, we employ the following commonly used verification strategy [5] with two stages. In the first stage, we formulate a direct problem by imposing the temperature $t_1^*(\tau)$ on the inner circumference of the tribo-couple. Together with the given temperature on the outer surface and the interface thermal conditions (8), this would allow us to compute the thermal field in the tribo-couple. Making use of the determined temperature, a solution of the thermoelasticity problem (3)–(6) is constructed analytically. The latter solution can then be used to derive an expression for the circumferential strain on the outer surface of the tribo-couple. In the second stage, we formulate the inverse problem, where condition (10) is used together with the circumferential strain computed on the previous stage in order to restore the temperature on the inner surface by making use of the proposed algorithm. By comparing the solution of the inverse problem with the temperature $t_1^*(\tau)$ imposed when formulating the direct problem on stage 1, we can draw a conclusion about the efficiency of the algorithm. When solving the inverse problem in this stage, we also introduce some random small errors in the distribution of the circumferential strain in order to verify the stability of the algorithm.

By following this strategy, let us first consider the direct heat-conduction and thermoelasticity problems by imposing the following boundary temperatures

$$t_1^*(\tau) = T_0 + B(1 - \cos 2\tau),\ t_2^*(\tau) = T_0 \tag{19}$$

and pressures $p_1(\tau) = \mathrm{C}\mathrm{H}(\tau)$ and $p_2(\tau) = 0$, where $B, C = \text{const}$, $\mathrm{H}(\tau)$ is the Heaviside step function, to determine the circumferential strain distribution on the outer surface $\rho = k_2$. Then, we can approximate the constructed strain within certain accuracy and use it as the input data for the inverse problem to determine the temperature $t_1^{(l)}, l = 1, \ldots, m$, on the inner surface $\rho = k_1$. By comparing the computed values $t_1^{(l)}, l = 1, \ldots, m$, with the actual $t_1(\tau)$, $\tau \in [0, \tau_m]$, imposed in (19), we can evaluate the accuracy of the proposed solution algorithm for the considered inverse problem of thermoelasticity.

The distribution of the dimensionless circumferential strain $\varepsilon(\tau) = \varepsilon_{\varphi\varphi}^{(2)}(k_2, \tau) \times 10^4$ on the outer surface $\rho = k_2$ is shown in Figure 2a. The strain was computed from the direct problem under the thermal loading (19) for the following parameters $R_0 = 5.0 \times 10^{-2}$ [m], $R_1 = 3.5 \times 10^{-2}$ [m], $R_2 = 6.0 \times 10^{-2}$ [m], $n_1 = 10^{-3}$ [m/GPa], $n_2 = 10^{-4}$ [m/GPa], $R = 5.0 \times 10^{-3}$ [m^2 × K/Wt], $R_* = 1.1 \times 10^{-3}$ [m^2 × K/Wt]; $\sigma_* = 10^2$ [MPa], $T_0 = 20$ [K], $B = 200$ [K], $C = 10^2$, $f = 0.25$, $\omega = 1.22$ [rad/s], and $\tau_m = 2.5$.

Now we can use the computed strain as the input data for solving the inverse problem in order to reconstruct the thermal loading on the inner circumference of cylinder 1. It is also important to analyze the effect of small errors in the input data (which can be induced by the errors in the stain measurement, etc.). For modeling of such errors, let us substitute the strain distribution at the discrete time moments τ_i with the values $\tilde{\varepsilon}(\tau_i)$ computed by the formula $\tilde{\varepsilon}(\tau_i) = \varepsilon(\tau_i)(1 + \theta_i \times 10^{-2})$, where θ_i are arbitrary numbers from the interval $[-1, 1]$ with the uniform distribution law and represent $\tilde{\varepsilon}(\tau)$ by a linear spline. This means that the input data are encountered with an arbitrary error falling within 1%.

In Figure 2b, the open circles denote the time distribution of the temperature $t_1^{(i)}$, $i = \overline{1, 250}$ on the inner surface of the cylinder 1, found by solving the inverse thermoelasiticy problem with the computational step $h = 0.01$. It is shown that the maximum relative error of the computed values in comparison to the corresponding values imposed in the direct problem (19) falls within 1.8%, which confirms the stability of the proposed solution algorithm with respect to the small errors in the input data. Due to the fact that the solutions to well-posed direct problems are stable with respect to small errors in the input data, the error in computing the thermal stresses, strains, and displacements by using the thermal loading (19) of the one computed by solving the inverse problem can be dismissed.

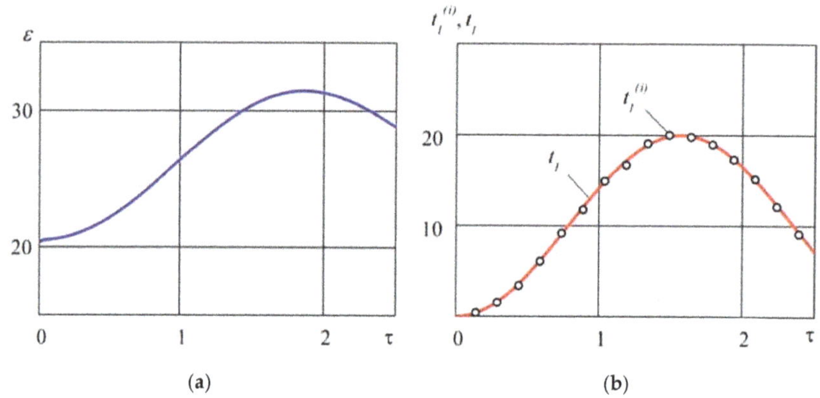

Figure 2. The circumferential elastic strain $\varepsilon(\tau) = \varepsilon^{(2)}_{\varphi\varphi}(k_2, \tau) \times 10^4$ (**a**) computed by the temperature $t_1(\tau)$ given in (19) by solving the direct problem versus the dimensionless time τ; the dimensionless temperature $t_1^{(i)}$ on the inner circumference (**b**) as given by formula (19) (solid lines) and computed by solving the inverse problem (open circles).

5. Conclusions

A problem on the determination of temperature and thermal-stress fields in a cylindrical tribo-couple with frictional heating on the interface is formulated for the case when the thermal loading on one of its circumferences is unknown. The additional information about the transient variation of the circumferential strain on the surface where the thermal loading is known was used as a supplementary condition for the formulated inverse thermoelasticity problem governed by a Volterra integral equation of the first kind. Due to the fact that the kernel of this integral equation $K_1(\tau - \eta)$ takes only positive values on the interval $\eta \in [0, \tau]$, monotonically increases for the entire range of variables, and suffers a root singularity at the point $\eta = \tau$, this integral equation can be regarded as one of Abel kind. The presence of the integrable singularity in this kernel at $\eta = \tau$ implies that there is no delay in the maximum thermal response of the circumferential strain $\varphi_*(\tau)$ to a change in the variation profile of temperature $t_1(\tau)$ at $\eta = \tau$ in view of the integral dependence of this strain on the temperature within the cylindrical layers of the tribo-couple. This feature of kernel $K_1(\tau - \eta)$ ensures the conditional correctness of the inverse problem. The correctness condition in this case was derived in the form of the fitting condition for the, circumferential strain on the periphery of the tribo-couple, and the pressures applied to its surfaces at the initial moment of time.

It is worth noting that the analogous kernels within the framework of inverse heat-conduction problems solely exhibit quite different features, which, in the final count, makes these problems ill-posed [5,6].

Another advantage of the proposed technique is that the system of algebraic Equation (17), which is the discrete analog for Equation (16), was represented in the form (18). This, in view of the appearance of its matrix $\mathbf{Q}_2$ ensures the stability of its solution with respect to small errors in the input data. An algorithm for solving the formulated inverse problem is suggested on the basis of the linear spline approximation technique. The efficiency of the algorithm was verified by solving the direct problem under the given thermal loading in order to determine the circumferential strain, which was then used as the input data for the inverse problem on the reconstruction of thermal loading.

These key features of the proposed algorithm may serve for benefit of setting up technological and experimental cylindrical tribo-systems expecting incomplete information about thermal loading for engineering applications and the wear analysis [28,29].

Author Contributions: Conceptualization, R.K.; methodology, A.Y.; validation and formal analysis, Y.T.; computation, E.H. All authors have read and agreed to the published version of the manuscript.

Funding: This research was partially supported by the budget program of Ukraine "Support for the Development of Priority Research Areas" (CPCEC 6451230) under the grant No. 0120U100499.

Institutional Review Board Statement: Not applicable.

Informed Consent Statement: Not applicable.

Data Availability Statement: Data Sharing is not applicable.

Conflicts of Interest: The authors declare no conflict of interest. The funders had no role in the design of the study; in the collection, analyses, or interpretation of data; in the writing of the manuscript, or in the decision to publish the results.

References

1. Tokovyy, Y.V.; Ma, C.-C. *The Direct Integration Method for Elastic Analysis of Nonhomogeneous Solids*; Cambridge Scholars Pub.: Cambridge, UK, 2021; 342p.
2. Dwivedi, S.D.; Vishwakarma, M.; Soni, A. Advances and researches on non destructive testing: A review. *Mater. Today Proc.* **2018**, 53690–53698. [CrossRef]
3. Wang, B.; Zhong, S.; Lee, T.-L.; Fancey, K.S.; Mi, J. Non-destructive testing and evaluation of composite materials/structures: A state-of-the-art review. *Adv. Mech. Eng.* **2000**, *12*, 1–28. [CrossRef]
4. Tokovyy, Y.V.; Ma, C.-C. Elastic analysis of inhomogeneous solids: History and development in brief. *J. Mech.* **2019**, *35*, 613–626. [CrossRef]
5. Alifanov, O.M. *Inverse Heat Transfer Problems*; Springer: Berlin, Germany, 1994; 348p.
6. Beck, J.V.; Blackwell, B.; Clair, C.R. *Inverse Heat Conduction: Ill-Posed Problems*; Wiley: New York, NY, USA, 1985; 308p.
7. Ootao, Y. Inverse problem of thermal deformation in a cylinder. In *Encyclopedia of Thermal Stresses*; Hetnarski, R.B., Ed.; Springer: Dordrecht, The Netherland, 2014; Volume 5, pp. 2578–2585. [CrossRef]
8. Nowacki, W. *Thermoelasticity*; Pergamon Press: New York, NY, USA, 1962; 628p.
9. Hetnarski, R.B.; Eslami, M.R. *Thermal Stresses—Advanced Theory and Applications*; Springer: Dordrecht, The Netherland, 2009; 560p.
10. Kalynyak, B.M.; Tokovyy, Y.V.; Yasinskyy, A.V. Direct and inverse problems of thermomechanics concerning the optimization and identification of the thermal stressed state of deformed solids. *J. Math. Sci.* **2019**, *236*, 21–34. [CrossRef]
11. Chang, W.J.; Lee, H.L.; Yang, Y.C. Estimation of heat flux and thermal stresses in functionally graded hollow circular cylinders. *J. Stresses* **2011**, *34*, 740–755. [CrossRef]
12. Golbahar Haghighi, M.R.; Malekzadeh, P.; Afshari, M. Inverse estimation of heat flux and pressure in functionally graded cylinders with finite length. *Compos. Struct.* **2015**, *121*, 1–15. [CrossRef]
13. Grysa, K.; Maciag, A. Solving direct and inverse thermoelasticity problems by means of Trefftz base functions for finite element method. *J. Stresses* **2011**, *34*, 378–393. [CrossRef]
14. Kushnir, R.M.; Yasinskyy, A.V.; Tokovyy, Y.V. Reconstruction of thermal loading of a functionally-graded hollow sphere by the surface displacements. *Math. Methods Phys. Mech. Fields* **2020**, *63*, 149–160. (In Ukrainian)
15. Segall, A.E.; Engels, D.; Drapaca, C. Inverse determination of thermal boundary conditions from transient surface temperatures and strains in slabs and tubes. *Mater. Manuf. Process.* **2012**, *27*, 860–868. [CrossRef]
16. Yasinskii, A.V. Identification of thermal and thermostressed states of a two-layer cylinder from surface displacements. *Int. Appl. Mech.* **2008**, *44*, 34–40. [CrossRef]
17. Grylitskiy, D.V.; Pyryev, Y.A.; Mandzyk, Y.I. Quasistatic thermoelastic contact problem for infinite two layer circular cylinder under friction heating. *J. Therm. Stresses* **1997**, *20*, 47–65. [CrossRef]
18. Yevtushenko, A.; Kuciej, M.; Och, E. Theoretical nonlinear model of frictional heat generation in braking. *Heat Transf. Res.* **2019**, *50*, 1007–1022. [CrossRef]
19. Yevtushenko, A.; Kuciej, M.; Topczewska, K. Frictional heating during braking of the C/C composite disc. *Materials* **2020**, *13*, 2691. [CrossRef] [PubMed]
20. Yasinskyy, A.; Tokova, L. Inverse problem on the identification of temperature and thermal stresses in an FGM hollow cylinder by the surface displacements. *J. Therm. Stresses* **2017**, *40*, 1471–1483. [CrossRef]
21. Yasins'kyi, A.V. Reconstruction of temperature fields and thermal stresses for given displacements in the case of friction contact of layers. *Mater. Sci.* **2002**, *38*, 814–823. [CrossRef]
22. Yasins'kyi, A.V. Inverse problem of evaluation on the coefficient of friction of layers according to the data of measurements of the surface displacements. *Mater. Sci.* **2003**, *39*, 704–711. [CrossRef]
23. Eslami, M.R.; Hetnarski, R.B.; Ignaczak, J.; Noda, N.; Sumi, N.; Tanigawa, Y. *Theory of Thermal Stresses. Explanations, Problems and Solutions*; Springer: Dordrecht, The Netherlands, 2013; 789p.
24. Farlow, S.J. *Partial Differential Equations for Scientists and Engineers*; Dover Publ.: New York, NY, USA, 1993; 414p.
25. Corduneanu, C. *Integral Equations and Applications*; Cambridge Univ. Press: Cambridge, UK, 2008; 366p.

26. Hudramovich, V.S.; Hart, E.L.; Marchenko, O.A. Reinforcing inclusion effect on the stress concentration within the spherical shell having an elliptical opening under uniform internal pressure. *Strength Mater.* **2020**, *52*, 832–842. [CrossRef]
27. Delves, L.M.; Mohamed, J.L. *Computational Methods for Integral Equations*; Cambridge Univ. Press: Cambridge, UK, 2008; 388p.
28. Blau, P.J. *Tribosystem Analysis: A Practical Apprach to the Diagnosis of Wear Problems*; CRC Press: Boca Raton, FL, USA, 2016; 194p.
29. Dykha, A.; Padgurskas, J.; Musial, J.; Matiukh, S. *Wear Models and Diagnostics of Cylindrical Sliding Tribosystem*; Foundation of Mechatronics Development: Bydgoszcz, Poland, 2020; 198p.

materials

MDPI

Article

Effect of Frictional Slipping on the Strength of Ribbon-Reinforced Composite

Yosyf Piskozub [1,*] and Heorhiy Sulym [2]

1 Department of Applied Mathematics and Physics, Ukrainian Academy of Printing, Pidgolosko 19, 79020 L'viv, Ukraine

2 Department of Mechanics and Applied Computer Science, Faculty of Mechanical Engineering, Bialystok University of Technology, Wiejska 45c, 15351 Bialystok, Poland; h.sulym@pb.edu.pl

* Correspondence: piskozub@pancha.lviv.ua or pjozef57@gmail.com

Abstract: A numerical–analytical approach to the problem of determining the stress–strain state of bimaterial structures with interphase ribbon-like deformable inhomogeneities under combined force and dislocation loading has been proposed. The possibility of delamination along a part of the interface between the inclusion and the matrix, where sliding with dry friction occurs, is envisaged. A structurally modular method of jump functions is constructed to solve the problems arising when nonlinear geometrical or physical properties of a thin inclusion are taken into account. A complete system of equations is constructed to determine the unknowns of the problem. The condition for the appearance of slip zones at the inclusion–matrix interface is formulated. A convergent iterative algorithm for analytical and numerical determination of the friction-slip zones is developed. The influence of loading parameters and the friction coefficient on the development of these zones is investigated.

Keywords: ribbon-like reinforcement; composite; thin inclusion; bimaterial; nonperfect contact; friction; jump functions

MSC: 00A06; 30E20; 45F15; 74M15; 74M25

Citation: Piskozub, Y.; Sulym, H. Effect of Frictional Slipping on the Strength of Ribbon-Reinforced Composite. *Materials* **2021**, *14*, 4928. https://doi.org/10.3390/ma14174928

Academic Editor: Michele Bacciocchi

Received: 1 July 2021
Accepted: 26 August 2021
Published: 30 August 2021

1. Introduction

The theory and practice of the design and use of progressive composite materials with flat reinforcement provides indisputable evidence that their tensile strength in the transversal direction, in the case of unidirectional ribbon reinforcement, is between 50 and 75% of the strength in the longitudinal direction; the use of fibers typically yields only between 2 and 15% [1–7]. Reviews and monographs [3,4,7,8] note the advantages of flat reinforcement, which improves manufacturability and the mechanical properties of the composite, increases the reinforcement factor and the resistance to leakage failure, and reduces the statistical variation of the designed properties. It is a testament to the big prospective application of composites with ribbon-like reinforcement. The use of external ribbon-like reinforcement in steel-reinforced concrete makes it possible to save between 15 and 45% of metal in comparison with reinforced concrete and pure metal structures. In addition, thin ribbon-like elements are a common phenomenon in micro- and nanostructures [8–10].

In composite mechanics, two key issues can be distinguished: the determination of (1) the average effective properties of the composite as a whole, and (2) the elastic or plastic deformation processes and the possible failure of composite constituents, including contact loss at the matrix-filler interface. In the process of the exploitation of composites, the phenomenon of cracking and delamination is frequent; i.e., reinforcing heterogeneities can be in both ideal and non-ideal contact with the basic material, including at the interface of the media. Consideration of friction in the study of contact phenomena is one of the most pressing problems for mechanical engineering and materials science in the analysis of phenomena and processes occurring in the moving elements of machines, at various technological operations [11–20]. Thus, friction may be accompanied by electrical, thermal, vibrational, and chemical processes that dampen internal dynamic processes, significantly affecting the intensity in the wear and tear of materials and, consequently, the reliability and durability of the structural elements made of them [12,13,16,19]. The influence of friction can be both negative and positive.

From the point of view of structural integrity mechanics, friction between crack faces at their relative displacement is useful in most cases, since it causes dissipation of internal strain energy; thus, it reduces stress concentration, and reduces or even eliminates alternating plastic deformation under an alternating load. It is also known that the development of the residual stress field contributes to the adaptation of the material to operational loads. The compression of composite materials arising under the action of frictional forces improves the redistribution of shear stresses, even in the case of macroscopic failure of the reinforcement-matrix interface.

The negative consequences of friction are mainly the wear and tear on the contact surfaces, as well as thermal radiation. At an excessive intensity, the latter can sometimes cause unpredictable changes in the mechanical, physical, and chemical properties of the material and the distribution of physical fields, influencing diffusion processes, in particular hydrogen diffusion, and the development of fracture phenomena.

Most progressive multiscale methods [4,9,10] recognize a priori that the overall behavior of a composite is highly dependent on the local details. For example, local imperfections in the spacing and direction of fibers or ribbons can negatively affect the overall bearing capacity of a structure consisting of a composite material.

However, it is impossible to draw reliable conclusions about this, especially to optimize the stress–strain state of the structure using direct numerical methods such as the finite element method. Therefore, the development of prediction methods that can adequately reflect the complex mechanical behavior near such inhomogeneities is a challenge and requires the application of either analytical tools or numerical–analytical methods.

The need to take into account the aforementioned geometric nonlinearity (a priori unknown contact spots) significantly complicates the process of problem-solving and requires the use of various approximate methods, even for bodies of simple geometry [4,6]. However, these methods do not always allow for the correct consideration of thin-walled heterogeneity, nor do they guarantee the accuracy of a solution for load optimization in future applications. During the operation of materials that have the structure that is being considered in this work, especially when loaded by concentrated force and dislocation factors, undesirable critical states of adhesion loss between the constituent parts may arise. This can be avoided by locating the loading points in the so-called "safety zone" when the shear stresses on the contact surfaces at each of their points do not exceed the critical value.

It should be noted that the vast majority of research studying the structural stability of bodies with thin inhomogeneities does not cover the entire possible range of force or dislocation loading of structures. Additionally, it does not provide a full opportunity to determine the critical load and so-called "safety zone" for applications in order to optimize the properties of structures for certain types of loading.

This work aims to develop a numerical–analytical method to study structures with ribbon-like deformable elements, with possible frictional contact between the constituent elements; moreover, it aims to study the mechanical effects of loading by force and dislocation factors on its strength.

2. Formulation of the Problem

Consider an unbounded isotropic bulk consisting of two half-spaces with the elastic constants E_k, ν_k, G_k $(k = 1,2)$, pressed to the interface by normal stresses $\sigma_{yy}^{\infty} < 0$, and under the action of uniformly distributed at infinity stresses σ_{xxk}^{∞} $(k = 1,2)$. The external longitudinal shear load is determined by the stresses σ_{yz}^{∞} and σ_{xzk}^{∞} uniformly distributed at infinity, concentrated intensity forces Q_k, and screw dislocations with the Burger's vector component b_k at points $\varsigma_{*k} \in S_k$ $(k = 1, 2)$, oriented along the axis in such a way that their action causes a quasi-static antiplane SSS in the body. To ensure the straightness of the material interface at infinity, the stresses must satisfy the conditions $\sigma^{\infty}_{xz2}G_1 = \sigma^{\infty}_{xz1}G_2$, $\frac{\nu_2\sigma^{\infty}_{yy}-(1-\nu_2)\sigma^{\infty}_{xx2}}{G_2} = \frac{\nu_1\sigma^{\infty}_{yy}-(1-\nu_1)\sigma^{\infty}_{xx1}}{G_1}$.

We will study the SSS of the body section with a plane xOy perpendicular to the direction Oz of its longitudinal displacement. The plane sections of half-spaces perpendicular to this axis form two half-planes S_k $(k = 1,2)$, and the abscissa axis corresponds to the interface $L \sim x$ between them (Figure 1).

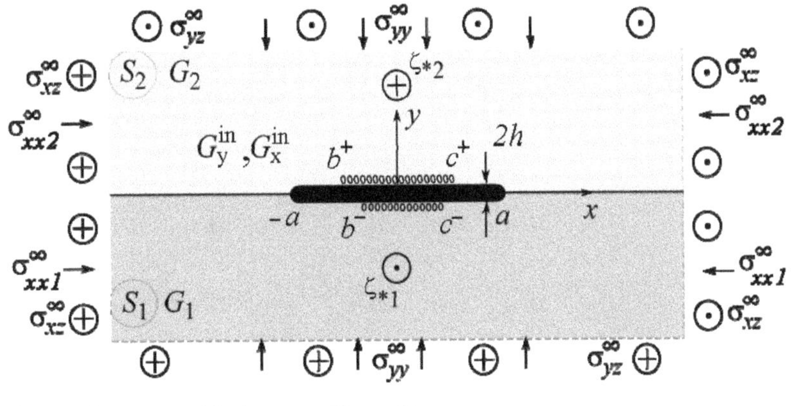

Figure 1. Geometry and load pattern of the problem.

Along the segment $L' = [-a; a]$ is a thin inclusion of thickness $2h$ $(h \ll a)$ (Figure 1), of which the upper and lower banks may come in contact with the matrix non-ideally on the intervals $L''^{\pm} = [b^{\pm}; c^{\pm}]$ $(|b^{\pm}| \le |a|, |c^{\pm}| \le |a|)$, respectively. G^{in}_x and G^{in}_y are the shear moduli of the inclusion material. The upper index "in" denotes the values describing the inclusion material's SSS.

The contact between the half-spaces along the line $L \backslash L'$ and at the inclusion–matrix interface at the sections $L' \backslash L''$ is also mechanically ideal:

$$w(x,+0) = w(x,-0),\ \sigma_{yz2}(x,+0) = \sigma_{yz1}(x,-0),\ x \in L\backslash L' \quad (1)$$

$$\begin{gathered} w(x,-h) = w^{in}(x,-h),\ \sigma^{in}_{yz}(x,-h) = \sigma_{yz1}(x,-h),\ x \in L'\backslash L''^{-}, \\ w(x,h) = w^{in}(x,h),\ \sigma^{in}_{yz}(x,h) = \sigma_{yz2}(x,h),\ x \in L'\backslash L''^{+}. \end{gathered} \quad (2)$$

At the contact areas $L''^{\pm}$, we assume stick-slip contact conditions [18], wherein mutual slippage of contacting body surfaces can start, causing heat release, energy dissipation, wear [11,16–18], etc., and that all points of $L''^{\pm}$, tangential stresses (friction forces) are equal to:

$$\sigma^{in}_{yz}(x,\pm h) = \sigma_{yz2}(x,\pm h) = -\mathrm{sgn}\left(w^{in}(x,\pm h) - w(x,\pm h)\right)\tau^{\max}_{yz}(x), \quad (3)$$

where $\tau^{\max}_{yz}(x) = -\alpha\sigma_{yy}(x)$ $(\sigma_{yy} < 0)$, α is the sliding friction coefficient. Outside the area $L''^{\pm}$, in the absence of slip on the inclusion surface, the tangential stresses may not exceed the allowable maximum

$$|\sigma_{yz}(x,\pm h)| \le \tau^{\max}_{yz}(x)\ (\sigma_{yy} < 0) \quad (4)$$

and there is no mutual displacement of contact surfaces (displacement jump). The sign (direction of action) of the tangential stresses is chosen depending on the sign of the displacement difference $w^{in}(x,\pm h) - w(x,\pm h)$ at the point in question at L''.

In the case of normal pressure in the first approximation, we obtain:

$$\tau_{yz}^{\max}(x) = -\alpha\sigma_{yy}^{\infty} \tag{5}$$

It is not difficult to obtain a more exact expression by constructing the solution of the corresponding plane problem. The application of friction law in classical form (5) gives an opportunity, of course, to simplify boundary conditions for the main problem, but the choice of complex friction models [11–14,17], considering wear, will not fundamentally complicate the solution process.

3. Materials and Methods

In general, the formulated problem contains three component modules: an internal problem (the stress–strain state in the inclusion), an external problem (the stress–strain state in the matrix), and the contact conditions (1)–(4), which relate them.

The stress–strain state in a thin inclusion (the internal problem) is described by an appropriate mathematical model. Due to the small thickness of the inclusion, it is possible to construct approximate relations between the components of the stress tensor and displacement vector on the opposite sides of the inclusion, which adequately describe its SSS. For example, a rather general model of a physically nonlinear thin inclusion is built in [6]:

$$\begin{cases} G_x^{in}(\sigma_{xz}^{in})\left\langle \frac{\partial w^{in}}{\partial x}\right\rangle_h(x) - 2\sigma_{xz}^{in}(-a) - \frac{1}{h}\int\limits_{-a}^{x}\left[\sigma_{yz}^{in}\right]_h(\xi)d\xi = 0, \\ G_y^{in}(\sigma_{yz}^{in})\left[w^{in}\right]_h(x) + h\left\langle\sigma_{yz}^{in}\right\rangle_h(x) = 0, \end{cases} \tag{6}$$

where G_x^{in} and G_y^{in} are the variable shear moduli of the inclusion material. Taking them to be constant, we obtain a special case of Hooke's law. Hereinafter, the following notations are used: $[\varphi]_h = \varphi(x,-h) - \varphi(x,+h)$ and $\langle\varphi\rangle_h = \varphi(x,-h) + \varphi(x,+h)$; the indexes "+" and "−" correspond to the limit values of functions at the upper and lower edges of the line L.

To solve the external problem, it is convenient to apply the method referenced in [6,7], which uses the well-known jump function method (JFM). According to its paradigm, a thin inclusion in the matrix is modeled by jumps of the components of the stresses and displacement vectors on the line L':

$$\begin{gathered} [\sigma_{yz}]_h \cong \sigma_{yz}^- - \sigma_{yz}^+ = f_3(x), \\ \left[\frac{\partial w}{\partial x}\right]_h \cong \frac{\partial w^-}{\partial x} - \frac{\partial w^+}{\partial x} = \left[\frac{\sigma_{xz}}{G}\right]_h \equiv \frac{\sigma_{xz}^-}{G_1} - \frac{\sigma_{xz}^+}{G_2} = f_6(x),\ x \in L'; \end{gathered} \tag{7}$$

$$f_3(x) = f_6(x) = 0,\ \text{if } x \notin L'. \tag{8}$$

Furthermore, we can obtain the dependences according to which components of the stress tensor and the derivatives of the displacement vector inside an unbounded plane S obtain the form:

$$\begin{gathered} \sigma_{yzk}(\varsigma) + i\sigma_{xzk}(\varsigma) = \sigma_{yzk}^0(\varsigma) + i\sigma_{xzk}^0(\varsigma) - \frac{G_1G_2}{G_1+G_2}\frac{1}{\pi}\int\limits_{L'}\frac{f_6(\xi)d\xi}{\xi-\varsigma} + \\ + i\frac{G_k}{G_1+G_2}\frac{1}{\pi}\int\limits_{L'}\frac{f_3(\xi)d\xi}{\xi-\varsigma}\ (\varsigma = x+iy,\ \varsigma \in S_k;\ r = 3,\ 6;\ k = 1,\ 2), \end{gathered} \tag{9}$$

and their boundary values on the upper and lower banks of the line L are the following:

$$\begin{gathered} \sigma_{yzk}^{\pm}(x,\pm h) = \mp\frac{G_k}{G_1+G_2}f_3(x) - \frac{G_1G_2}{G_1+G_2}\frac{1}{\pi}\int\limits_{L'}\frac{f_6(\xi)d\xi}{\xi-x} + \sigma_{yz}^{0\pm}(x,\pm h), \\ \sigma_{xzk}^{\pm}(x,\pm h) = \mp\frac{G_k}{G_1+G_2}f_6(x) + \frac{G_1G_2}{G_1+G_2}\frac{1}{\pi}\int\limits_{L'}\frac{f_3(\xi)d\xi}{\xi-x} + \sigma_{xz}^{0\pm}(x,\pm h). \end{gathered} \tag{10}$$

The values, further marked with the index "0" on top, correspond to the bulk's SSS model without inhomogeneities (inclusions, cracks, etc.) under the corresponding external load (homogeneous solution). Hereinafter, the notations [6] are used:

$$\begin{gathered}\sigma^0_{yzk}(\varsigma) + i\sigma^0_{xzk}(\varsigma) = \tau + i\{\tau_k + D_k(\varsigma) + \tfrac{G_k - G_j}{G_1 + G_2}\overline{D}_k(\varsigma) + \tfrac{2G_k}{G_1+G_2}D_j(\varsigma)\}, \\ D_k(\varsigma) = -\tfrac{Q_k + iG_k b_k}{2\pi(\varsigma - \varsigma_{*k})}\ (\varsigma \in S_k,\ k = 1, 2;\ j = 3-k). \end{gathered} \tag{11}$$

Additional balance conditions must be imposed on the solution of the external problem:

$$\begin{gathered}\int_{-a}^{a} f_3(\xi)d\xi = 2h(\sigma^{in}_{xz}(a) - \sigma^{in}_{xz}(-a)), \\ \int_{-a}^{a} f_6(\xi)d\xi = [w](a) - [w](-a). \end{gathered} \tag{12}$$

Equation (10) can be used directly to determine the critical values of the load applied to the structure at which slippage will begin. At some point(s) $L''^{\pm}$ when the maximum allowable slippage $\tau^{\max}_{yz}(x)$ is reached:

$$\sigma_{yzk}(x, \pm h) = \tau^{\max}_{yz}(x), \tag{13}$$

that is, once the expressions for the jumps f_r $(r = 3, 6)$ are known, the applied load can be investigated and its critical values determined.

It is possible to apply the classical JFM to the solution of the obtained system of Equations (1)–(12), which provides the substitution of (10) into (6) using (1)–(4), and obtains the resulting system of singular integral equations (SSIE) to determine the unknown f_r $(r = 3, 6)$ and the stresses in the matrix using (9) and inside the inclusion using (1)–(4). However, such a scheme for solving the problem works well in the case of an ideal contact of the structure components, simple geometry, and linear constitutive properties of the inclusion material. In the case of a non-ideal contact, with the a priori unknown dimensions of the slip zones $(c^{\pm} - b^{\pm})$, rather complicated algorithms for solving the SSIEs are required, which do not always guarantee calculation accuracy.

We propose a different approach to solving such a problem, which can be called a structurally modified JFM. The idea is to combine all equations into a global system without substituting the boundary conditions (1)–(4) into the model Equation (6), and limit values of matrix components of the stress–strain relations (9). Furthermore, it is convenient to solve this system of equations by any numerical–analytical method, for example, by the collocation method. Submitting the system of Equations (1)–(12) in discrete form in the set of collocation points $\left(x_n,\ n = \overline{1, N}\right)$, we obtain the system of $6N$ linear algebraic equations (SLAE) for the determination of $6N$ unknowns $\sigma^{in}_{yz}(x_n, \pm h)$, $\frac{\partial w^{in}}{\partial x}(x_n, \pm h)$, $f_r(x_n)$ $\left(r = 3, 6;\ n = \overline{1, N}\right)$. Of course, the number of unknowns for solving the problem increases, which is not crucial with modern computational capabilities. However, the construction of SLAEs is considerably simplified, as its modularity allows for making independent changes in separate modules with significantly less effort than constructing and solving new classes of problems with considerably more complicated parameters.

4. Numerical Results and Discussion

The accuracy of the solution in the case of non-ideal contact is very sensitive to the correct determination of the position and size of the slip zones. Within the framework of the proposed structural–modular MFS, we apply an iterative approach for this at each point of the interval $L''^{\pm}$: (1) gradually with growth, we apply a small load to check condition (13) of the beginning of the slippage process; (2) as soon as condition (13) is satisfied at certain points x_n, we assign values $\tau^{\max}_{yz}$ to values $\sigma^{in}_{yz}(x_n, \pm h)$, $\sigma_{yzk}(x_n, \pm h)$ in all these points and re-solve the SLAE; (3) we check whether constraint (4) is satisfied everywhere; if not, we assign values $\tau^{\max}_{yz}$ again to values $\sigma^{in}_{yz}(x_n, \pm h)$, $\sigma_{yzk}(x_n, \pm h)$ in those points x_n where (4) is not satisfied. The process is repeated until condition (4) is satisfied in all points $\left(x_n,\ n = \overline{1,N}\right)$. It is proven that such an iterative algorithm is convergent under monotonically increasing non-contrast loading. Despite the increased number of SLAE equations, the calculations showed that a relative error of 0.1% of the results can be achieved already at 21 collocation points for no more than seven iteration steps in the worst case of a very stiff inclusion. The obtained results were validated by comparing them with known partial solutions for an interfacial crack and an interfacial thin rigid inclusion, as well as a thin elastic interfacial inclusion at its ideal contact with the matrix [6,7,15,18].

Figures 2 and 3 show the results of the study of the "safety zone" for the intensities and coordinates of concentrated forces $\left(\widetilde{Q}_2 = -\widetilde{Q}_1 = \widetilde{Q},\ \widetilde{\varsigma}_{*k} = \varsigma_{*k}/a = \widetilde{x}_{*k} + i\widetilde{y}_{*k},\ \widetilde{\varsigma}_{*2} = \overline{\widetilde{\varsigma}}_{*1},\ \widetilde{y}_{*k} = \pm i\widetilde{d}\right)$, where $\widetilde{Q} = Q/\pi a G_{av}, \widetilde{x}_{*2} = x_{*2}/a,\ \widetilde{d} = d/a,\ G_{av} = \{\sqrt{G_1G_2},\ \max(G_1,\ G_2),\ Q/\pi a\}$. The "safety zone" for a concentrated factor of a certain value will be understood as the coordinates of its application at which slippage does not yet start at any point of the boundary L'; i.e., condition (4) is not fulfilled. We define the boundary of the "safety zone" from the following criterion: condition (4) starts to be fulfilled at least at one point. It can be argued that an inclusion harder than the matrix changes the form of the "safety zone" much less than a softer one. The particular case of no inclusion $\widetilde{G}_y^{in} = \widetilde{G}_k \left(\widetilde{G}_y^{in} = G_y^{in}/G_{av};\ \widetilde{G}_k = G_k/G_{av}\right)$ shows the coincidence of the results with those obtained in [6,7,18]. Additionally, note the expected trends of linear dependence of the growth of the critical value $\widetilde{Q}^*$ on the increase in the value $\widetilde{\tau}_{yz}^{\max\pm} = \frac{\tau_{yz}^{\max\pm}}{G_{av}}$, as well as on the increase in the distance $\widetilde{d}$ of the points of application of the concentrated force. These effects are especially appreciable for the "soft" inclusion, when $\widetilde{G}_y^{in} \ll \widetilde{G}_k$.

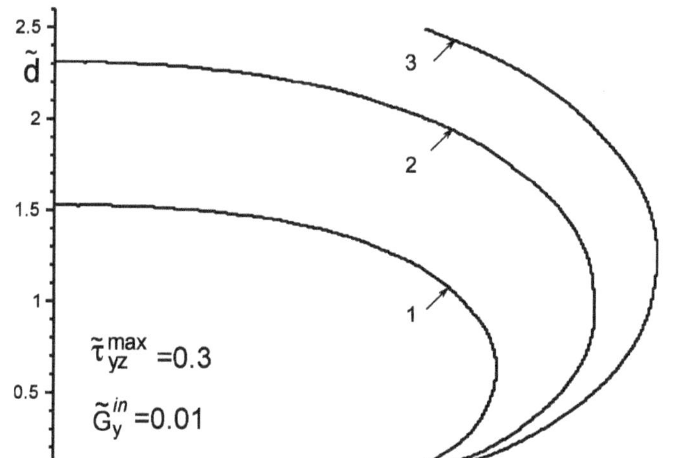

Figure 2. The boundary of the "safety zone" when loading the structure with a softer than the matrix inclusion by a concentrated force: 1—$\widetilde{Q}^* = 0.5$; 2—$\widetilde{Q}^* = 0.75$; 3—$\widetilde{Q}^* = 0.9$.

Figures 4–9 show the effect of slip on the stress $\widetilde{\sigma}_{yz} = \sigma_{yz}/G_{av}$ distribution along with the inclusion–matrix interface, as well as the growth of the slip zone size and its intensity $\widetilde{w}_{sl} = w_{sl}/a$ depending on the problem parameters. It is noteworthy that for a softer than the matrix inclusion, the slippage appears and grows faster than for a more rigid inclusion with the same problem parameters (Figures 4, 6 and 7). Decreasing the distance of the application points from the inclusion axis, as well as increasing the load intensity, is expected to increase the slip area and its magnitude (Figures 5–7 and 9). However, the displacement of the coordinates of the force application points along the inclusion axis to its apex (Figure 8), as well as the distance from this axis (Figure 5), reduces the slip intensity. The loading by an applied screw dislocation with a Burger's vector $\widetilde{b}_2 = b_2 G_2 / G_{av}$ in p. $x_{*2} = 0$; $y_{*2} = a$ leads to the appearance of two slip zones antisymmetric for the vertical axis of inclusion (Figure 9).

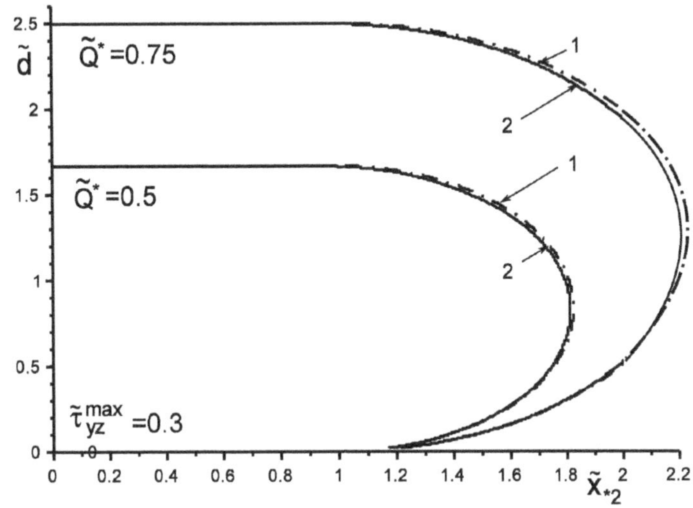

Figure 3. The boundary of the "safety zone" when loading with the concentrated force $\widetilde{Q}^*$ of the structure without ($1—\widetilde{G}_y^{in} = 1$) and with harder than matrix inclusion ($2—\widetilde{G}_y^{in} = 10$).

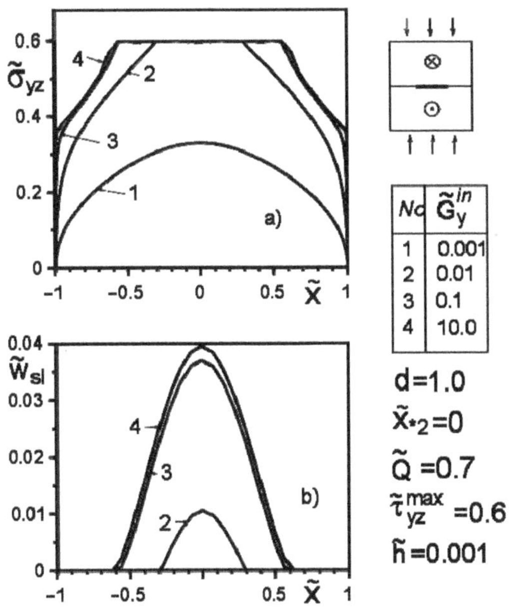

Figure 4. Stress distribution along with the inclusion–matrix boundary, (**a**) and the size of the slip zone (**b**) depending on the ratio $\widetilde{G}_y^{in} / \widetilde{G}_k$

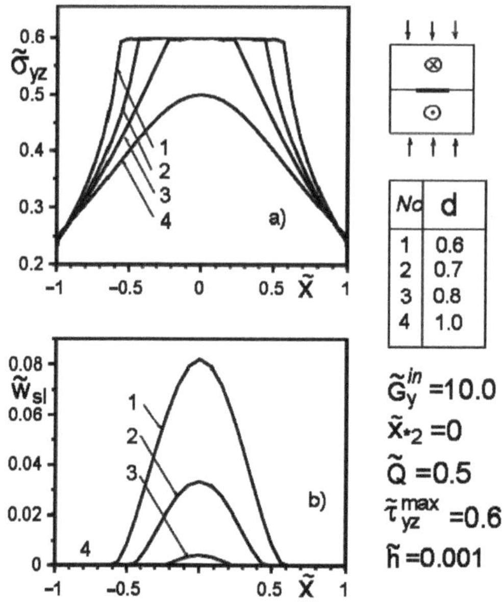

Figure 5. Stress distribution along with the inclusion–matrix boundary (**a**) and the value of the slip zone (**b**) for inclusion harder than matrix, depending on the distance of the force application point from its axis.

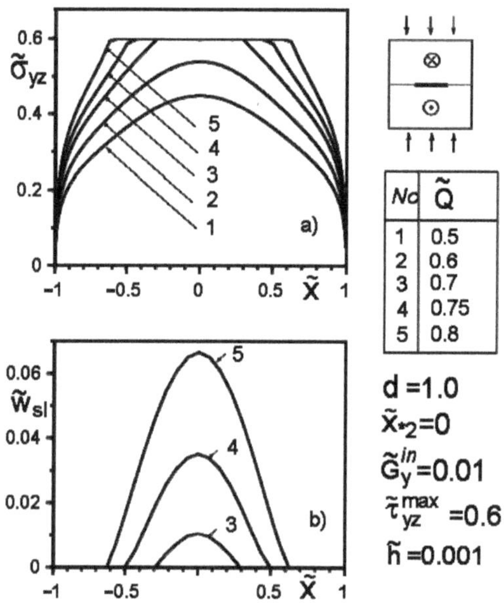

Figure 6. Stress distribution along with the inclusion–matrix boundary (**a**) and the size of the slip zone (**b**) for a softer than matrix inclusion as a function of force intensity growth.

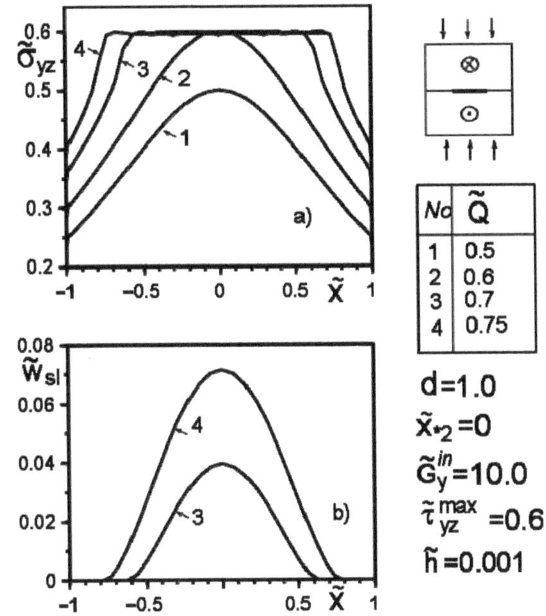

Figure 7. Stress distribution along with the inclusion–matrix boundary (**a**) and the value of the slip zone (**b**) for a harder than matrix inclusion as a function of force intensity growth.

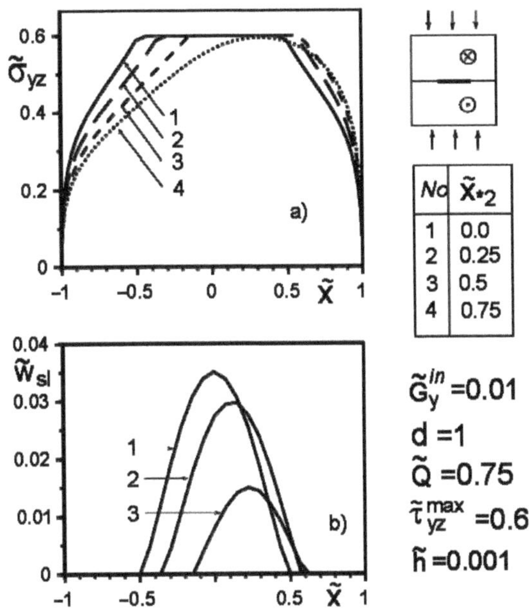

Figure 8. Stress distribution along with the inclusion–matrix boundary (**a**) and the value of the slip zone (**b**) for a softer inclusion than matrix depending on the change in the coordinates of the force application points along the inclusion axis.

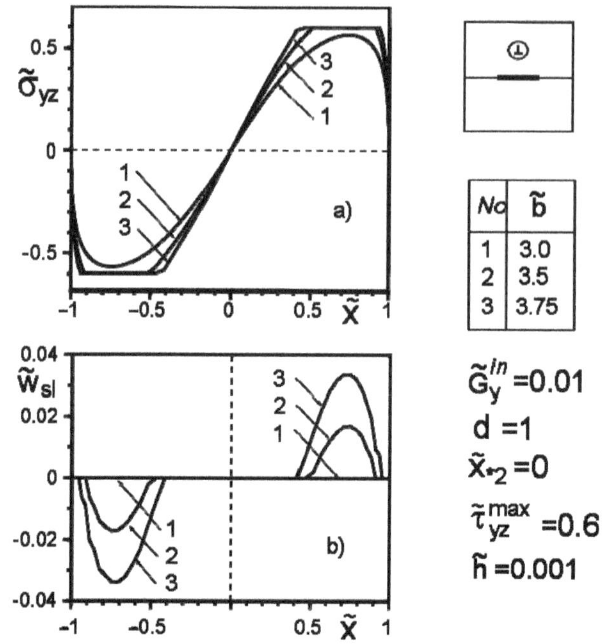

Figure 9. Stress distribution along with the inclusion–matrix boundary (**a**) and the value of the slip zone (**b**) for a softer than matrix inclusion under screw dislocation loading.

5. Conclusions

The proposed numerical–analytical approach to study the effects of mechanical contact imperfections and friction effects of a bimaterial structure with an interfacial thin ribbon-like deformable inclusion made it possible to obtain some important new results. Firstly, the a priori unknown configuration of the sliding friction zone was determined under different types of loads.

It is revealed that with the friction coefficient, type, and place of load application unchanged, the slippage will start at a lower load and the slippage zone will be larger in the case of a harder than the matrix inclusion. It can be argued that the shape and size of the "safety zone" changes much less in the case of a harder than matrix inclusion than in the case of a softer one.

The calculations confirmed the linear attenuation of the sizes of the safety and slip zones from a decrease in the intensity or distance from the inclusion of the concentrated factors. This is especially noticeable for the softer inclusion than the matrix inclusion.

The distribution of stresses in the inclusion vicinity, calculated to optimize the load regime of the investigated structure, makes it possible to conclude that the greatest influence on its stress–strain state is in the inclusion vicinity; accordingly, the appearance of slippage between the components has the case of applying concentrated forces in the point above the inclusion center.

The obtained results and the proposed method can be used and developed to study the value of the energy dissipated on the inclusion, heat release due to friction, development of wear processes, and the interaction of band inhomogeneities in composite structures with subsequent determination of their effective mechanical characteristics, optimization of operating load modes and the like.

Author Contributions: Conceptualization, Y.P. and H.S.; methodology, Y.P. and H.S.; software, Y.P.; validation, Y.P. and H.S.; formal analysis, Y.P.; investigation, Y.P. and H.S.; resources, H.S.;

data curation, Y.P.; writing—original draft preparation, Y.P.; writing—review and editing, Y.P.; visualization, Y.P.; supervision, H.S.; project administration, Y.P.; funding acquisition, H.S. All authors have read and agreed to the published version of the manuscript.

Funding: This study was performed within the framework of project financing through the program of the Minister of Science and Higher Education of Poland named "Regional Initiative of Excellence" in 2019–2022, project number 011/RID/2018/19, amount of financing 12,000,000 PLN.

Institutional Review Board Statement: Not applicable.

Informed Consent Statement: Not applicable.

Data Availability Statement: The data presented in this study are openly available at DOI 10.1007/s11003-018-0114-2, reference number [6].

Conflicts of Interest: The authors declare no conflict of interest.

Nomenclature

SSS	stress–strain state;
SSIE	system of singular integral equations;
SLAE	system of linear algebraic equations;
x, y, z	Cartesian coordinates;
f_r	jump functions;
E_k, ν_k, G_k	elastic modulus of the material;
S_k	half-planes (sections of the body);
$a, h, b^{\pm}, c^{\pm}$	dimensions of the inclusion and slip zones;
$w, \sigma_{xz}, \sigma_{yz}, \sigma_{xx}, \sigma_{yy}$	displacement, stresses (components of SSS);
$L' = [-a; a]$	line, modelling the presence of thin inclusion;
Q_k, b_k	magnitudes of concentrated forces and screw dislocations;
$\sigma_{yz}^{\infty}, \sigma_{xzk}^{\infty}, \sigma_{yy}^{\infty}, \sigma_{xxk}^{\infty}$	uniformly distributed in infinity shear stresses;
α	coefficient of the sliding friction;
Special denotations	
$[\varphi]_h = \varphi(x,-h) - \varphi(x,+h)$,	$\langle\varphi\rangle_h = \varphi(x,-h) + \varphi(x,+h)$;
superscripts "+" and "−"	denotes boundary values of functions on the upper and the lower with respect to width inclusion borders accordingly;
superscript "in"	marks the values corresponding to inclusion;
superscript "°"	marks the values in the corresponding problem without any inclusion;
superscript "~"	marks the terms that become dimensionless;
subscript "k"	denotes the terms corresponding to half-plains.

References

1. Comninou, M.; Schmueser, D.; Dundurs, J. Frictional slip between a layer and a substrate caused by a normal load. *Int. J. Eng. Sci.* **1980**, *18*, 131–137. [CrossRef]
2. Kalker, J.J. A survey of the mechanics of contact between solid bodies. *Z. Angew. Math. Mech.* **1977**, *57*, T3–T17.
3. Kachanov, M.; Sevostianov, I. *Micromechanics of Materials, with Applications/Solid Mechanics and Its Applications*; Springer: Berlin, Germany, 2018; Volume 249, p. 712.
4. Nemat-Nasser, S.; Hori, M. *Micromechanics: Overall Properties of Heterogeneous Materials*; North-Holland Series in Applied Mathematics and Mechanics; Elsevier: Amsterdam, The Netherlands, 1993; Volume 37, p. 687.
5. Sekine, H. Mechanics of debonding along the surfaces of dispersed flat inclusions in composite materials (A model of debonding along the surface of a flat inclusion). *Trans. ASME J. Appl. Mech.* **1982**, *48A*, 1415–1420.
6. Sulym, H.T.; Piskozub, I.Z. Nonlinear deformation of a thin interface inclusion. *Mater. Sci.* **2018**, *53*, 600–608. [CrossRef]
7. Sulym, H.T. *Bases of Mathematical Theory of Thermo-elastic Equilibrium of Solids Containing Thin Inclusions*; Research and Publishing Center of NTSh: L'viv, Ukraine, 2007; p. 716. (In Ukrainian)
8. Davim, J.P.; Charitidis, C.A. (Eds.) *Nanocomposites. Materials, Manufacturing and Engineering*; Walter de Gruyter GmbH: Berlin, Germany; Boston, MA, USA, 2013; p. 211.
9. Wang, J.; Karihaloo, B.L.E.; Duan, H.L. Nano-mechanics or how to extend continuum mechanics to nano-scale. *Bull. Pol. Acad. Sci. Tech. Sci.* **2007**, *55*, 133–140.
10. Wang, Y.; Huang, Z.M. Analytical Micromechanics Models for Elastoplastic Behavior of Long Fibrous Composites: A Critical Review and Comparative Study. *Materials* **2018**, *11*, 1919. [CrossRef] [PubMed]

11. Bogdanovich, P.N.; Tkachuk, D.V. Thermal and Thermomechanical Phenomena in Sliding Contact. *J. Frict. Wear* **2009**, *30*, 153–163. [CrossRef]
12. Evtushenko, A.A.; Kutsei, M. Effect of pressure evolution during braking on temperature of a pad-disk tribosystem. *J. Frict. Wear* **2010**, *31*, 317–325. [CrossRef]
13. Goryacheva, I.G. *Contact Mechanics in Tribology;* Series: Solid Mechanics and Its Applications; Springer: Berlin, Germany, 1998; Volume 61, p. 346.
14. Hills, D.A.; Nowell, D.; Sackfield, A. *Mechanics of Elastic Contact;* Butterworth-Heinemann: Oxford, UK, 1993; p. 238.
15. Kharun, I.V.; Loboda, V.V. A set of interface cracks with contact zones in combined tension-shear field. *Acta Mech.* **2003**, *166*, 43–56. [CrossRef]
16. Pyriev, S.Y.; Yevtushenko, A.A.; Sulym, G.T. Thermomechanical Wear during Quasistationary Heat Generation by Friction. *J. Frict. Wear* **2012**, *33*, 315–321. [CrossRef]
17. Sosnovskiy, L.A. *Tribo-fatigue. Wear-fatigue Damage and Its Prediction (Foundations of Engineering Mechanics);* Series: Foundations of Engineering Mechanics; Springer: Berlin, Germany, 2005.
18. Sulym, H.T.; Pasternak, Y.M.; Piskozub, J.Z.; Piskozub, L.G. Longitudinal shear of a bimaterial with frictional sliding contact in the interfacial crack. *J. Theor. Appl. Mech.* **2015**, *54*, 529–539.
19. Sun, C.T.; Qian, W. A treatment of interfacial cracks in the presence of friction. *Int. J. Fract.* **1998**, *94*, 371–382. [CrossRef]
20. Belhocine, A.; Oday, I.A. Thermomechanical Model for the Analysis of Disc Brake Using the Finite Element Method in Frictional Contact. *Multiscale Sci. Eng.* **2020**, *2*, 27–41. [CrossRef]

materials MDPI

Article

Effect of the Transverse Functional Gradient of the Thin Interfacial Inclusion Material on the Stress Distribution of the Bimaterial under Longitudinal Shear

Yosyf Piskozub [1,*], Liubov Piskozub [2] and Heorhiy Sulym [3]

1 Department of Applied Mathematics, Faculty of Computer Science and Telecommunications, Cracow University of Technology, Warszawska Str. 24, 31-155 Cracow, Poland
2 Department of Applied Mathematics and Physics, Ukrainian Academy of Printing, Pidgolosko Str. 19, 79020 Lviv, Ukraine
3 Department of Mechanics and Applied Computer Science, Faculty of Mechanical Engineering, Bialystok University of Technology, Wiejska 45C, 15-351 Bialystok, Poland
* Correspondence: yosyf.piskozub@pk.edu.pl; Tel.: +48-574-560-665 or +38-073-222-4246

Abstract: The effect of a functional gradient in the cross-section material (FGM) of a thin ribbon-like interfacial deformable inclusion on the stress–strain state of a piecewise homogeneous linear–elastic matrix under longitudinal shear conditions is considered. Based on the equations of elasticity theory, a mathematical model of such an FGM inclusion is constructed. An analytic–numerical analysis of the stress fields for some typical cases of the continuous functional gradient dependence of the mechanical properties of the inclusion material is performed. It is proposed to apply the constructed solutions to select the functional gradient properties of the inclusion material to optimize the stress–strain state in its vicinity under the given stresses. The derived equations are suitable with minor modifications for the description of micro-, meso- and nanoscale inclusions. Moreover, the conclusions and calculation results are easily transferable to similar problems of thermal conductivity and thermoelasticity with possible frictional heat dissipation.

Keywords: functionally graded material; thin inclusion; composites; nonperfect contact; frictional heating; crack; stress intensity factor

Citation: Piskozub, Y.; Piskozub, L.; Sulym, H. Effect of the Transverse Functional Gradient of the Thin Interfacial Inclusion Material on the Stress Distribution of the Bimaterial under Longitudinal Shear. *Materials* **2022**, *15*, 8591. https://doi.org/10.3390/ma15238591

Academic Editor: Francisco J. G. Silva

Received: 3 November 2022
Accepted: 28 November 2022
Published: 2 December 2022

1. Introduction

Many structural materials contain numerous thin inhomogeneities in the form of inclusions of different origins [1–7]. Quite often, these inclusions are used as elements to reinforce the structural parts of machines and structures or as fillers in composite materials or coatings [8–15]. The use of nanocomposites with specific properties in engineering and technology has significantly shifted the interest from the study of objects at the macro level (10^0–10^{-1} m) and micro level (10^{-3}–10^{-6} m) to the nano level (10^{-9} m) [16–20].

One of the typical examples of composite materials is the structure with thin ribbon inclusions. Structural elements made with the use of FGM have proven to be rather effective in practice [21–23]. In this way, it is possible to achieve a significant improvement in their mechanical, rheological, thermal, or other properties or the formation of protective thin layers [18,24–27].

The mathematical modeling of nanostructures requires the construction of more complex constitutional laws in comparison with the macro level [28,29]. Therefore, it is important to construct methods for studying the stress–strain state of such structures. To model a thin inclusion, there are mainly two basic approaches using analytical methods. The first one is based on the use of Eshelby's analytical solution [21,30,31] for an ellipsoidal inclusion, in which a limiting transition with a decrease in one of the characteristic dimensions of the inclusion is performed. However, its application to thin interphase inclusions is impossible. The second one is based on the principle of the conjugation of continua of different measurability [32–37] and the method of jump functions [28,29,38–41]. According to this method, the inclusions are replaced by a certain surface (in the two-dimensional case, a line) of the discontinuity of the physical–mechanical fields, which describes the perturbing effect of a thin inclusion. Successful attempts were made in [17,33,42–45] to apply it to consider the influence of various physical and contact nonlinearities in the antiplane problem of elasticity theory for two compressed half-spaces with interfacial defects. The frictional slip with possible heat generation for contacting bodies [34,44,46–49] and the boundary element approach [50–52] were also considered here. Inhomogeneity of the mechanical properties of structural materials can be both designed for a specific purpose (FGM) and a consequence of technological processes of obtaining new materials and their processing (FSW, ball-burnishing process, etc.) [53,54]. Such factors cause additional complexity in the constitutive relations for the mathematical modeling of the behavior of such materials. However, the case of a thin inclusion of an inhomogeneous material has not been practically studied.

The process of improving the mathematical models of FGM [11,25,26,55–58] is complicated by the complex geometry of structural elements and the consideration of imperfections in the contact of their components. This is especially important to ensure their qualitative design, both in terms of mechanical strength [11,59–63] and in terms of the consideration of thermal, magnetic, and piezoelectric load factors [13,21,24,44,53,54,58,64,65].

The works [17,31,35,39] have been devoted to the consideration of surface energy and stresses in nanocomposites. Consideration of the heterogeneity of inclusions' properties at the micro and nano scale is particularly important because the heterogeneity density of matter (discrepancy variation in mechanical and other properties) of matter bodies with a decrease in their scale usually increases, and the impact of this heterogeneity increases even further.

This publication aims to develop the method of jump functions and to construct a convenient structured and highly versatile approach to study the longitudinal displacement and thermal heating of bodies with thin inclusions of arbitrary physical nature, including those made from FGM.

2. Formulation of the Problem

Consider an unbounded isotropic structure consisting of two half-spaces with elastic moduli G_k $(k = 1, 2)$, which is subjected to an external longitudinal shear load determined by uniformly distributed infinity stresses σ_{yz}^{∞} and σ_{yz}^{∞}, concentrated forces of intensity Q_k and screw dislocations with vector Burgers component b_k at points $\varsigma_{*k} \in S_k (k = 1,\ 2)$. The stresses must satisfy the conditions $\sigma_{xz2}^{\infty} G_1 = \sigma_{xz1}^{\infty} G_2$ at infinity to ensure the straightness of the interface.

Let us investigate the stress–strain state (SSS) of the body section with a plane xOy perpendicular to the direction Oz of its longitudinal displacement (external problem). The plane sections of half-spaces perpendicular to this axis form two half-planes S_k $(k = 1, 2)$ and the abscissa axis corresponds to the interface $L \sim x$ between them (Figure 1). At the interface of half-spaces (plane xOz), there is a tunnel section $L' = [-a;\ a]$ in the direction of the shear axis z, in which a certain object of general thickness $2h$ $(h \ll a)$ is inserted.

According to the paradigm of the method of jump functions [36], the presence of a thin inclusion in the bulk is modeled by jumps in the components of stress and displacement vectors in [38,40,41]:

$$\begin{aligned} &[\sigma_{yz}]_h \cong \sigma_{yz}^{-} - \sigma_{yz}^{+} = f_3(x),\ x \in L' \\ &\left[\frac{\partial w}{\partial x}\right]_h \cong \frac{\partial w^-}{\partial x} - \frac{\partial w^+}{\partial x} = \left[\frac{\sigma_{xz}}{G}\right]_h \equiv \frac{\sigma_{xz}^-}{G_1} - \frac{\sigma_{xz}^+}{G_2} = f_6(x); \end{aligned} \tag{1}$$

$$f_3(x) = f_6(x) = 0, \text{ if } x \notin L'. \tag{2}$$

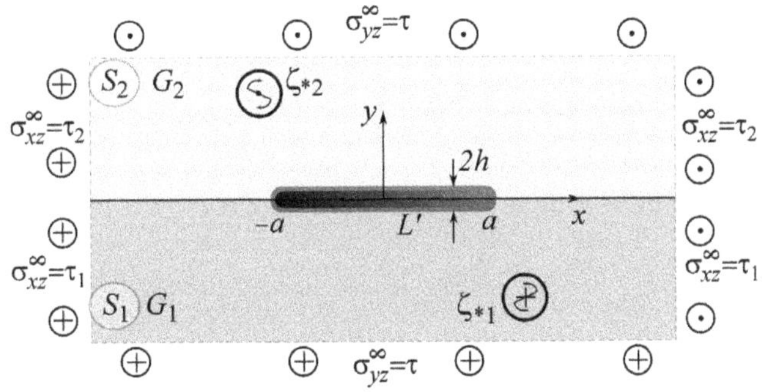

Figure 1. The loading and geometric scheme of the problem.

It is hereinafter marked $[\bullet]_h = \bullet(x,-h) - \bullet(x,+h)$, $\langle\bullet\rangle_h = \bullet(x,-h) + \bullet(x,+h)$; superscripts "+" and "−" correspond to the boundary values of the functions on the upper and lower banks of the line L.

The mathematical model of a thin inclusion is given as complicated conditions of imperfect contact between opposite matrix surfaces adjacent to the inclusion (internal problem) [28,38–40,65]. The general model of a thin, physically nonlinear inclusion is presented in [28,29,38], where the methods of modeling thin objects involve the integration of the defining relations describing the physical and mechanical properties of the material of the inclusion, with the subsequent consideration of the smallness of one of the linear dimensions of the inclusion.

Let us consider a similar model for a thin inclusion, assuming that the mechanical properties of the inclusion material are coordinate-dependent. This will allow us to model the inclusions of a functionally graded material:

$$\begin{cases} G_x^{in}(x)\left\langle \frac{\partial w^{in}}{\partial x}\right\rangle_h(x) - 2\sigma_{xz}^{in}(-a) - \frac{1}{h}\int\limits_{-a}^{x}\left[\sigma_{yz}^{in}\right]_h(\xi)d\xi = 0, \\ G_y^{in}(x)\left[w^{in}\right]_h(x) + h\left\langle\sigma_{yz}^{in}\right\rangle_h(x) = 0. \end{cases} \tag{3}$$

Here, $G_x^{in}(x)$, $G_y^{in}(x)$ are the variable shear moduli of the inclusion's material. As a special case, considering their values to be constant, we obtain Hooke's law. The upper index "*in*" denotes the terms describing the inclusion material's SSS components.

Contact between matrix components and the inclusion at L' and between the bimaterial structure components along a line $L\backslash L'$ is supposed to be mechanically perfect,

$$\begin{array}{c} w(x,\pm h) = w^{in}(x,\pm h), \quad \sigma_{yz}^{in}(x,\pm h) = \sigma_{yzk}(x,\pm h),\ x \in L', \\ w(x,+0) = w(x,-0), \\ \sigma_{yz2}(x,+0) = \sigma_{yz1}(x,-0), \qquad x \in L\backslash L', \end{array} \tag{4}$$

or frictional in some areas $x \in L_f \subset L'$, as was considered in works [41,46–48],

$$\sigma_{yz}^{in}(x,\pm h) = \sigma_{yzk}(x,\pm h) = -\mathrm{sgn}\left[w^{in}\right]_h \tau_{yz}^{\max}. \tag{5}$$

Here, $\tau_{yzK}^{\max}$ is the limit value of shear stresses, at which the slippage begins. In this case, however, additional iterative methods should be applied to determine the area of the slip zones depending on the specific types of external loading of the composite [41].

3. Materials and Methods

Expressions for the components of the stress tensor and the derivatives of displacements on the line L of the infinite plane $S = S_1 \cup S_2$, as well as inside the latter, can be obtained by applying the results of [37] to the solution of the external problem

$$\begin{gathered}\sigma_{yz}(z) + i\sigma_{xz}(z) = \sigma^0_{yz}(z) + i\sigma^0_{xz}(z) + \\ + ip_k g_3(z) - Cg_6(z) \quad (z \in S_k;\ r = 3, 6;\ k = 1, 2);\end{gathered} \tag{6}$$

$$\begin{gathered}\sigma^{\pm}_{yzk}(x) = \mp p_k f_3(x) - Cg_6(x) + \sigma^{0\pm}_{yz}(x), \\ \sigma^{\pm}_{xzk}(x) = \mp Cf_6(x) + p_k g_3(x) + \sigma^{0\pm}_{xz}(x),\end{gathered} \tag{7}$$

where the notation [28,41] is introduced:

$$\begin{gathered}g_r(z) \equiv \frac{1}{\pi}\int\limits_{L'} \frac{f_r(x)dx}{x-z}, \quad s_r(x) \equiv \int\limits_{-a}^{x} f_r(x)dx, \\ p = \frac{1}{G_1+G_2},\ p_k = G_k p,\ C = G_1 G_2 p.\end{gathered} \tag{8}$$

The superscript "+" corresponds to $k = 2$ and "–" corresponds to $k = 1$. Values marked with superscript "0" correspond to values in a continuous medium without inclusions under the same external load (homogeneous solution) [28,41].

Using (7), (8) and boundary conditions (4), it is easy to obtain from model (3) a system of singular integral equations:

$$\begin{gathered}\begin{cases}(p_2 - p_1)f_6(x) + 2pg_3(x) - \frac{s_3(x)}{hG^{in}_x(x)} = F_3(x), \\ (p_2 - p_1)f_3(x) + 2Cg_6(x) - \frac{G^{in}_y(x)s_6(x)}{h} = F_6(x),\end{cases} \\ F_3(x) = \frac{2}{G^{in}_x}\sigma^{in}_{xz}(-a) - \left(\sigma^0_{xz2}(x)/G_2 + \sigma^0_{xz1}(x)/G_1\right), \\ F_6(x) = \left\langle \sigma^0_{yz} \right\rangle(x) - G^{in}_y\left\langle \frac{\sigma^0_{yzk}(x)}{G_k} \right\rangle - \frac{G^{in}_y}{h}\begin{bmatrix}0\\w\end{bmatrix}(-a).\end{gathered} \tag{9}$$

Balance conditions on the power balance and unambiguity of displacements while moving around the thin defect must be added to the solution of the external problem:

$$\begin{gathered}\int\limits_{-a}^{a} f_3(\xi)d\xi = 2h\left(\sigma^{in}_{xz}(a) - \sigma^{in}_{xz}(-a)\right), \\ \int\limits_{-a}^{a} f_6(\xi)d\xi = [w](a) - [w](-a).\end{gathered} \tag{10}$$

Solving (9) and (10) using the methods in [29,38,41], it is easy to obtain a system of linear algebraic equations with unknown coefficients of the decomposition of the jump functions $f_r(x)$ into a series by orthogonal Jacobi or Chebyshev polynomials.

An important aspect of the study of the strength of such structures is the improvement of their strength criteria. In fracture mechanics, it is acceptable to use the stress intensity factor to describe the behavior of the SSS in the vicinity of the crack tip [42,45,61–63,66]. This is not sufficient for the case of a thin deformable inclusion. In [45], the authors obtained the two-term asymptotical expressions for the distribution of SSS in the vicinity of the thin inclusion tips using the introduced generalized stress intensity factors (GSIF):

$$K_{31} + iK_{32} = \lim_{r\to 0\,(\theta=0,\pi)} \sqrt{2\pi r}\left(\sigma_{yz} + i\sigma_{xz}\right). \tag{11}$$

Here, (r, θ) is a system of polar coordinates with the origin near the right or the left tip of the inclusion $z = \pm r\exp(i\theta) \pm a$.

Considering the well-known mathematical analogy [67], the obtained solutions to the antiplane problem can be regarded as solutions to the accordant heat conduction problem, if we take into account the correspondence of the values

$$w \sim T,\ \frac{\partial w}{\partial x} \sim \frac{\partial T}{\partial x},\ \frac{\partial w}{\partial y} \sim \frac{\partial T}{\partial y},\ q_x \sim \sigma_{xz},\ q_y \sim \sigma_{yz},$$
$$G_x \sim \lambda_x,\ G_y \sim \lambda_y,\ K_{31} \sim k_{qy},\ K_{32} \sim k_{qx}.$$

The terms are as follows: T—temperature, q_x, q_y—heat flows, λ_x, λ_y—thermal conductivity coefficients, k_{qy}, k_{qx}—heat flow intensity factors [40].

4. Numerical Results and Discussion

Since the main focus of this article is to investigate the effect of the functional gradient on the mechanical properties of the inclusion material, we will limit ourselves to one of the most representative variants of the structure loading: homogeneous longitudinal shear $\sigma_{yz}^{\infty} = \tau$ and $\sigma_{xzk}^{\infty} = \tau_k$ $(k = 1,2)$ at infinity. However, the calculations for loading by concentrated force factors or dislocations do not make any fundamental difference except for the necessity to consider the locality of their application [29,41].

The dependence $G_x(x)$, $G_y(x)$ on coordinate x for mathematical modeling can be defined as an arbitrary function (linear, exponential, power, periodic [58], etc.), which adequately reflects the desired practical properties of the material. To illustrate the method, let us consider one of the illustrative variants of the functional gradient of the inclusion material—the piecewise linear one:

$$G_x(x) = G_y(x) = \begin{cases} (G_{01} - G_0)\frac{x}{a} + G_{01}, & x \in [-a, 0]; \\ (G_{02} - G_{01})\frac{x}{a} + G_{01}, & x \in [0, a], \end{cases} \tag{12}$$

where G_0, G_{01}, G_{02}—some given constants.

To significantly reduce the number of calculations without loss of generality, it is convenient to use the following dimensionless quantities, marked by symbol "~" (tilde) on top:

$$\tilde{x} = x/a, \tilde{h} = h/a, \tilde{y} = y/a,$$
$$\tilde{G}_x^{in}(\tilde{x}) = G_x^{in}(x)/G_{gav}, \tilde{G}_y^{in}(\tilde{x}) = G_y^{in}(x)/G_{gav},$$
$$\tilde{\tau}_k = \tau_k/G_{gav},\ \tilde{\tau} = \tau/G_{gav},\ G_{gav} = \sqrt{G_1 G_2},$$
$$\tilde{G}_0 = G_0/G_{gav},\ \tilde{G}_{01} = G_{01}/G_{gav}, \tilde{G}_{02} = G_{02}/G_{gav},$$
$$\tilde{\sigma}_{xz}(\tilde{x}) = \sigma_{xz}(x)/G_{gav},\ \tilde{\sigma}_{yz}(\tilde{x}) = \sigma_{yz}(x)/G_{gav}.$$
$$\tilde{K}_{31} = \frac{K_{31}^{+}}{2\tilde{C}G_{gav}\sqrt{\pi a}},\ \tilde{K}_{32} = \frac{K_{32}^{+}}{2p_2 G_{gav}\sqrt{\pi a}},$$

where K_{31}^{+} K_{32}^{+} are the GSIFs near the tip $x = +a$ of the inclusion.

Figures 2–11 illustrate the dependence of the stress–strain behavior of the matrix in the inclusion vicinity on the variation in the parameters $\tilde{G}_0$, $\tilde{G}_{01}$, $\tilde{G}_{02}$, the values of which were chosen to reveal a qualitative picture of the FGM effect on the stress–strain parameters. It can be immediately concluded from Figures 2 and 3 that under the load $\tilde{\tau}$, the dimensionless $\tilde{K}_{31}^{+}$ are expected to decrease with the increasing shear moduli of any part of the inclusion, while at $\tilde{K}_{32}^{+}$ they appear to increase with increasing load $\tilde{\tau}_k$.

The effect of changes in the moduli $\tilde{G}_x(x)$, $\tilde{G}_y(x)$ on the stresses $\tilde{\sigma}_{yz}$, $\tilde{\sigma}_{xz}$ on the inclusion surface is more obvious if we choose a linear growth law for them along the inclusion axis (Figures 4–6). The magnitude of the surface stresses increases significantly in the stiffer part of the inclusion. The larger the stiffness gradient, the more significant the increase.

The choice of the piecewise linear law of moduli $\tilde{G}_x(x)$, $\tilde{G}_y(x)$ change in the Formulae (14) as $\tilde{G}_{01} = \tilde{G}_{02}$ (variant 1) or $\tilde{G}_0 = \tilde{G}_{01}$ (variant 2) has a more contrasting effect on the surface stresses $\tilde{\sigma}_{yz}$, $\tilde{\sigma}_{xz}$, especially in the vicinity of the gradient breaking point $x = 0$ (Figures 7 and 8). Moreover, variant 2 of the functional dependence of the inclusion material moduli leads to partial unloading in the softer part of the inclusion near the breaking point $x = 0$ (Figure 8).

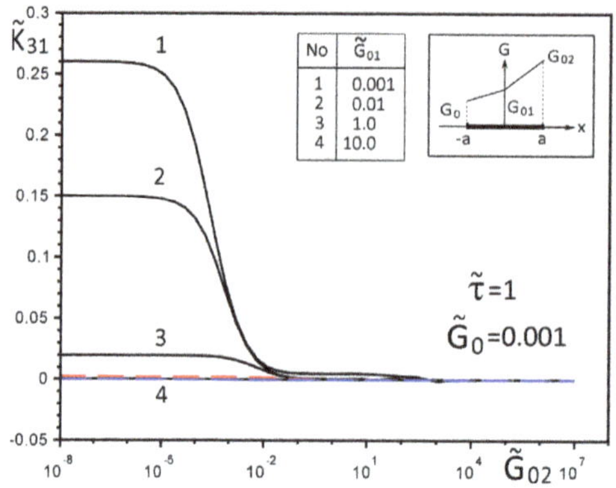

Figure 2. Influence of the parameters $\widetilde{G}_0$, $\widetilde{G}_{01}$, $\widetilde{G}_{02}$ on the GSIF $\widetilde{K}_{31}^{+}$ under the load, uniformly distributed on infinity stress $\sigma_{yz}^{\infty} = \tau$.

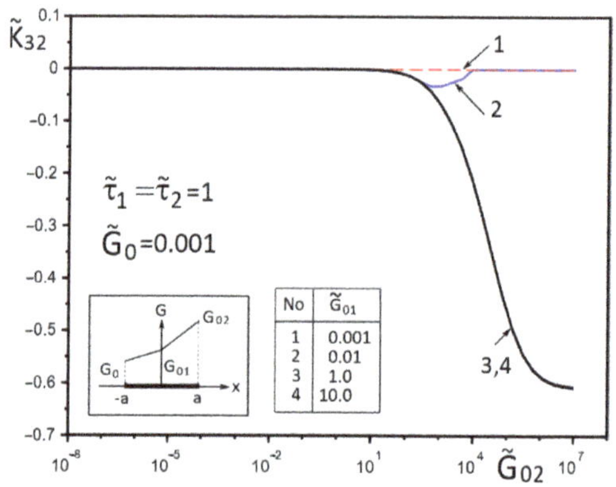

Figure 3. Influence of the parameters $\widetilde{G}_0$, $\widetilde{G}_{01}$, $\widetilde{G}_{02}$ on the GSIF $\widetilde{K}_{32}^{+}$ under the load, uniformly distributed on infinity stress $\sigma_{xzk}^{\infty} = \tau_k \, (k = 1, 2)$.

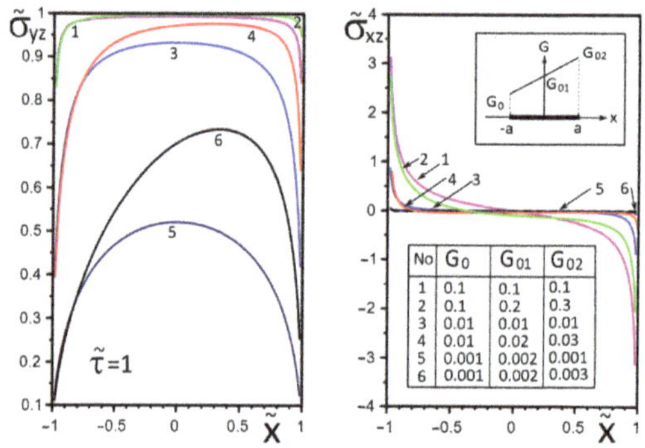

Figure 4. Stress distribution along with the upper interface (inclusion–matrix half-space S_2) with a linear distribution of material stiffness.

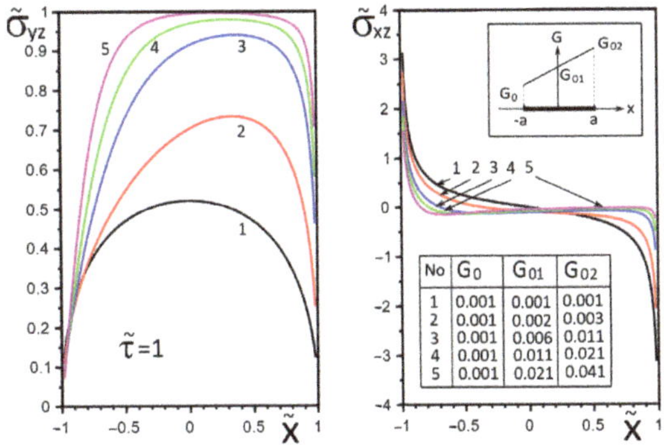

Figure 5. Stress distribution along with the upper interface (inclusion–matrix half-space S_2) with a linear distribution of material stiffness.

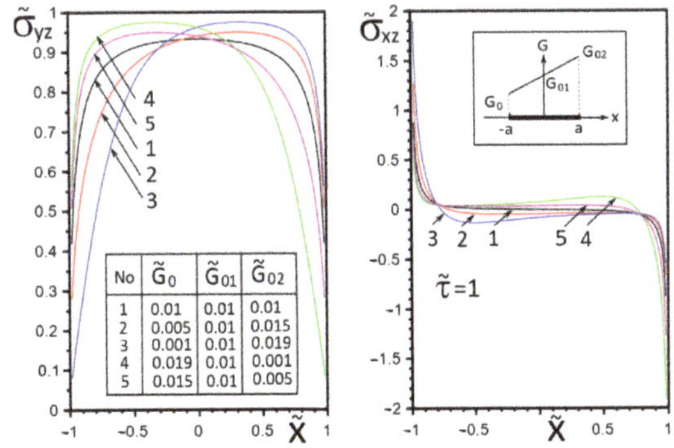

Figure 6. Stress distribution along with the upper interface (inclusion–matrix half-space S_2) with a linear distribution of material stiffness.

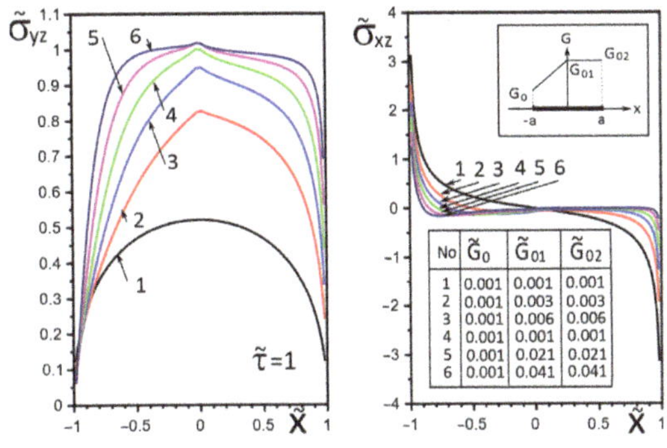

Figure 7. Stress distribution along with the upper interface (inclusion—matrix half-space S_2) with a piecewise linear distribution of material stiffness.

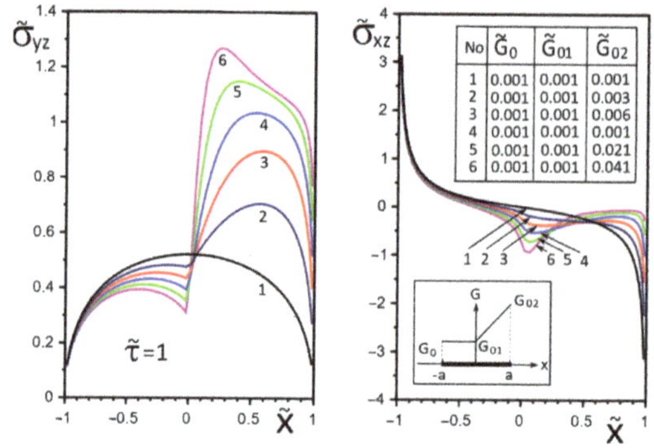

Figure 8. Stress distribution along with the upper interface (inclusion—matrix half-space S_2) with a piecewise linear distribution of material stiffness.

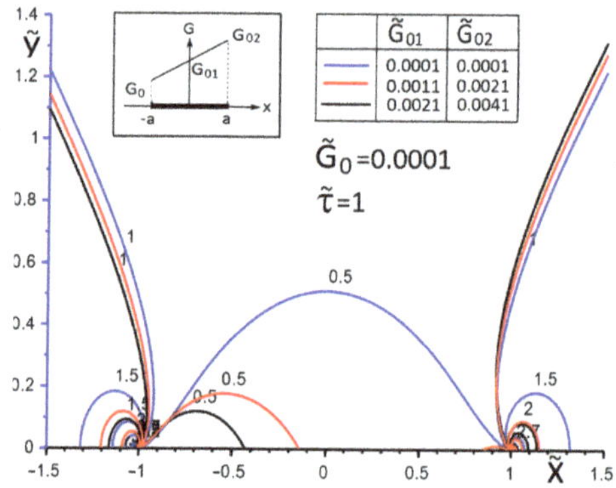

Figure 9. Stress distribution in the matrix at the vicinity of the inclusion with a linear distribution of material stiffness.

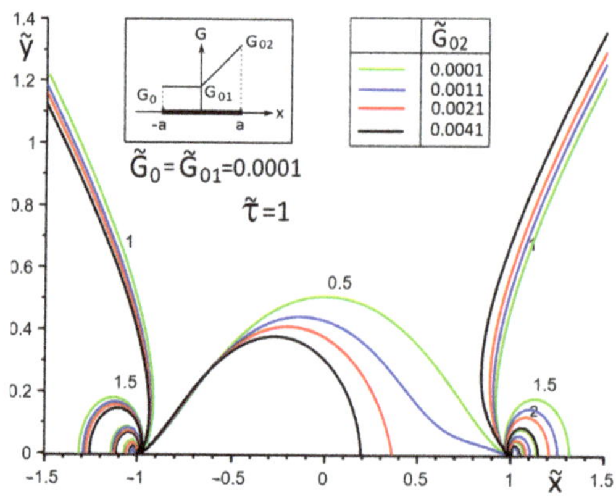

Figure 10. Stress distribution in the matrix at the vicinity of the inclusion with the piecewise linear distribution of material stiffness.

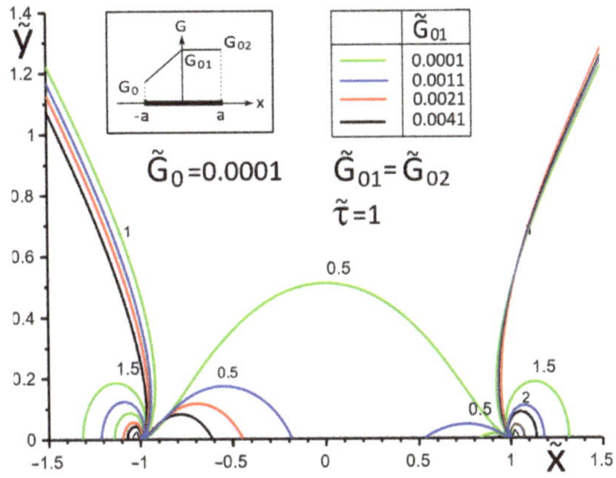

Figure 11. Stress distribution in the matrix at the vicinity of the inclusion with the piecewise linear distribution of material stiffness.

Figures 9–11 illustrate the changes in the stress field in the matrix in the inclusion vicinity under different variants of the law of functional change of the inclusion material moduli. The trends towards a decrease in the stress magnitudes in the vicinity of the stiffer parts of the inclusion are visible.

5. Conclusions

The proposed sufficiently simple and mathematically correct methodology made it possible for us to construct, for the first time, a mathematical model of a deformable thin linear interfacial inclusion with essentially inhomogeneous linear mechanical properties. Such a model can be used to simulate a thin inclusion from a functionally graded material and to solve the corresponding problems of defining the stress–strain field of the corresponding micro- or nanostructures by efficient analytical–numerical methods (the jump function method and its modifications), without the need to involve purely numerical approaches (in particular, FEM).

The calculations of the stress–strain field components for simple test cases of the functional dependence of the shear moduli of inclusion material have demonstrated the expected qualitative picture of their effect on the variation in the FGM parameters. In particular, (1) the stress magnitude increases significantly in the vicinity of the inclusion regions with increased stiffness; (2) the combination of the inclusion materials from parts with piecewise linear mechanical characteristics may lead to partial unloading of the inclusion and matrix in their softer part in the vicinity of the breaking point of the gradient dependence of the inclusion material parameters; (3) the contrast of the stress field changes of the inclusion and matrix is proportional to the increase in the gradient dependence.

The conclusions and calculation results are easily transferable to analogous problems of thermal conductivity and thermoelasticity with possible frictional heat generation and can be used for recommendations on the optimal operating parameters of structures.

The discussed conclusions can be useful in designing the functionally gradient mechanical properties of the material of inclusions and in the optimization of engineering structures to increase their strength and service life. The proposed method is effective for solving a wide class of problems of deformation of solids with thin deformable inclusions of finite length and can be used for SSS calculation for different FGM inclusions.

Author Contributions: Conceptualization, Y.P. and H.S.; methodology, Y.P. and H.S.; software, Y.P. and L.P.; validation, Y.P. and H.S.; formal analysis, Y.P.; investigation, Y.P. and H.S.; resources, H.S.; data curation, Y.P.; writing—original draft preparation, Y.P.; writing—review and editing, Y.P.;

visualization, Y.P. and L.P.; supervision, H.S.; project administration, H.S.; funding acquisition, H.S. All authors have read and agreed to the published version of the manuscript.

Funding: This investigation was performed within the framework of research project No. 2017/27/B/ST8/01249, funded by the National Science Centre, Poland, and with project financing through the program of the Minister of Education and Science of Poland named "Regional Initiative of Excellence" in 2019–2022, project No. 011/RID/2018/19; amount of financing: 12,000,000 PLN.

Institutional Review Board Statement: Not applicable.

Informed Consent Statement: Not applicable.

Data Availability Statement: The data presented in this study are openly available at https://ua1lib.org/book/665574/5c937e (accessed on 9 January 2022), reference number [37]; https://doi.org/10.3390/ma15041435 (accessed on 9 January 2022), reference number [38]; and https://doi.org/10.3390/ma14174928 (accessed on 9 January 2022), reference number [41].

Conflicts of Interest: The authors declare no conflict of interest.

Nomenclature

FGM	functionally gradient material;
GSIF	generalized stress intensity factor;
SSS	stress–strain state
x, y, z	Cartesian coordinates;
f_r	jump functions;
E_k, ν_k, G_k, G_x^{inK}, G_y^{inK}	elastic properties of the materials;
S_k	half-planes (sections of the solid);
w, σ_{xz}, σ_{yz}, σ_{xx}, σ_{yy}	displacement, stresses (components of SSS);
$L' = [-a; a]$	line, modeling the presence of thin inclusion;
Q_k, b_k	magnitudes of concentrated forces and screw dislocations;
σ_{yz}^{∞}, σ_{xzk}^{∞}	uniformly distributed in infinity shear stresses.

References

1. Mura, T.; Barnett, D.M. Micromechanics of Defects in Solids. *J. Appl. Mech.* **1983**, *50*, 477. [CrossRef]
2. Wang, Y.; Huang, Z. Analytical Micromechanics Models for Elastoplastic Behavior of Long Fibrous Composites: A Critical Review and Comparative Study. *Materials* **2018**, *11*, 1919. [CrossRef] [PubMed]
3. Wang, J.; Karihaloo, B.L.E.; Duan, H.L. Nano-mechanics or how to extend continuum mechanics to nanoscale. *Bull. Pol. Acad. Sci. Tech. Sci.* **2007**, *55*, 133–140.
4. Zhou, K.; Hoh, H.J.; Wang, X.; Keer, L.M.; Pang, J.H.; Song, B.; Wang, Q.J. A review of recent works on inclusions. *Mech. Mater.* **2013**, *60*, 144–158. [CrossRef]
5. Mencik, J. *Mechanics of Components with Treated or Coated Solids*; Kluwer Academic Publishing: Dordrecht, The Netherlands, 1996; 360p.
6. Nemat-Nasser, S.; Hori, M. *Micromechanics: Overall Properties of Heterogeneous Materials*; North-Holland Series in Applied Mathematics and Mechanics; Elsevier: Amsterdam, The Netherlands, 1993; Volume 37, 687p.
7. Pasternak, I.; Sulym, H.; Ilchuk, N. Interaction of Physicomechanical Fields in Bodies with Thin Structural Inhomogeneities: A Survey. *J. Math. Sci.* **2021**, *253*, 63–83. [CrossRef]
8. Chen, J. Determination of thermal stress intensity factors for an interface crack in a graded orthotropic coating-substrate structure. *Int. J. Fract.* **2005**, *133*, 303–328. [CrossRef]
9. Johnson, W.S.; Masters, J.; O'Brien, T.K.; Naik, R. Simplified Micromechanical Equations for Thermal Residual Stress Analysis of Coated Fiber Composites. *J. Compos. Technol. Res.* **1992**, *14*, 182–186. [CrossRef]
10. Shevchuk, V.A. Modeling and Computation of Heat Transfer in a System "Body-Multilayer Coating". *Heat Transf. Res.* **2006**, *37*, 412–423. [CrossRef]
11. Kashtalyan, M.; Menshykova, M. Three-dimensional analysis of a functionally graded coating/substrate system of finite thickness. *Philos. Trans. R. Soc. Lond. Ser. A Math. Phys. Eng. Sci.* **2008**, *366*, 1821–1826. [CrossRef]
12. Bao, G.; Wang, L. Multiple cracking in functionally graded ceramic/metal coatings. *Int. J. Solids Struct.* **1995**, *32*, 2853–2871. [CrossRef]
13. Zhao, J.; Silberschmidt, V.V. Microstructure-based damage and fracture modelling of alumina coatings. *Comput. Mater. Sci.* **2005**, *32*, 620–628. [CrossRef]
14. Hashin, Z. Thin interphase/imperfect interface in elasticity with application to coated fiber composites. *J. Mech. Phys. Solids* **2002**, *50*, 2509–2537. [CrossRef]

15. Kulchytsky-Zhyhailo, R.; Matysiak, S.J.; Bajkowski, A.S. Semi-analytical solution of three-dimensional thermoelastic problem for half-space with gradient coating. *J. Therm. Stress.* **2018**, *41*, 1169–1181. [CrossRef]
16. Davim, P.; Constantinos, A. (Eds.) *Nanocomposites. Materials, Manufacturing and Engineering*; Charitidis, Walter de Gruyter GmbH: Berlin, Germany; Boston, MA, USA, 2013; 211p.
17. Kim, C.I.; Schiavone, P.; Ru, C.-Q. The Effects of Surface Elasticity on an Elastic Solid with Mode-III Crack: Complete Solution. *Trans. ASME J. Appl. Mech.* **2009**, *77*, 021011. [CrossRef]
18. Uchida, Y. Properties of functionally graded materials, Manufactured by progressive lamination method for applications. *Aichi Inst. Technol. Res. Rep.* **2004**, *39–B*, 39–51.
19. Ichikawa, K. (Ed.) *Functionally Graded Materials in the 21ST Century, A Workshop on Trends and Forecasts*; Kluver Academic Publishers: Boston, MA, USA; Dordrecht, The Netherlands; London, UK, 2000; 264p.
20. Miyamoto, Y.; Kaysser, W.A.; Rabin, B.H.; Kawasaki, A.; Ford, R.G. *Functionally Graded Materials: Design, Processing and Applications*; Materials Technology Series; Springer: Berlin/Heidelberg, Germany, 1999; pp. 247–313. [CrossRef]
21. Duan, H.; Wang, J.; Huang, Z.; Karihaloo, B. Eshelby formalism for nano-inhomogeneities. *Proc. R. Soc. A Math. Phys. Eng. Sci.* **2005**, *461*, 3335–3353. [CrossRef]
22. Elperin, T.; Rudin, G. Thermal stresses in a coating–substrate assembly caused by internal heat source. *J. Therm. Stress.* **2016**, *39*, 90–102. [CrossRef]
23. Zhang, N.; Khan, T.; Guo, H.; Shi, S.; Zhong, W.; Zhang, W. Functionally Graded Materials: An Overview of Stability, Buckling, and Free Vibration Analysis. *Adv. Mater. Sci. Eng.* **2019**, *2019*, 1354150. [CrossRef]
24. Li, Y.; Feng, Z.; Hao, L.; Huang, L.; Xin, C.; Wang, Y.; Bilotti, E.; Essa, K.; Zhang, H.; Li, Z.; et al. A Review on Functionally Graded Materials and Structures via Additive Manufacturing: From Multi-Scale Design to Versatile Functional Properties. *Adv. Mater. Technol.* **2020**, *5*, 1900981. [CrossRef]
25. Bishop, A.; Lin, C.Y.; Navaratnam, M.; Rawlings, R.D.; McShane, H.B. A functionally gradient material produced by a powder metallurgical process. *J. Mater. Sci. Lett.* **1993**, *12*, 1516–1518. [CrossRef]
26. Kieback, B.; Neubrand, A.; Riedel, H. Processing techniques for functionally graded materials. *Mater. Sci. Eng. A* **2003**, *362*, 81–105. [CrossRef]
27. Mistler, R.E. High strength alumina substrates produced by a multiple-layer casting technique. *Am. Ceram. Soc. Bull.* **1973**, *52*, 850–854.
28. Sulym, H.T.; Piskozub, I.Z. Nonlinear Deformation of a Thin Interface Inclusion. *Mater. Sci.* **2018**, *53*, 600–608. [CrossRef]
29. Sulym, H.; Piskozub, Y.; Polanski, J. Antiplane Deformation of a Bimaterial with Thin Interfacial Nonlinear Elastic Inclusion. *Acta Mech. Autom.* **2018**, *12*, 190–195. [CrossRef]
30. Eshelby, J.D. The determination of the elastic field of an ellipsoidal inclusion, and related problems. *Proc. Soc. Lond. A Math. Phys. Eng. Sci.* **1957**, *241*, 376–396. [CrossRef]
31. Sharma, P.; Ganti, S. Size-Dependent Eshelby's Tensor for Embedded Nano-Inclusions Incorporating Surface/Interface Energies. *ASME J. Appl. Mech.* **2004**, *71*, 663–671. [CrossRef]
32. Benveniste, Y.; Baum, G. An interface model of a graded three-dimensional anisotropic curved interphase. *Proc. R. Soc. A Math. Phys. Eng. Sci.* **2006**, *463*, 419–434. [CrossRef]
33. Benveniste, Y.; Miloh, T. Imperfect soft and stiff interfaces in two-dimensional elasticity. *Mech. Mater.* **2001**, *33*, 309–323. [CrossRef]
34. Martynyak, R.M.; Serednytska, K.I. *Contact Problems of Thermoelasticity for Interface Cracks in Bimaterials*; Rastr-7: Lviv, Ukraine, 2017; 168p.
35. Gurtin, M.E.; Murdoch, A.I. A continuum theory of elastic material surfaces. *Arch. Ration. Mech. Anal.* **1975**, *57*, 291–323. [CrossRef]
36. Steigmann, D.J.; Ogden, R.W. Elastic surface—Substrate interactions. *Proc. R. Soc. A Math. Phys. Eng. Sci.* **1999**, *455*, 437–474. [CrossRef]
37. Sulym, H.T. *Bases of Mathematical Theory of Thermo-elastic Equilibrium of Solids Containing Thin Inclusions*; Research and Publishing Center of NTSh: L'viv, Ukraine, 2007; 716p. Available online: https://ua1lib.org/book/665574/5c937e (accessed on 19 January 2022). (In Ukrainian)
38. Hutsaylyuk, V.; Piskozub, Y.; Piskozub, L.; Sulym, H. Deformation and Strength Parameters of a Composite Structure with a Thin Multilayer Ribbon-like Inclusion. *Materials* **2022**, *15*, 1435. [CrossRef] [PubMed]
39. Piskozub, J.Z. Effect of surface stresses on the tensely deformed state of thin interface microinclusion. *Math. Model. Comput.* **2021**, *8*, 69–77. [CrossRef]
40. Sulim, G.T.; Piskozub, J.Z. Thermoelastic equilibrium of piecewise homogeneous solids with thin inclusions. *J. Eng. Math. Spec. Issue Thermomechanics* **2008**, *61*, 315–337. [CrossRef]
41. Piskozub, Y.; Sulym, H. Effect of Frictional Slipping on the Strength of Ribbon-Reinforced Composite. *Materials* **2021**, *14*, 4928. [CrossRef] [PubMed]
42. Evtushenko, A.A.; Sulim, G.T. Stress concentration near a cavity filled with a liquid. *Sov. Mater. Sci.* **1981**, *16*, 546–549. [CrossRef]
43. Lee, Y.-D.; Erdogan, F. Interface cracking of FGM coatings under steady-state heat flow. *Eng. Fract. Mech.* **1998**, *59*, 361–380. [CrossRef]
44. Martynyak, R.M.; Kryshtafovych, A. Friction contact of two elastic half-planes with local recesses in boundary. *J. Frict. Wear* **2000**, *21*, 6–15.

45. Piskozub, Y.; Sulym, H. Asymptotics of stresses in the vicinity of a thin elastic interphase inclusion. *Mater. Sci.* **1996**, *32*, 421–432. [CrossRef]
46. Sulym, H.; Piskozub, L.; Piskozub, Y.; Pasternak, I. Antiplane Deformation of a Bimaterial Containing an Interfacial Crack with the Account of Friction I. Single Loading. *Acta Mech. Autom.* **2015**, *9*, 115–121. [CrossRef]
47. Sulym, H.; Piskozub, L.; Piskozub, Y.; Pasternak, I. Antiplane Deformation of a Bimaterial Containing an Interfacial Crack with the Account of Friction 2. Repeating and Cyclic Loading. *Acta Mech. Autom.* **2015**, *9*, 178–184. [CrossRef]
48. Sulym, H.; Pasternak, I.; Piskozub, L. Longitudinal shear of a bi-material with fractional sliding contact in the interfacial crack. *J. Theor. Appl. Mech.* **2016**, *54*, 529–539. [CrossRef]
49. Yevtushenko, A.; Rozniakowska, M.; Kuciej, M. Transient temperature processes in composite strip and homogeneous foundation. *Int. Commun. Heat Mass Transf.* **2007**, *34*, 1108–1118. [CrossRef]
50. Pasternak, I.M.; Pasternak, R.M.; Sulym, H.T. 2D boundary element analysis of defective thermoelectroelastic bimaterial with thermally imperfect but mechanically and electrically perfect interface. *Eng. Anal. Bound. Elem.* **2015**, *61*, 194–206. [CrossRef]
51. Pasternak, Y.M.; Sulym, H.T.; Piskozub, L.G. Model tonkogo vklyucheniya v usloviyah yego idealnogo i neidealnogo kontaktnogo vzaimodeystviya s okruzhayuschim materialom. In Proceedings of the Trudy VI Mezhdunar. Simp. po Tribofatike MSTF 2010, Minsk, Belarus, 25 October–1 November 2010; BGU: Minsk, Belarus, 2010; Volume 2, Chapter 2. pp. 399–404.
52. Chen, X.; Liu, Q. Thermal stress analysis of multi-layer thin films and coatings by an advanced boundary element method. *Comput. Model. Eng. Sci.* **2001**, *2*, 337–349.
53. Rodríguez, A.; Calleja, A.; de Lacalle, L.N.L.; Pereira, O.; González, H.; Urbikain, G.; Laye, J. Burnishing of FSW Aluminum Al–Cu–Li Components. *Metals* **2019**, *9*, 260. [CrossRef]
54. Sánchez, A.J.; Egea, A.; Rodríguez, A.; Celentano, D.; Calleja, A.; López de Lacalle, L.N. Joining metrics enhancement when combining FSW and ball-burnishing in a 2050 aluminum alloy. *Surf. Coat. Technol.* **2019**, *367*, 327–335. [CrossRef]
55. Bouhamed, A.; Jrad, H.; Mars, J.; Wali, M.; Gamaoun, F.; Dammak, F. Homogenization of elasto-plastic functionally graded material based on representative volume element: Application to incremental forming process. *Int. J. Mech. Sci.* **2019**, *160*, 412–420. [CrossRef]
56. Chen, Y.Z. Study of multiply-layered cylinders made of functionally graded materials using the transfer matrix method. *J. Mech. Mater. Struct.* **2011**, *6*, 641–657. [CrossRef]
57. Jin, Z.-H. An asymptotic solution of temperature field in a strip of a functionally graded material. *Int. Commun. Heat Mass Transf.* **2002**, *29*, 887–895. [CrossRef]
58. Majak, J.; Mikola, M.; Pohlak, M.; Eerme, M.; Karunanidhi, R. Modelling FGM materials. An accurate function approxi-mation algorithms. In Proceedings of the Modern Materials and Manufacturing (MMM 2021), Tallinn, Estonia, 27–29 April 2021; IOP Conference Series: Materials Science and Engineering. 2021; Volume 1140.
59. Takahashi, M.; Itoh, Y.; Kashiwaya, H. Fabrication and evaluation of W/Cu gradient material by sintering and infiltration technique. In Proceedings of the First International Symposium on Functionally Gradient Materials FGM, Tokyo, Japan, 8–9 October 1990; Functionally Gradient Materials Forum, The Society of Non-Traditional Technology. 1990; pp. 129–134.
60. El-Borgi, S.; Erdogan, F.; Ben Hatira, F. Stress intensity factors for an interface crack between a functionally graded coating and a homogeneous substrate. *Int. J. Fract.* **2003**, *123*, 139–162. [CrossRef]
61. Erdogan, F.; Wu, B.H. Crack problem in FGM layers under thermal stresses. *J. Therm. Stress.* **1996**, *19*, 237–265. [CrossRef]
62. Wang, B.L.; Mai, Y.W.; Noda, N. Fracture mechanics analysis model for functionally graded materials with arbitrarily dis-tributed properties. *Int. J. Fract.* **2002**, *116*, 161–177. [CrossRef]
63. Yildirim, B.; Dag, S.; Erdogan, F. Three dimensional fracture analysis of FGM coatings under thermomechanical loading. *Int. J. Fract.* **2005**, *132*, 371–397. [CrossRef]
64. Guo, L.-C.; Noda, N.; Ishihara, M. Thermal Stress Intensity Factors for a Normal Surface Crack in a Functionally Graded Coating Structure. *J. Therm. Stress.* **2007**, *31*, 149–164. [CrossRef]
65. Sulym, H.T.; Piskozub, Y.Z. Conditions of contact interaction (a survey). *Mat. Metod. Fiz. Mekhanichni Polya* **2004**, *47*, 110–125.
66. Ding, S.-H.; Li, X. Thermal stress intensity factors for an interface crack in a functionally graded layered structures. *Arch. Appl. Mech.* **2010**, *81*, 943–955. [CrossRef]
67. Kit, G.S. On Analogy between Longitudinal Shear and Stationary Thermal Conductivity of Bodies with Inclusions and Cracks. *Dokl. USSR Ser. A* **1977**, *4*, 336–340.

MDPI
St. Alban-Anlage 66
4052 Basel
Switzerland
Tel. +41 61 683 77 34
Fax +41 61 302 89 18
www.mdpi.com

Materials Editorial Office
E-mail: materials@mdpi.com
www.mdpi.com/journal/materials

www.ingramcontent.com/pod-product-compliance
Lightning Source LLC
Chambersburg PA
CBHW041017270926
51850CB00024B/1380
* 9 7 8 3 0 3 6 5 8 4 6 3 8 *